The Critical Point

The Critical Point

A historical introduction to the modern theory of critical phenomena

CYRIL DOMB

Bar-Ilan University, Ramat-Gan, Israel

CRC Press
Taylor & Francis Group
Boca Raton London New York

CRC Press is an imprint of the
Taylor & Francis Group, an **informa** business
A TAYLOR & FRANCIS BOOK

CRC Press
Taylor & Francis Group
6000 Broken Sound Parkway NW, Suite 300
Boca Raton, FL 33487-2742

First issued in paperback 2019

ISBN-13: 978-0-7484-0435-3 (hbk)
ISBN-13: 978-0-367-40138-2 (pbk)

British Library Cataloguing in Publication Data
A catalogue record for this book is available from the British Library.

Library of Congress Cataloging Publication Data are available

Cover design by Amanda Burragry

Typeset in Times 10/12 by Santype International Limited, Salisbury, Wilts

Visit the Taylor & Francis Web site at
http://www.taylorandfrancis.com

and the CRC Press Web site at
http://www.crcpress.com

Contents

Preface

This monograph has its origin in a course of graduate lectures on 'Critical Phenomena' which I gave for many years in an Intercollegiate Programme at London University in England, and subsequently at Bar-Ilan University in Israel. The lectures were designed for the general condensed matter physicist, the aim being to provide background and current information on the exciting developments which were taking place. The course started in 1960, and had to be expanded from time to time to take account of important new ideas.

When the renormalization group (RG) burst on the scene in the 1970s it provided a wealth of new opportunities for exploration, and these attracted much new talent and ability into the field. Some took the view that much of the previous work had been made obsolete, and that the ideas of the RG should be introduced at the beginning of the course. This was not the attitude which I adopted. The RG did not provide exact solutions; it made daring and drastic approximations which required justification, and independent alternative assessments of critical behaviour were therefore of vital importance. To me it seemed that the RG had filled the large missing gaps in the previous treatment, and that it enhanced rather than detracted from the value of the work which had preceded it. I therefore did not change my course but added a final section which explained the RG approach, outlined its conclusions and assessed their significance.

In fact there are two alternative methods of teaching a scientific topic which has experienced revolutionary progress. The first ignores history and formulates the subject from first principles in the most logical manner. The second describes experimental data and theoretical ideas as they developed, even though the path is less direct and may involve *ad hoc* conjectures and blind alleyways. Using the second approach the present monograph aims to present a historical account of the development of a coherent theory of critical phenomena. It can also perhaps serve as a role model in the history of science illustrating the collaboration between experiment and theory, the formulation and application of new concepts, the exploitation of abstract theoretical models, and the basic importance of pursuing accurate theoretical calculations.

I have benefited particularly from the specialized review articles in the volumes on *Phase Transitions and Critical Phenomena* which I have been privileged to edit first with the late Mel Green and subsequently with Joel Lebowitz. I hope that this monograph will serve as an elementary introduction to these more specialized and sophisticated reviews.

The important question of notation presented significant problems. Our discussion extends to fields with well-established conventional notations which it is unwise to contravene. Hence inevitably the same symbols are used to denote two or even three different properties, sometimes even in the same chapter: m represents magnetization, but also mass; s represents entropy, but also spin; ϵ represents refractive index, energy of interaction and $(4 - d)$ $(d = \text{dimension})$ in the Wilson ϵ-expansion. It is hoped that once the reader has been alerted to the situation the chances of confusion will be substantially reduced. For additional guidance a list of symbols and what they represent is appended to each chapter.

I owe a debt of gratitude to a number of my associates with whom I have discussed the subjects in the book: first and foremost to my students who later became colleagues, most notably Martin Sykes, Michael Fisher, David Gaunt, Geoff Joyce and Dennis Rapaport; and to experts on the RG, particularly Amnon Aharony and David Mukamel who helped to clear up my difficulties. Bryan Coles encourages me to embark on writing the book, and gave me steadfast support without which I could never have completed the project. Taylor & Francis showed exemplary patience as I failed to meet one deadline after another.

Finally, I must acknowledge the help of academic institutions with which I have been associated: King's College, London, where most of my own contributions and those of my collaborators in this field were initiated; Sussex University where I spent a fruitful sabbatical leave writing much of Chapters 5 and 6; and finally Bar-Ilan University where the rest of the writing took place, and where I have had the rewarding experience of helping to build up an excellent physics department when at an advanced stage of my academic career.

CYRIL DOMB
June 1995
Jack & Pearl Resnick Institute of Advanced Technology
Bar-Ilan University, Ramat-Gan, Israel

Foreword

About the Author and the Subject

It is both a privilege and a pleasure to offer a Foreword to this book on *The Critical Point* by Cyril Domb. It is a privilege because there are few opportunities, indeed, for presenting an introduction to a book by a world authority and true pioneer who was there in the earliest days of the revolutionary developments that he describes, that he himself helped launch and then continued to guide. From a vigorous and active retirement [1], Cyril Domb can now look back nearly 50 years to when he first heard Lars Onsager talk on 'Transition Points' at the conference on 'Fundamental Particles and Low Temperature Physics' in Cambridge on 23 July 1946 [2]. Onsager described, in characteristically Delphic terms, the exact solution for the free energy of the two-dimensional rectangular Ising model [3] and the new method of derivation, involving spinor algebra, which he had just developed with Bruria Kaufman [4]. As Professor Domb shows in this text, it was that work and its subsequent extension to the correlation functions and long-range order [5, 6] that led to the 'Onsager Revolution' in our understanding of critical points in fluids, in magnets, in alloys, in superfluids, and in many other systems. The classical theories, going back to van der Waals, Maxwell, Curie and Weiss, and refined by Bethe, Bragg and Williams, Landau, and others, had to be overturned and rebuilt. The 'Reconciliation' came first through the concept of non-trivial critical exponents, later welded into a powerful (but still phenomenological) theory of *scaling* and *universality*. That, in turn, through the striking insights and imagination of Kenneth G. Wilson, finally gained its foundation and justification in the *renormalization group* theory of critical phenomena. This is the dramatic story that Cyril Domb lived through and thought through: he tells it here, for the student and non-specialist reader, with the practised hand of a devoted and experienced scholar and teacher.

By way of a Foreword perhaps no more need be written; but I would like to say more about the author and to explain why it is a personal pleasure to contribute this Foreword: that is because it provides the opportunity to pay tribute to a teacher and mentor who opened for me areas of science which proved fertile ground for many of my own researches. Indeed, the school of mathematical studies in statistical mechanics which Cyril Domb founded in 1954 in King's College, London – of which I am proud to have been a member – played a major role in exploring the manifold consequences of the Onsager breakthrough and moving the field beyond the famous and seductive classical or mean-field theories which had defined the previous eight decades. As a teacher and expositor, Cyril Domb brought new horizons and opportunities to his students and provided them with the tools and ideas to forge ahead. This book will surely do the same for those new to the field or for those wishing to revive or extend their acquaintance with it.

Cyril Domb is of that generation whose higher education was disrupted by World War II. In 1938 he left Hackney Downs School in London for Pembroke College, Cambridge, with a Major Open Scholarship. The war broke out for the United Kingdom in September 1939 but, wisely, students in science were allowed to complete their undergraduate studies. In June 1941, Cyril Domb left Cambridge with a degree in mathematics including, however, in the British applied mathematics tradition, courses in relativity, quantum mechanics and statistical mechanics [2]. His strong background in pure mathematics, however, stood him in good stead since, as he put it himself in his reminiscences of those days [2], 'a training in rigour', even when working on practical technical and engineering problems, 'enables one to decide with more confidence when it can be ignored'! In retrospect, the mathematical breadth and insight that Cyril Domb brought to statistical mechanics were crucial instruments in building on Onsager's results for the most basic two-dimensional models (later to be characterized as in the 'Ising universality class') in order to understand critical behaviour in real, three-dimensional systems in physics, chemistry, and other areas. An Appendix to this book points to some of those further directions: in particular, to percolation phenomena and macromolecular solutions, in which the author's pioneering work set the stage for significant modern advances.

A month after leaving Cambridge as a twenty-year old mathematician, Cyril Domb was plunged into research on radar at the Admiralty Signal Establishment. There he was exposed to other bright young men, in particular to Fred Hoyle (five years his senior), to Herman Bondi and to Tommy Gold. And, of course, he also encountered a host of problems demanding prompt and effective practical solutions. One of these solutions, for the intensity of radio signals propagating near the earth, was published with his mentor, M. H. L. Pryce, shortly after the war in the *Journal of the Institute of Electrical Engineers* [7].

On his return to Cambridge in 1946, Cyril Domb resumed his formal education as a graduate student under the nominal supervision of Fred Hoyle. However, he sought his own problems in his by-then-chosen field of statistical mechanics. As he relates [2], most of the standard texts and current research papers were discouraging: it appeared that 'only minor points' remained 'to be straightened out'. However, J. Frenkel's just-published book *Kinetic Theory of Liquids* [8] provided another, deeper, and more critical view, exposing a range of open problems in condensed matter science (and strongly belying the book's restrictive title!). In fact, Yakov Il'ich Frenkel had an understanding, seemingly rare in that era, of the central role to be played by simple models in seeking an understanding of complex physical systems [9]. This valuable insight, which in many ways characterizes the modern era of condensed matter theory, Philip W. Anderson being one of its most notable proponents, imbues Cyril Domb's work. And time has shown that it was, and still is, crucially important for progress in the study of critical-point behaviour. The significance of 'simplified physical models' is stated clearly, albeit modestly, in the Introduction to Cyril Domb's own seminal review of the field in 1960: *On the Theory of Cooperative Phenomena in Crystals* [10]. Although over 30 years old, this article is still valuable for serious students because of its thorough account of exact results for the bulk thermodynamic properties of a wide range of two-dimensional Ising models, a body of knowledge essentially complete by that time.

Furthermore, the 1960 article also expounds the fundamentals of Cyril Domb's own most important and singular contributions to the subject; namely, first, the idea that the systematic study of successive approximations to the solution of a difficult problem may teach one much more about the true answer than even the best of the individual approximations, and, second, the embodiment of this idea in methods for the extrapolation of long power-series expansions. Combined with serious attention to the mathematical problems of generating long enough series, this approach proved a particularly fruitful route to the truth; more than anything else – including even Onsager's exact results which, in the late 1950s, could still be discarded as interesting peculiarities restricted to two-dimensional systems [11] – Domb's methods provided the route for theory to escape from the tyrannical reign of the classical exponent values! The path was not easy and the strategy met with disparagement; but persistence, faith and, as often needed in research, some luck, particularly the fortunate collaboration with his gifted student Martin F. Sykes, won the battle: see, especially, the notable joint paper published in 1961 [12]. Even today, the generation and intelligent extrapolation of power series provides one of the more reliable and precise routes for the computation of critical-point parameters for old problems and for new ones [13, 14].

It is interesting to go back to Cyril Domb's thesis work, completed in 1948, to find the basic ideas already laid out and applied. Early in his studies he had rediscovered the transfer matrix approach for attacking the Ising model [2] – but in this he had been anticipated in 1941 by Kramers and Wannier, by Lassettre and Howe, and by Elliott Montroll (who later became a lifelong friend). However, Cyril Domb carried the power series expansion for the spontaneous magnetization of the square lattice Ising model, $M_0(T)$, to significantly higher order than previously known [15], specifically to order z^{18}, the coefficient being $-88\,048$, where, in terms of J, the basic coupling energy, one has $z = \exp(-2J/k_B T)$. On this basis he then proceeded to demonstrate that the spontaneous magnetization appeared to vanish at the critical point, T_c, in accord with the law

$$M_0(T) \sim (T_c - T)^\beta$$

where the exponent β (not so named until a decade later) was far smaller than the value $\beta = \frac{1}{2}$ dictated by the classical theories; rather one could most reasonably conclude [15] $\beta \lesssim 0.16$.

But was this bold surmise near the truth? The freshly minted 'Dr. Domb' moved to Oxford at the start of 1949 as an ICI Fellow. Later that year his colleague G. Stanley Rushbrooke, then a new Reader in Theoretical Physics, brought back the news from the 'STATPHYS 1' meeting in Florence that Onsager had announced the exact result for $M_0(T)$ [6]: it implied $\beta = \frac{1}{8} = 0.125$. (Onsager never published his derivation and the result became widely known only when C. N. Yang presented his independent calculation three years later [16].) Truly, the correct value for β did lie far below the classical prediction and was, in fact, only 20% smaller than Cyril Domb's series-based estimate! As the reader will learn from the text that follows, the precision and reliability of the extrapolation procedures improved [12, 13] and, indeed, many other critical exponents for two dimensions were accurately estimated prior to their analytic evaluation. But, despite the great progress initiated by the renormalization group and the ϵ (or dimensionality) expansion theories in 1972, we still lack an exact value for β (or for any other critical exponent) for the three-dimensional Ising model. There are thus, dear reader, still problems to solve!

I cannot close this Foreword without telling of Cyril Domb's successful efforts to carry the field forward in ways other than by his own many further seminal contributions to the research literature. Specifically, in 1971 he joined forces with M. S. Green, by then at Temple University. Six years earlier, Mel Green's *Conference on Phenomena in the Neighborhood of Critical Points* at the National Bureau of Standards in Washington [17] had been the first international meeting to recognize the breadth and depth of the field, bringing together a wide range of experimentalists and theorists including Peter Debye and George Uhlenbeck. Together, Domb and Green planned a series of volumes [18] designed to present a coherent picture of the progress achieved through the 1960s and to provide a standard reference for graduate students and researchers. Four volumes entitled *Phase Transitions and Critical Phenomena* were initially proposed [18]: Volume 1 was devoted to general, rigorous theorems and to exactly soluble models including an account by E. H. Lieb, F. Y. Wu and R. J. Baxter of their work which had, by then, extended Onsager's achievements in most startling ways. Volume 2 discussed thermodynamics and scaling and a host of novel topics: in particular, Benjamin Widom wrote on surface tensions – the first application of critical exponent relations and scaling concepts to properties beyond the bulk. Series expansions were featured in Volume 3: Cyril Domb himself contributed the first article; but, by then, an international cast was needed to review the subject, including G. A. Baker, Jr., M. Wortis, H. E. Stanley, and J. F. Nagle from the United States, D. D. Betts from Canada, and A. J. Guttmann from Australia.

As the field blossomed, so further volumes were assembled. Volume 5 started with a masterly overview, *Scaling, Universality, and Operator Algebras*, by Leo P. Kadanoff. By 1976 the renormalization group explosion had occurred and Volume 6 was designed to expound the basic theory and to review the applications already made: an introduction by Kenneth Wilson was followed by seven authoritative and definitive articles by the leading international researchers which still constitute a most valuable resource.

Sadly, Mel Green died prematurely in 1979. And Cyril Domb left King's College for Bar Ilan University in Israel; but happily he persuaded Joel Lebowitz, at Rutgers University, to join him as coeditor of the series[19]. It was then revived with Volume 7 in 1983. The first article, on *Defect-mediated Phase Transitions*, was written by David R. Nelson who had been a graduate student at Cornell when the renormalization-group wave first broke and swept the subject forward! Now, with Volume 16 recently published [19], the field of critical phenomena may reasonably be regarded as 'mature': but it is still full of life and that life has, indeed, spread far beyond its original source in a sharp focus on the peculiar properties of matter near fluid and magnetic critical points.

Cyril Domb, as I hope my account shows, has seen the play through all its major acts. He has consistently welcomed, encouraged, and explored the new ideas that continue to emerge. His perspective is historical, his experience direct and broad. One could not wish for a more expert and sympathetic guide to introduce one to a major triumph of theoretical and experimental science in our century.

MICHAEL E. FISHER
October 1995

References

1. A recent article by Cyril Domb is: *Common-Sense, Scientific Method and Accident Prevention*, Lecture to the Israel Physical Society, April 1995, published by the Nebenzahl Institute for Human Safety and Accident Prevention, Jerusalem College of Technology (Jerusalem 1995), pp. 1–13.
2. DOMB, C. (1990) Some Reminiscences about my Early Career, *Physica* A **168**, 1–21.

3. ONSAGER, L. (1994) Crystal Statistics. I. A two-dimensional model with an order-disorder transition, *Phys. Rev.* **65**, 117–149.
4. ONSAGER, L. and KAUFMAN, B. (1947) *Transition Points*, Physical Society Cambridge Conference Report, pp. 137–144.
5. KAUFMAN, B. and ONSAGER L. (1949) Crystal Statistics. II. Short-range order in a binary lattice, *Phys. Rev.* **76**, 1232–52.
6. ONSAGER, L. (1949) *Nuovo Cim. Suppl.* (9) **6**, 261.
7. DOMB, C. and PRYCE, M. H. L. (1947) The Calculation of Field Strengths over a Spherical Earth, *J. I. E. E.* **94**, 325–339.
8. FRENKEL, J. (1946) *Kinetic Theory of Liquids*, Oxford: Oxford University Press. Reprinted 1955 by Dover Publications, Inc., New York.
9. See the obituary of Frenkel by TAMM, I. E. (1962) *Sov. Phys. Uspekhi* **5** (2), Sept.–Oct. 1962 which I have cited on previous occasions.
10. DOMB, C. (1960) On the Theory of Cooperative Phenomena, *Adv. Phys., Phil. Mag. Suppl.* **9**, 149–361. The extensive tables of lattice configurational data should be especially noted.
11. See LANDAU, L. D. and LIFSHITZ, E. M. (1958) *Statistical Physics*, London: Pergamon Press, p. 438.
12. DOMB, C. and SYKES, M. F. (1961) Use of Series Expansions for the Ising Model Susceptibility and Excluded Volume Problem, *J. Math. Phys.* **2**, 63–67.
13. See, e.g., GUTTMANN, A. J. (1989) Asymptotic Analysis of Power-Series Expansions, in DOMB, C. and LEBOWITZ, J. L., Eds., *Phase Transitions and Critical Phenomena*, Vol. 13, London: Academic Press, pp. 1–234.
14. See, e.g., the references cited in FISHER, M. E. (1990) Low-dimensional Quantum Antiferromagnets: Criticality and Series Expansions at Zero Temperature, *Physica* A **168**, 22.
15. DOMB, C. (1949) Order-disorder Statistics II. A two-dimensional model, *Proc. Roy. Soc.* A **199**, 199–221.
16. YANG, C. N. (1952) The Spontaneous Magnetization of the Two-dimensional Ising Model, *Phys. Rev.* **85**, 808–816.
17. GREEN, M. S. and SENGERS, J. V., Eds. (1966) *Critical Phenomena: Proceedings of a Conference held in Washington D. C., April 1965*, NBS Misc. Publ. 273 Washington: National Bureau of Standards, pp. 1–242.
18. DOMB, C. and GREEN, M. S., Eds. (1972–1976) *Phase Transitions and Critical Phenomena*, Vols. 1–6, London: Academic Press.
19. DOMB, C. and LEBOWITZ, J. L., Eds. (1983–1994) *Phase Transitions and Critical Behavior*, Vols. 7–16, London: Academic Press.

1

Historical Survey

1.1 The Development of Statistical Mechanics

The idea that any piece of matter consists of a large number of elementary constituent particles is very old, and was the subject of considerable discussion among Greek philosophers. But such discussion would nowadays be termed metaphysical since no quantification was attempted, and no predictions were made which could be subjected to experimental test.

When Newton's laws of motion were established the door to numerical calculation was opened, and Daniel Bernoulli in 1738 introduced the first kinetic theory of gases in the modern sense (for a detailed historical analysis see Brush (1983)). He was able to account for Boyle's law, and to relate pressure and temperature to the kinetic energy of the molecules. Unfortunately there was little follow-up to Bernoulli's ideas, and well over 100 years elapsed without any major step forward.

When progress did take place in the latter half of the nineteenth century it was extremely rapid. In an obituary tribute to Rudolf Clausius, published in 1889, Josiah Willard Gibbs wrote:

> The origin of the kinetic theory of gases is lost in remote antiquity and its completion the most sanguine cannot hope to see. But a single generation has seen it advance from the stage of vague surmises to an extensive and well established body of doctrine. This is mainly the work of three men, Clausius, Maxwell, and Boltzmann, of whom Clausius was the earliest in the field and has been called by Maxwell the principal founder of the science. We may regard his paper (1875), 'Ueber die Art der Bewegung, welche wir Wärme nennen' as marking his definite entry into this field, although many points were incidentally discussed in earlier papers.

It seems reasonable, therefore, to regard the above paper by Clausius, 'The kind of motion we call heat', as the founding paper of statistical mechanics (for an English translation see Brush (1965)). In it he defined the gaseous state as one in which the space filled by the molecules of the gas is infinitesimal in comparison with the whole space occupied by the gas itself, and in which the influence of the molecular forces is infinitesimal. The specific heat of a gas was related to the kinetic energy of its molecules. The solid state was characterized as one in which the molecules vibrate about positions of equilibrium, and the liquid state as one in which the molecules no longer have any definite positions of equilibrium but remain in close association. In this way Clausius was able to explain in molecular terms a number of well-known physical phenomena. Finally he gave numerical estimates of the velocities of typical molecules.

The last values were criticized by a Dutch meteorologist, C. H. D. Buys-Ballot, who pointed out that if the molecules moved as fast as Clausius claimed, the mixing of gases by diffusion should be much faster than had been observed. To meet this criticism Clausius wrote a second paper in 1858 introducing molecular collisions into his model, with the new parameter of *mean free path* to characterize the distance travelled by molecules between successive collisions.

Surprisingly, Clausius assumed that the molecules all had equal velocities. It was James Clerk Maxwell, then a young professor at the University of Aberdeen, who pointed out in a lecture to the British Association in 1859 that the concepts of probability and statistics are those appropriate to a model involving molecular collisions. Even if the molecules started off with equal velocities, these would rapidly be changed by collisions until they formed a probability distribution. He proceeded to formulate this distribution subsequently known as the Maxwellian distribution. Although the arguments he advanced were not rigorous, the answer he gave was correct.

Maxwell, in the above lecture, used the mean free path to calculate the transport properties of a gas, the viscosity, heat conductivity, and diffusion rate. He continued to research on the kinetic theory of gases until his death in 1879, and in his lectures on this topic he emphasized the statistical nature of the calculations. In a lecture on 'Molecules' at the British Association Meeting at Bradford in 1873, Maxwell gave a figure for the number of molecules in a cubic centimetre of gas under standard conditions at 19 million million million (Loschmidt had been the first to make such an estimate in 1865 using Maxwell's theory to provide an approximation to the size of a molecule). Modern determinations differ only in replacing the 19 by 26.9. Maxwell then referred to the population statistician who ceases to focus attention on the properties of any one individual and concentrates instead on average properties. He divides the population into groups according to age, height, weight, hair colour or education, and is guided in his analysis by the mathematical theory of probability. But in applying this theory he is usually beset by two worries: is his population really large enough for the method to be valid, and is his population thoroughly homogeneous or is one section influenced by factors which do not apply to the remainder? Neither of these worries exist for the population of molecules.

Ludwig Boltzmann entered the field a few years after Maxwell, and in his first major paper in 1868 (and in a later paper in 1877) significantly extended the statistical basis of the theory. Suppose each molecule of a gas is allowed to have possible energies $\epsilon_1, \epsilon_2, \ldots, \epsilon_n$. A microstate of the gas is defined by specifying which of the molecules have energy ϵ_1, which have ϵ_2, and so on. The total energy E of the assembly of molecules is constant and can be distributed in a large number of different ways among the N molecules. It is convenient to assume that the ϵ_r and E are all multiples of a small unit ϵ, and Boltzmann was interested in the limit $\epsilon \to 0$, $N \to \infty$, E/N remaining finite. The important new basic assumption was that all microstates have the same a priori probability. A state in which a specific molecule has energy ϵ_i can be called a macrostate. The probability P_i of this macrostate is then proportional to the number of microstates in which the remaining energy $E - \epsilon_i$ is distributed among the other $(N - 1)$ molecules. This is a very large number and its asymptotic form can be determined.

Boltzmann was able to establish the distribution which carries his name

$$P_i \propto \exp(-\epsilon_i/kT), \tag{1.1}$$

where T = absolute temperature. For an ideal gas this gives the Maxwell distribution but it can also be extended to the case when an external force (such as gravity) is present. We have used a notation which can immediately be adapted to an assembly in which the molecules have quantized energy levels $\epsilon_1, \epsilon_2, \ldots, \epsilon_r$.

Boltzmann felt the need to justify his equal a priori probability assumption for microstates, and in 1871 he advanced what afterwards came to be known as the *ergodic hypothesis*: that a complex dynamical system with many degrees of freedom will over a long period spend the same amount of time in each microstate. The average value of any property of the system taken over such a long period will be equal to the average value taken over all microstates.

One of Boltzmann's major aims was to derive the second law of thermodynamics from the principles of mechanics. In pursuing this aim he was led to consider the kinetics of collisions in the approach to equilibrium, for which in 1872 he put forward the equation which bears his name. He also introduced the H function related to the velocity distribution of the molecules, which always decreases as a result of collisions, and achieves its minimum value in equilibrium. This function could be identified with minus the thermodynamic entropy, for an equilibrium state. But it provided a generalized definition of entropy for non-equilibrium states.

Turning back to the macrostate–microstate picture it was reasonable to assume that a typical irreversible process involving an increase of entropy corresponds to a transition from a less probable to a more probable macrostate. Hence, if W is the probability of the macrostate, Boltzmann conjectured the relation*

$$S = k \ln W. \tag{1.2}$$

The logarithmic term was required since entropies of independent systems are additive, whereas probabilities are multiplicative.

Boltzmann spent much time and effort subsequently in clarifying the statistical nature of the second law of thermodynamics (aided by Maxwell whose demon, 1871, provided conceptual help). He was also deeply involved in explaining the irreversibility paradox – how reversible microscopic equations can give rise to macroscopic irreversibility.

We have quoted Gibbs' assessment of the contributions of Clausius, Maxwell and Boltzmann. It was Gibbs himself who coined the term *statistical mechanics* and demonstrated how easily one could in principle relate the equilibrium thermodynamic behaviour of a macroscopic body to the properties of its microscopic constituents.

1.2 Gibbs Ensembles

Once the ergodic assumption had been made the dynamics of molecular collisions could be forgotten. In evaluating the equilibrium properties of an assembly of molecules one need only take an average over all possible microstates to derive equilibrium macroscopic behaviour. In 1884 Boltzmann had improved his earlier ideas on the calculation of the equilibrium behaviour of an assembly of N molecules with given total energy E.

In 1902 Gibbs introduced the concept of an *ensemble*, a collection of physical bodies following a statistical distribution which is chosen to simulate a particular thermodynamic situation. The above assembly used by Boltzmann was termed the *microcanonical ensemble*, and it simulates a thermodynamic body whose total energy is kept constant. The average then yields

$$\Omega = \sum_{\substack{\text{macrostates} \\ [n]}} \Omega_n = \exp(S/k) \tag{1.3}$$

$$S = k \ln \Omega. \tag{1.4}$$

Here Ω is used to denote the number of *complexions* or independent configurations of the assembly, and this definition is the same as (1.2) except for an arbitrary constant. If appropriate approximations are made for large numbers, one is led to a thermodynamic relation of the form

$$S = S(U,V,N) \tag{1.5}$$

* See §4.2 for further historical notes on this formula.

(where U = internal energy, V = volume), which is a fundamental thermodynamic relation (U being identified with E). We thus have

$$\frac{1}{T} = \frac{\partial(k \ln \Omega)}{\partial U}, \quad P = \frac{\partial(k \ln \Omega)}{\partial V}. \tag{1.6}$$

Gibbs then suggested taking an ensemble of macroscopic assemblies, the probability of energy E_n being proportional to $\exp(-\beta E_n)$. This has the important property that if two different ensembles are connected together with the same β, their equilibrium behaviour is unchanged by the connection. Hence, this ensemble simulates a thermodynamic body whose temperature is constant. The average yields

$$Z_N = \sum_n \exp(-\beta E_n) = \exp(-\psi/k). \tag{1.7}$$

Formation of the average of the energy E_n (which is to be identified with the internal energy U)

$$\langle E_n \rangle = \sum_n \exp(-\beta E_n/Z_N) = -\frac{\partial}{\partial \beta}(\ln Z_N)$$

and $(\partial/\partial V)(\ln Z_N)$, which is related to the pressure, and comparison with thermodynamics enables β to be identified with $1/kT$, and ψ with $-F/T$ where F is the free energy of Helmholtz. Thus, one is led to the fundamental thermodynamic relation

$$F = -kT \ln Z_N = F(T,V,N). \tag{1.8}$$

Gibbs termed this the *canonical ensemble.* It can be used much more widely than the microcanonical ensemble which is restricted to non-interacting systems, like ideal gases. In principle it provides a mechanism for calculating the equilibrium thermodynamic behaviour of any macroscopic body – solid, liquid, gas, plasma, etc. – in terms of the properties of its constituent molecules. The requirement is a knowledge of the different possible energies E_n of the macroscopic body, corresponding to all the different possible arrangements of the constituent molecules. In terms of the canonical ensemble we have

$$\begin{aligned} U &= -\frac{\partial}{\partial \beta}(\ln Z_N) = T^2 \frac{\partial}{\partial T}(F/T) \\ S &= -\frac{\partial}{\partial T}(kT \ln Z_N), \quad P = -kT \frac{\partial}{\partial V}(\ln Z_N). \end{aligned} \tag{1.9}$$

It was easy to show that fluctuations of quantities like the energy are negligible for a macroscopic assembly. Z_N was subsequently termed the partition function (from the German *Zustandssumme* or sum over states).

In dealing with the thermodynamics of heterogeneous substances Gibbs had found it useful to introduce the *chemical potential* μ_i of each species i which plays a similar role in controlling chemical equilibrium to that of the temperature in controlling thermal equilibrium. Just as it is often convenient to use the temperature as an independent thermodynamic variable rather than the energy, so it is often convenient to use the chemical potential of a given species as a variable rather than its concentration. Gibbs introduced the *grand canonical ensemble* to simulate a thermodynamic body with constant chemical potentials, the number of assemblies with ν_1 molecules of species 1, ν_2 of species 2, ..., ν_r of species r being proportional to $\exp(\mu_1\nu_1 + \mu_2\nu_2 + \cdots + \mu_r\nu_r - \beta E_n)$. Performing the averaging, one is led to

$$\zeta = \sum_{\nu_1, \nu_2, \ldots, \nu_r, n} \exp(\mu_1\nu_1 + \mu_2\nu_2 + \cdots + \mu_r\nu_r - \beta E_n) = \exp(\Phi/k) \tag{1.10}$$

where formation of $\langle E_n \rangle$, $\langle \nu_i \rangle$ and $(\partial/\partial V)(\ln \zeta)$ and comparison with thermodynamics leads to the identification

$$\Phi = \frac{PV}{T} = \Phi(T,V,\mu_i) \tag{1.11}$$

which is again a fundamental thermodynamic relation.

In terms of the grand partition function, ζ, the thermodynamic relations are

$$\nu_i = kT\frac{\partial}{\partial \mu_i}(\ln \zeta), \quad P = kT\frac{\partial}{\partial V}(\ln \zeta), \quad S = \frac{\partial}{\partial T}(kT \ln \zeta). \tag{1.12}$$

The above ensembles defined by Gibbs provide a complete prescription for the transition from the microscopic to the macroscopic properties of a physical or chemical body. It is possible to define other ensembles, e.g. constant pressure ensembles; for lattice models with different species it is convenient to sum over the *concentrations* keeping the total number of molecules constant. In each case the Gibbs procedure must be followed to determine the precise thermodynamic function to which the ensemble leads.

We have formulated Gibbs' ideas in terms of quantum systems with discrete energy levels. Gibbs himself was, of course, concerned with classical systems with continuum energy levels, and he used integrals in place of sums. For example, for the canonical ensemble the partition function would be defined by

$$Z_N = \int \exp(-\beta E)\, d\mathbf{x}_1 \cdots d\mathbf{x}_N\, d\mathbf{p}_1 \cdots d\mathbf{p}_N \tag{1.13}$$

instead of (1.7), the integral being taken over all possible positions $\mathbf{x}_i$ and moments $\mathbf{p}_i$ of the N molecules. For an assembly of molecules with central intermolecular potentials $\phi(|\mathbf{x}_i - \mathbf{x}_j|)$ between molecules at $\mathbf{x}_i$ and $\mathbf{x}_j$

$$E = \frac{1}{2m}\sum_{i=1}^{N}\mathbf{p}_i^2 + \sum_{\substack{\langle i,j\rangle \\ \text{pairs}}}\phi(|\mathbf{x}_i - \mathbf{x}_j|). \tag{1.14}$$

We then find for this classical partition function,

$$Z_N = (2\pi mkT)^{3N/2}\int \exp\{-\beta[\Phi(\mathbf{x}_1, \mathbf{x}_2, \ldots, \mathbf{x}_N)]\}\, d\mathbf{x}_1 \cdots d\mathbf{x}_N \tag{1.15}$$

where

$$\Phi(\mathbf{x}_1, \mathbf{x}_2, \ldots, \mathbf{x}_N) = \sum_{\substack{\langle i,j\rangle \\ \text{pairs}}}\phi(|\mathbf{x}_i - \mathbf{x}_j|). \tag{1.16}$$

For many quantum systems a classical approximation is adequate. It is shown in standard texts on statistical mechanics (e.g. Rushbrooke 1949) that the only change required in (1.15) to link up properly with quantum statistical mechanics is to divide (1.15) by a factor h^{3N}:

$$Z_N = \left(\frac{2\pi mkT}{h^2}\right)^{3N/2}\int \exp\{-\beta[\Phi(\mathbf{x}_1, \mathbf{x}_2, \ldots, \mathbf{x}_N)]\}\, d\mathbf{x}_1 \cdots d\mathbf{x}_N. \tag{1.17}$$

Finally if the energy levels of a quantum system are not known explicitly, but the dynamics are governed by a Hamiltonian operator $\mathscr{H}_N$, it is convenient to write (1.7) in the alternative form

$$Z_N = \langle \exp(-\beta \mathscr{H}_N)\rangle \tag{1.18}$$

where $\langle\ \rangle$ represent the trace of the operator.

1.3 Non-interacting and Interacting Systems

Our discussions in this monograph will be confined to equilibrium properties, and accepting the ergodic assumption, the basic task (following Gibbs) is to calculate the partition function for a particular ensemble; an exact solution is the best we can achieve but this is possible only in a limited number of cases, and more generally we must endeavour to find suitable approximations.

If the thermodynamic properties of matter in equilibrium are surveyed, it will be found that they fall into two groups: those which are smooth and continuous, and those which have sharp discontinuities. As examples of the first group we may cite the properties of ideal or nearly ideal gases (energy, entropy, specific heat, equation of state), of ideal or nearly ideal solids, of ideal or nearly ideal mixtures of gases or solids; paramagnetism and diamagnetism; properties of electrons and phonons in normal metals. The second group is usually associated with phase transitions of various types: liquid–vapour equilibrium and the critical point, the melting of solids, phase separation in liquid or solid mixtures or solutions, order–disorder transitions in alloys, ferromagnetism, antiferromagnetism, superconductivity, λ-point anomalies (e.g. liquid helium).

Standard statistical mechanics can deal with the first group with relative ease. For the thermodynamic properties of an ideal gas of N molecules, it is reasonable to assume in the first instance that the interactions between molecules can be ignored as far as the construction of the partition function is concerned, and it can then be shown that the partition function Z_N for the canonical ensemble is given by

$$Z_N = Z_{\text{gas}}^N, \tag{1.19}$$

where

$$Z_{\text{gas}} = \sum_r \exp(-\beta\epsilon_r) \tag{1.20}$$

and ϵ_r are the energy levels of the molecules of the gas.

It is often possible to simplify (1.20) further since the energy of a molecule is made up of translational, vibrational, rotational and electronic contributions

$$\epsilon_r = \epsilon_{\text{trans}} + \epsilon_{\text{vib}} + \epsilon_{\text{rot}} + \epsilon_{\text{elect}} \tag{1.21}$$

with little interaction between them, and hence to a reasonable approximation

$$Z_{\text{gas}} = Z_{\text{trans}} Z_{\text{vib}} Z_{\text{rot}} Z_{\text{elect}}. \tag{1.22}$$

But even if approximation (1.21) is not accepted, the only problem to be solved is the determination of the energy levels ϵ_r of the molecules and then the partition function (1.20) can be constructed. Elementary mathematical considerations show that the thermodynamic behaviour resulting from (1.19) and (1.20) will then be continuous.

In general the discontinuities described in properties of the second group above originate in intermolecular interactions, and the problem of how to take such interactions into account in calculating a partition function will be central to our discussions. Some of the mathematics involved in a proper treatment of interactions is specialized and sophisticated. But it is our contention that the *results* and *significance* of such calculations can be understood by the general condensed matter physicist even if the detailed mathematical techniques are beyond him.

A slightly non-ideal gas interaction between molecules can be taken into account by perturbation theory, but as long as only a finite number of terms are considered the continuity of the thermodynamic properties will not be destroyed. Discontinuous behaviour can be introduced only by taking the perturbation series to infinity so that the mathematical character of the partition function is changed. In fact the problem to be tackled in dealing with phase transitions is the 'strong interaction' problem in which the interactions can no longer be treated as a small perturbation, but play a dominant role in the calculations and in the resulting physical properties.

We will be largely concerned with behaviour near the critical points of fluids, magnets and solutions, and with various types of λ-point transition. The past four decades have seen great progress in the theoretical understanding of how this behaviour depends on the nature of the intermolecular interactions, e.g. their symmetry and range. We shall endeavour to describe this progress and bring the reader up to the frontiers of current research.

In the remaining sections of this chapter we shall survey briefly the major developments in the exploration of critical behaviour from 1869 to the present day. This should provide a bird's eye view of the background to, and emergence of, the modern theory. The treatment in this chapter will be rather cursory, but the topics raised will be discussed at greater length later in the book.

1.4 The Classical Period

The term classical is used in the sense of well established (see Uhlenbeck 1965) and not with any relationship to quantum theory. The period starts in 1869 with the introduction of the term 'critical point' by Thomas Andrews in his Bakerian lecture to the Royal Society. Andrews had used the term before in 1863 (Brush 1983) but the exposition of the true nature of the critical point of a fluid took place in the Bakerian lecture. The period is characterized by the interplay of experiment and theory, and the introduction of concepts and calculations which seemed adequate to provide a qualitative and quantitative description of the phenomena under consideration. Developments which took place in different areas of physics seemed to have many features in common, and led eventually to the idea of a unified description of all phenomena involving critical points or λ-points.

The classical period terminated in 1944 with the publication by Onsager of the exact solution of a two-dimensional interacting model, which gave results in basic disagreement with the classical calculations. As a result the classical theories were discredited in relation to quantitative predictions of critical behaviour. But the concepts which they had introduced retained their usefulness, and played their part in the reconstruction of a wider and more embracing theory; eventually the classical calculations found a place in the wider framework.

1.4.1 Liquid–Gas Critical Point

The relation between gases and liquids was the subject of much attention during the first half of the nineteenth century. It was realized quite early that the liquid phase ceased to exist above a certain temperature, but the general assumption was that it disappeared into the gaseous phase. Thus Faraday talked of the 'disliquefying point', while Mendeleev used the term 'absolute boiling temperature' for the point at which the latent heat of evaporation becomes zero. For an excellent survey of the early history see Rowlinson (1969).

In addition to careful and accurate experiments on carbon dioxide described in the Bakerian lecture of 1869, Thomas Andrews introduced a new concept of symmetry between the liquid and gaseous phases; the two phases merged at the critical point into one fluid phase, 'but if any one should ask whether it is now in the gaseous or liquid state, the question does not, I believe, admit of a positive reply'. He emphasized this feature in the title of his lecture, 'On the Continuity of the Gaseous and Liquid States of Matter', and pointed out how it was possible, by a suitable choice of path, to pass from the liquid to the gaseous phase without any discontinuity. Figure 1.1 represents the isothermals of carbon dioxide taken from Andrews' Bakerian lecture (1869). The difference of behaviour below and above the critical temperature is manifest. Andrews estimated the critical temperature as 30.92 °C.

Only four years elapsed before van der Waals used the newly developing ideas on the kinetic theory of gases to give a plausible theoretical explanation of Andrews' experimental data: van der Waals assumed that a gas is made up

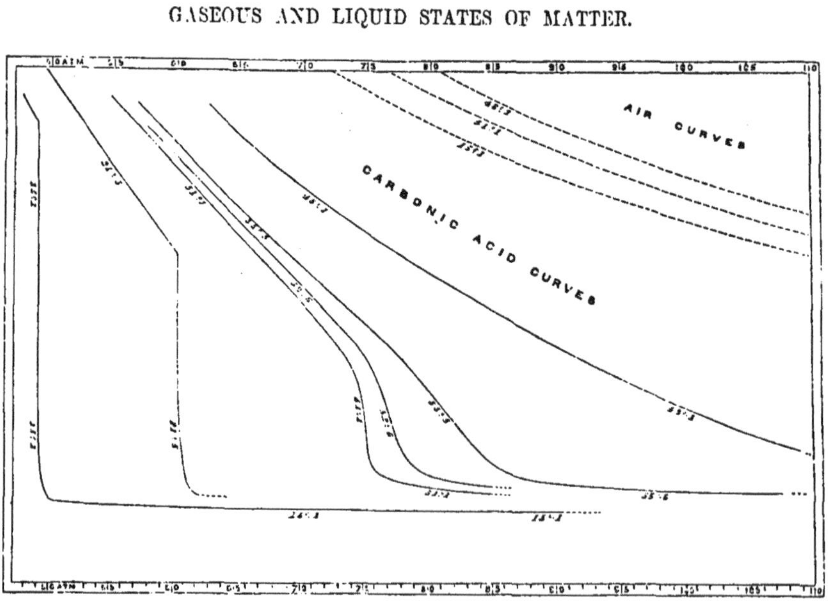

Figure 1.1 Isothermals of carbon dioxide taken from Andrews' Bakerian lecture to The Royal Society (1869). Andrews' estimate for the critical temperature was 30.92 °C.

of molecules with a hard core and a long-range mutual attraction. The range of the attractive forces was assumed to be long compared with the mean free path, and they give rise to a negative 'internal pressure'. To calculate its value van der Waals made use of the virial theorem which Clausius had introduced in 1870 only a few years before and which states that

$$\left\langle \sum_i \tfrac{1}{2} m v_i^2 \right\rangle = -\frac{1}{2} \left\langle \sum_i (x_i X_i + y_i Y_i + z_i Z_i) \right\rangle \tag{1.23}$$

where v_i is the velocity, (x_i, y_i, z_i) coordinates and (X_i, Y_i, Z_i) the force on the ith molecule. He found that

$$P_{\text{internal}} = -a/v^2 \tag{1.24}$$

where $v = V/N$ is the volume per molecule. For the hard core he made the simplest assumption that the available volume is reduced from v to $v - b$. Hence the equation he put forward was

$$P = P_{\text{internal}} + RT/(v - b) \tag{1.25}$$

$$(P + a/v^2)(v - b) = RT. \tag{1.26}$$

Van der Waals was a doctoral student, and the above ideas were put forward in his thesis in Dutch. Their importance was quickly recognized by Maxwell who reviewed the thesis in *Nature* in 1874, and in a lecture to the Chemical Society in 1875 where he wrote as follows

> The molecular theory of the continuity of the liquid and gaseous states forms the subject of an exceedingly ingenious thesis by Mr. Johannes Diderick van der Waals, a graduate of Leyden. There are certain points in which I think he has fallen into mathematical errors, and his final result is certainly not a complete expression for the interaction of real molecules, but his attack on this difficult question is so able and so brave, that it cannot fail to give a notable impulse to molecular science. It has certainly directed the attention of more than one inquirer to the study of the Low-Dutch language in which it is written.

It was in this lecture that Maxwell put forward his famous 'equal-area' construction (Figure 1.2) which completes the van der Waals treatment of liquid–gas equilibrium.

Boltzmann too was impressed by van der Waals' results, and although he said that 'unfortunately, van der Waals had to abandon mathematical rigour at a certain point in order to carry out his calculation', he was nevertheless impressed by the practical value of his theory.

It is significant that both Maxwell and Boltzmann were reasonable happy with the attractive a/v^2 term; their main anxiety was the repulsive term.

The new concept of 'internal pressure' which van der Waals introduced was to bear fruit 30 years later in a completely different area of physics.

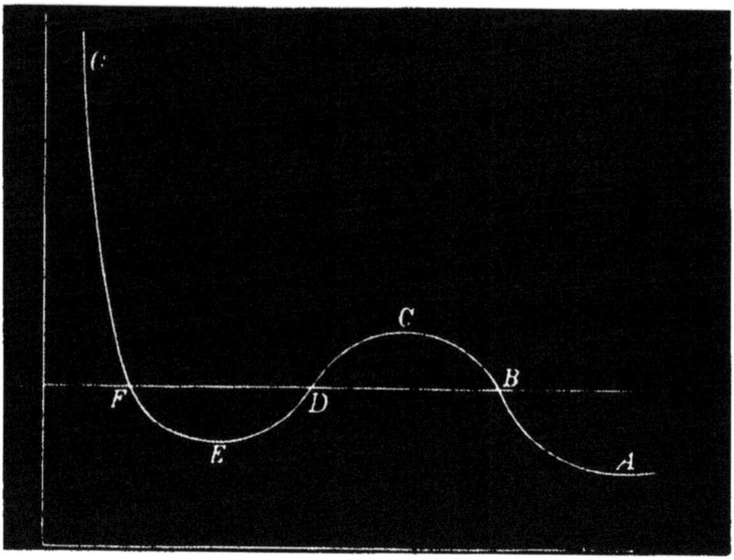

Figure 1.2 Maxwell's 'equal-area' construction to determine the horizontal part of the isotherm in van der Waals' equation. Area FED = area DCB. (From a lecture reproduced in *Nature* in 1875.)

1.4.2 Curie Point of a Ferromagnet

The fact that a magnet loses its magnetic power at high temperatures was noted by Gilbert in his famous treatise *De Magnete* in 1600. A more detailed quantitative investigation started in the nineteenth century involving Faraday in the 1830s, Barrett in 1874, Bauer in 1880 and Hopkinson in 1889 who was the first to introduce the term 'critical temperature' for 'the temperature at which the magnetism disappears'. But the definitive paper for magnets comparable with that of Andrews for fluids was written by Curie in 1895 before he started his more famous investigations of radioactivity.

One of the most interesting new ideas put forward in this paper is the analogy between magnets and fluids. Taking pressure P as the analogue of magnetic field, and density $\rho = N/V$ as the analogue of magnetization, Curie points out the close similarity between the P–ρ and M–H isothermals (Figure 1.3). The paramagnetic state at high temperatures corresponds to the gaseous phase, and the ferromagnetic state at low temperatures to the liquid phase. Curie says that this analogy could be used to suggest new and useful experiments, and he poses the question whether there exists a precisely defined critical point with associated critical constants for a ferromagnet analogous to a fluid.

It was this analogy which led Pierre Weiss in 1907 to postulate his *molecular field* hypothesis, in which he assumed that the mutual interactions between

P. CURIE — 1895

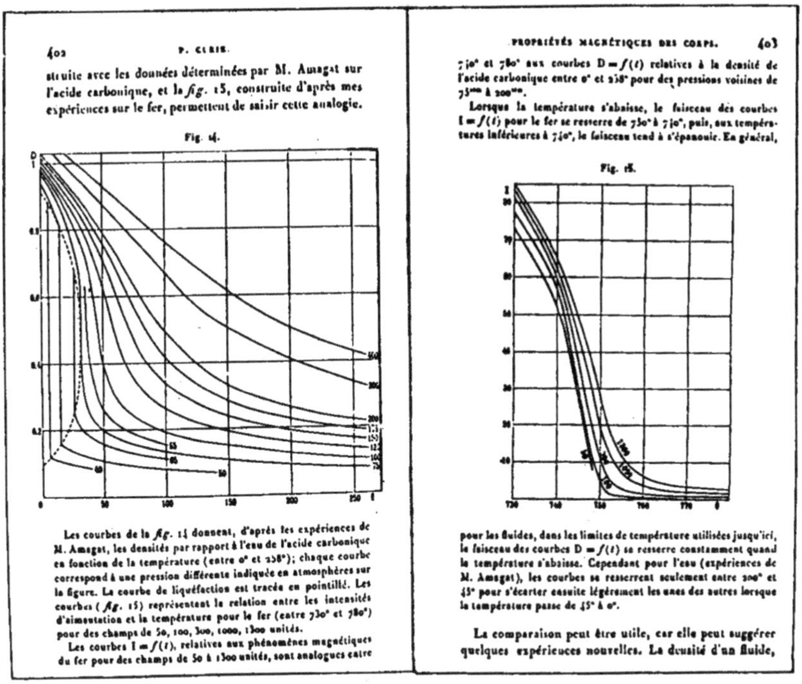

402 P. CURIE.

struite avec les données déterminées par M. Amagat sur l'acide carbonique, et la *fig.* 15, construite d'après mes expériences sur le fer, permettent de saisir cette analogie.

Fig. 14.

Les courbes de la *fig.* 14 donnent, d'après les expériences de M. Amagat, les densités par rapport à l'eau de l'acide carbonique en fonction de la température (entre 0° et 258°); chaque courbe correspond à une pression différente indiquée en atmosphères sur la figure. La courbe de liquéfaction est tracée en pointillé. Les courbes (*fig.* 15) représentent la relation entre les intensités d'aimantation et la température pour le fer (entre 730° et 780°) pour des champs de 50, 100, 300, 1000, 1300 unités.

Les courbes I = *f*(*t*), relatives aux phénomènes magnétiques du fer pour des champs de 50 à 1300 unités, sont analogues entre

PROPRIÉTÉS MAGNÉTIQUES DES CORPS. 403

740° et 780° aux courbes D = *f*(*t*) relatives à la densité de l'acide carbonique entre 0° et 258° pour des pressions voisines de 75atm à 200atm.

Lorsque la température s'abaisse, le faisceau des courbes I = *f*(*t*) pour le fer se resserre de 750° à 740°, puis, aux températures inférieures à 740°, le faisceau tend à s'épanouir. En général,

Fig. 15.

pour les fluides, dans les limites de température utilisées jusqu'ici, le faisceau des courbes D = *f*(*t*) se resserre constamment quand la température s'abaisse. Cependant pour l'eau (expériences de M. Amagat), les courbes se resserrent seulement entre 200° et 45° pour s'écarter ensuite légèrement les unes des autres lorsque la température passe de 45° à 0°.

La comparaison peut être utile, car elle peut suggérer quelques expériences nouvelles. La densité d'un fluide,

Figure 1.3 The analogy between fluids and magnets noted by Curie in 1895. On the left-hand side are Amagat's density–temperature curves for CO_2 at different pressures, and on the right-hand side Curie's own measurements of magnetization–temperature curves for iron at different magnetic fields.

the molecules could be replaced by a uniform field nM proportional to the magnetization in the same direction. He states, 'One may give a nM the name *Internal Field* to mark the analogy with the internal pressure of van der Waals'.

For the magnetic susceptibility of a gas Curie had discovered experimentally the inverse temperature dependence

$$\chi = C/T \tag{1.27}$$

and Langevin had applied the techniques of statistical mechanics to explain this result theoretically. For an ideal gas of molecules, each having a magnetic moment m, Langevin (1905) derived the magnetic equation of state (analogous to $PV = RT$ for an ideal gas)

$$M = L(mH/kT) \tag{1.28}$$

where $L(x)$ is the Langevin function $\coth x - 1/x$.

For a ferromagnet Weiss put forward the simple modification

$$M = L\left(\frac{m(H + H_{\text{int}})}{kT}\right) \quad (H_{\text{int}} = nM) \tag{1.29}$$

and this led to far-reaching conclusions. There is indeed a sharply defined critical point analogous to that for a fluid; Weiss and Kamerlingh Onnes (1910) later termed this the *Curie point* in memory of Curie who had been killed in an accident in 1906. Below the Curie temperature there is a non-zero spontaneous magnetization; in the paramagnetic state above the Curie temperature T_c, Curie's relation (1.27) for the susceptibility is modified to

$$\chi = C/(T - T_c) \tag{1.30}$$

and this usually called the Curie–Weiss law.

The magnet–fluid analogy which proved so fruitful seems to have been forgotten for more than 30 years until it was rediscovered in the lattice–gas model by Cernuschi and Eyring in 1939. It was fruitful again in subsequent developments after 1944.

1.4.3 *Microscopic Critical Behaviour of Fluids*

The experiments of Andrews were concerned with the thermodynamic properties of bulk fluids. A new experimental phenomenon which excited interest at the beginning of the twentieth century was critical opalescence,* in which a colourless transparent fluid suddenly becomes opaque in a narrow region of temperature near T_c; the onset of opacity is accompanied by dramatic changes of colour. It was clear that a considerable increase in light scattering takes place near the critical point, and this was explained by Smoluchowski (1908) and Einstein (1910) as arising from large fluctuations in density as follows.

Einstein showed how Boltzmann's formula (1.2) relating the probability of any state of a system to its entropy could be used to calculate fluctuations of thermodynamic quantities. For the density he derived the formula

$$\langle \Delta\rho^2 \rangle = \frac{\rho^2 kT}{V} K_T \tag{1.31}$$

where K_T is the isothermal compressibility,

$$K_T = \frac{1}{\rho}\left(\frac{\partial \rho}{\partial P}\right)_T = -\frac{1}{V}\left(\frac{\partial V}{\partial P}\right)_T. \tag{1.32}$$

* For a detailed history see §4.1.

Assuming that a change in density is accompanied by a change of refractive index (ε) according to the Lorentz–Lorenz law

$$\frac{\varepsilon - 1}{\varepsilon + 2} = A\rho \tag{1.33}$$

and that scattering takes place randomly, Einstein came to the conclusion that the scattering of light of wavelength λ should be proportional to K_T/λ^4. Since van der Waals' equation leads to an infinite value of K_T at T_c, this seemed to provide a satisfactory explanation of critical opalescence.

It was remarkably perceptive of Ornstein and Zernike (1914, 1916) to note an inconsistency in the treatment arising from the assumption that the fluctuations in all elements of volume are independent of one another. In fact they pointed out that there must be a correlation between different elements which increases indefinitely in range as the critical point is approached.

To treat this correlation Ornstein and Zernike introduced a basic new function, afterwards called the 'pair distribution function', which has played a central role in the theory of liquids ever since. Let $\nu(d\mathbf{r})$ be a random variable representing the number of molecules in a volume $d\mathbf{r}$ centred at $\mathbf{r}$. Since $d\mathbf{r}$ is very small, the probability of occupation by more than one particle being of order $d\mathbf{r}^2$ can be neglected. Hence we can write

$$\langle \nu(d\mathbf{r}) \rangle = n_1(\mathbf{r})\, d\mathbf{r} \tag{1.34}$$

where $n_1(\mathbf{r})$ is the density which they took to be a constant ρ. Similarly for the correlation between particles at points $\mathbf{r}_1,\mathbf{r}_2$,

$$\langle \nu(d\mathbf{r}_1)\nu(d\mathbf{r}_2) \rangle = n_2(\mathbf{r}_1,\mathbf{r}_2)\, d\mathbf{r}_1\, d\mathbf{r}_2 \tag{1.35}$$

and for a homogeneous isotropic fluid $n_2(\mathbf{r}_1,\mathbf{r}_2)$ is of the form $n_2(r)$ ($r = |\mathbf{r}_1 - \mathbf{r}_2|$).

Ornstein and Zernike introduced the function

$$g(r) = n_2(r)/\rho^2 \tag{1.36}$$

which approaches 1 at large distances where the correlation becomes negligible. They used the fluctuation relation (1.31) to derive a fundamental identity between K_T and $g(r)$. But they also obtained an integral equation for $g(r)$ which they were able to solve explicitly.

The basic physical idea behind Ornstein and Zernike's treatment was to differentiate between the direct influence of molecular interactions which should be short ranged and was represented by a function $f(r)$, and the correlation between densities, represented by $g(r)$ above, which should become long ranged as the critical temperature is approached. The integral equation relates $g(r)$ and $f(r)$.

The original paper of Ornstein and Zernike makes difficult reading, and certain aspects of their treatment are obscure. (The whole subject has been beautifully clarified in a classic review paper by M. E. Fisher (1964).) But if I were asked to nominate the contribution in the classical period which gave greatest insight into the nature of critical behaviour I would choose that of Ornstein and Zernike.

Their calculations suggested that the correlations fall off asymptotically as

$$\frac{1}{r}\exp(-\kappa r) \tag{1.37}$$

where the value of κ can be determined from van der Waals' equation ($\kappa \sim (T - T_c)^{1/2}$). Their detailed conclusions about light scattering differed from those of Einstein: near the critical temperature the λ^{-4} Rayleigh dependence on wavelength ceases to be valid, and there is a 'whitening' of the scattered light. By temperature T_c the wavelength dependence has become of the form λ^{-2}.

1.4.4 *Critical Behaviour of Binary Alloys*

The early years of the twentieth century saw the development of X-ray diffraction as a powerful tool in the investigation of crystal structures, and compounds such as NaCl were found to have a regular ordered structure. However, the ionic bond between Na and Cl is so strong that no significant disordering occurs when the temperature is raised; the crystal melts before it disorders.

In 1919* Tammann suggested that similar ordering might occur in metallic alloys, and this was demonstrated experimentally a few years later (Bain 1923, Johansson and Linde 1925) by the existence of superlattice lines in the X-ray diffraction pattern of copper–gold alloys. But for such systems there were soon indications that significant disordering takes place with increase in temperature, and that this is accompanied by an anomalous specific heat.

The mathematical description of the disordering process is usually associated with the names of Bragg and Williams. In their classic paper† in 1934 they introduced a parameter S to characterize the *degree of order*, and used Boltzmann's principle to calculate its behaviour as a function of temperature. They found a pattern closely analogous to the Weiss theory of ferromagnetism, with S falling rapidly to zero at a critical temperature T_c, and remaining zero for $T > T_c$. In fact for a binary alloy with equal concentrations of constituents the equation derived for S was

$$S = \tanh(ST_c/T) \tag{1.38}$$

* I am indebted to Dr L. Muldawer of Temple University for a detailed account of the early history.

† This was W. L. Bragg's Bakerian lecture to the Royal Society.

which is of the same form as (1.28) with $H = 0$ and tanh x replacing the Langevin function $L(x)$.

In 1935 Bragg and Williams wrote a second paper elaborating their ideas. Because of the analogy with ferromagnetism they called T_c the *Curie temperature* of the alloy; S was now termed *long-distance order* to differentiate it from another parameter which had been introduced by Bethe to characterize the *short-range order* which persists above T_c. But they also apologized in this paper for having ignored the work of other investigators, notably Gorsky (1928), Borelius (1934) and Dehlinger (1930, 1932, 1933), who had been thinking along similar lines.

Some years ago shortly before Sir Lawrence Bragg's death, I had an opportunity to speak to him about the background to the above two papers. Sir Lawrence told me that in 1933 he had given a seminar in Manchester describing qualitatively how he thought ordering takes place in binary alloys. E. J. Williams was in the audience, and at the end of the seminar he presented Sir Lawrence with pencilled notes on a sheet of paper, which, he claimed, gave numerical substance to the qualitative ideas. Sir Lawrence was naturally impressed, but suggested that if the mathematics was really so simple, someone must surely have done it before. But no one in the audience knew of any such calculations in the literature, and Bragg and Williams proceeded to write a paper which was presented to the Royal Society.

A few days after the corrected proofs of the paper had been returned to the Royal Society, Sir Lawrence on clearing up his desk was shocked to find a preprint by Borelius developing ideas very similar to those which he had just sent off. He told me that he had been very embarrassed in subsequent years to find that all the credit for the development had been given to him and Williams; he would be pleased if a more correct perspective could be introduced.

Sir Lawrence added that the reason for their obtaining the credit was probably because the Bragg–Williams papers described the ideas more clearly than those of any of the other investigators; anyone looking at the literature today will readily confirm that this is so. The important new concept of long-range order is clearly described in the second paper; and a significant distinction is drawn between long-range order and long-range forces. It is stated clearly that short-range forces can give rise to long-range order, a conclusion exactly parallel to that of Ornstein and Zernike for fluids described above of which Bragg and Williams were apparently unaware.

1.4.5 *Landau's Theory of λ-Point Transitions: Universality*

It was L. D. Landau (1937) who first attempted to provide a unified description for all transitions of the type we have been considering. In addition to the phenomena described above, experimental evidence was accumulating about specific heat anomalies in liquid helium, ammonium chloride and a number of other substances which were termed λ-point transitions, and the superconducting transition was also in this category. Ehrenfest (1933) had introduced a thermodynamic classification of higher-order transitions, but there were difficulties associated with his treatment (see e.g. Pippard 1957), and we shall avoid describing λ-point transitions as second-order phase transitions although this terminology is common.

Landau generalized the ideas introduced in the theory of alloys to all systems manifesting λ-point transitions. He suggested that for every such system one must identify an *order parameter* analogous to the long-range order in an alloy, which would be zero on the high-temperature side of the transition and non-zero on the low-temperature side. He emphasized the important role which symmetry plays in phase transitions, and suggested that the important features of behaviour in the vicinity of a λ-point could be determined by expanding the free energy in a power series as a function of the order parameter η. In the ferromagnetic transition the order parameter is the spontaneous magnetization; in the fluid, the density difference between liquid and gas. Landau argued that for reasons of symmetry the form of the expansion of the Gibbs function will be

$$\Phi(P,T,\eta) = \Phi_0(P,T) + A(P,T)\eta^2 + B(P,T)\eta^4 + \cdots \tag{1.39}$$

and the Curie temperature corresponds to

$$A(P,T) = 0, \quad B(P,T) > 0. \tag{1.40}$$

Above the Curie temperature $A > 0$ and the solution corresponding to a minimum of Φ is $\eta = 0$; this is the phase of higher symmetry. Below the Curie temperature the minimum of Φ corresponds to a non-zero value of η given by

$$\eta^2 = -A/2B \tag{1.41}$$

and this is the phase of lower symmetry.

Landau's theory enabled one to understand why all the systems discussed previously had the same essential pattern of critical behaviour; even though equations of state like (1.26) and (1.29) look very different, they both conform to (1.40). Critical exponents are the same for all λ-point transitions, and in later terminology the transitions would be described as universal.

The behaviour of typical thermodynamic quantities for a ferromagnet near T_c was as follows $\left(\tau = \dfrac{T}{T_c} - 1\right)$:

$$\begin{array}{lll} \text{Spontaneous magnetization} & M_0 \sim (-\tau)^{1/2} & (\tau < 0) \\ \text{Initial susceptibility} & \chi_0 \sim \tau^{-1} & (\tau > 0) \\ \text{Critical isotherm} & H \sim M^3 & (\tau = 0) \\ \text{Derivative of susceptibility} & \mathrm{d}\chi_0/\mathrm{d}H \sim \tau^{-4} & (\tau > 0). \end{array} \tag{1.42}$$

The specific heat tended to one value C_- on approaching the transition from a lower temperature, and to another C_+ from above. The transcription to appropriate thermodynamic variables can be readily made for other systems.

Landau's paper contained important additional ideas which will be discussed in §4.9, and his treatment of critical fluctuations, which gives the Ornstein–Zernike result (1.37), will be considered in §4.9.1. On the Landau picture it is reasonable to assume that the behaviour (1.37) is also universal.

1.5 The Onsager Revolution

An extremely simple atomic model of ferromagnetism was suggested by Lenz to his student Ising in 1925. Each spin in a lattice can orient parallel or antiparallel to an external field, and interacts only with its nearest neighbours. Does the resultant lattice show a non-zero spontaneous magnetization when the field is reduced to zero? Ising was able to solve the problem only in one dimension, where there is no spontaneous magnetization at any non-zero temperature. In 1928 Heisenberg introduced his vector-coupled model which seemed to have a sounder quantum mechanical basis, and this model commanded the attention of most physicists in the field. But the Ising model continued to attract the interest of theoreticians.

In 1936 Peierls showed that the two-dimensional Ising model did indeed have a spontaneous magnetization. In 1944 Onsager published an exact solution of the partition function of this model for the simple quadratic lattice in zero field. The result was a shattering blow to classical theory. The specific heat (Figure 1.4) was not discontinuous as required by (1.42) but logarithmically infinite (this had been conjectured previously by Kramers and Wannier (1941). But more important, the partition function was non-analytic at T_c so that an expansion of the type used by Landau was completely invalid.

In a subsequent paper with B. Kaufman in 1949, Onsager calculated the correlations but they did not fit in with the results of the Ornstein–Zernike treatment. Finally he was able to calculate the spontaneous magnetization (1949) (later calculated independently by C. N. Yang (1952)) and this was of the form $(-\tau)^{1/8}$ in the critical region, very different from $(-\tau)^{1/2}$ of the classical Weiss theory (1.42) (Figure 1.5).

Experimental evidence on critical behaviour also began to accumulate in conflict with classical predictions. Annake Levelt-Sengers (1974) has pointed out that such evidence for fluids was clearly recognized by Verschaffelt around 1900, but his work did not attract the attention it deserved. In 1945 Guggenheim undertook a critical analysis of experimental data on the coexistence curves of a number of gases. According to van der Waals' theory these gases should obey a law of corresponding states (see Chapter 2), i.e. if reduced units T/T_c, ρ/ρ_c are used (where ρ_c, T_c are the density and temperature at the critical point) the coexistence curves for the different gases should fall on a single universal curve. Guggenheim found that the data provided good support for such a law of corresponding states (Figure 1.6), but the form of the curve was cubic, $\Delta\rho \sim (-\tau)^{1/3}$, rather than quadratic as required by van der Waals' theory.

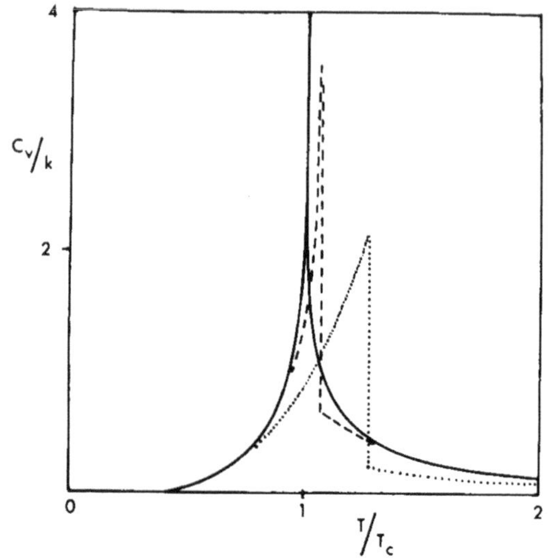

Figure 1.4 Onsager's classic calculation (1944) of the specific heat of the spin $\frac{1}{2}$ Ising model on a simple quadratic lattice. The solid curve represents the true specific heat which is logarithmically infinite. The dotted curve represents the Bethe closed-form approximation, and the dashed curve an improved closed-form approximation due to Kramers and Wannier.

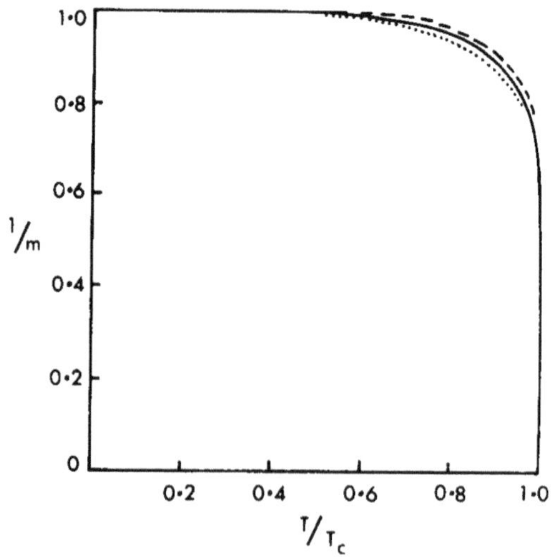

Figure 1.5 Spontaneous magnetization of the two-dimensional Ising model (solid curve, SQ lattice; dashed curve, honeycomb lattice; dotted curve, triangular lattice). The magnetization drops to zero very steeply at T_c ($\sim(T_c - T)^{1/8}$); closed-form approximations give $(T_c - T)^{1/2}$.

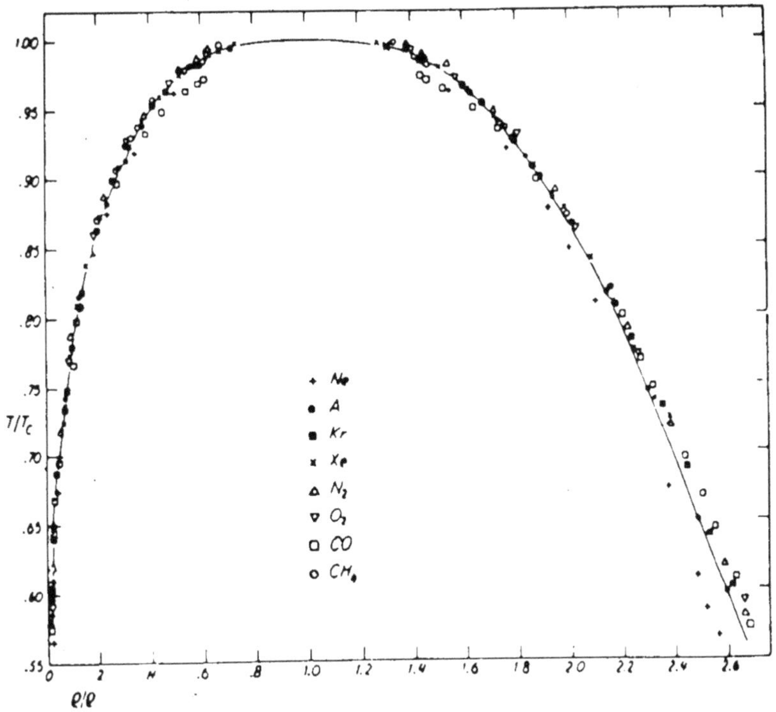

Figure 1.6 Reduced densities of coexisting liquid and gas phases for a number of simple molecular fluids (Guggenheim 1945). The experimental points support a law of corresponding states, but the universal curve is cubic rather than quadratic as required by van der Waals' theory.

In 1954 Habgood and Schneider published precise measurements of the isotherms of xenon near the critical point. They found that the critical isotherm is much flatter than the cubic curve predicted by van der Waals (Figure 1.7), and they suggested that the third and fourth derivatives of P with respect to ρ are zero at the critical point.

Onsager's calculations on the Ising model disagreed with classical theory but also disagreed with experiment. The latter was not unreasonable since the calculations applied only in two dimensions; it was therefore important to obtain theoretical results for three-dimensional systems. But the exact techniques of Onsager were specific to two dimensions, and to the partition function and its derivative in zero field.

During the two decades which followed the publication of Onsager's solution, the critical properties of model systems had to be established on an individual basis – each model and each property entailed a separate calculation. The most useful tool turned out to be the generation of lengthy perturbation

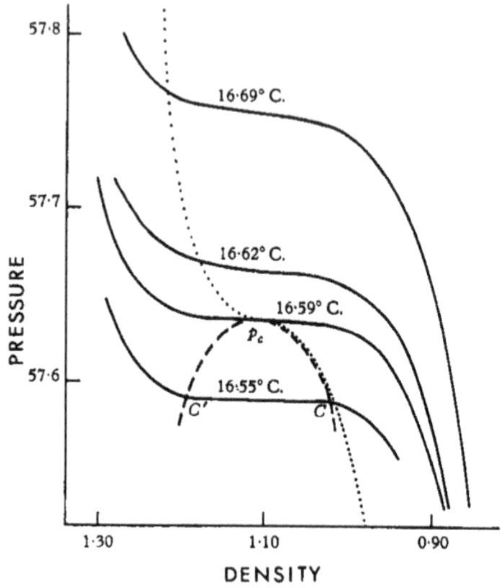

Figure 1.7 Isotherms of xenon near the critical point (Habgood and Schneider 1954). The dashed line marks the region of coexistent phases. The dotted line is the critical isotherm according to van der Waals' equation to be contrasted with the measured 16.59 °C isothermal.

series expansions at high and low temperatures, and the present writer and his research group at King's College, London (particularly M. F. Sykes, M. E. Fisher and their collaborators) were involved in this development.

There were a number of novel features in the series approach (see e.g. Domb and Green 1974). The actual generation of terms required a close familiarity with graph theory, and the programming of sophisticated graphical enumeration problems on computers. Then there were problems of interpretation, and here the Onsager solution served as a valuable guide: it was usual to assume branch point singularities, i.e. that the critical behaviour of specific heat, magnetic susceptibility, etc., is of the form $(1 - T_c/T)^{-\theta}$, and when the terms were consistent in sign, the asymptotic behaviour of the coefficients could be related directly to critical points, exponents and amplitudes. But in many important cases, particularly for three-dimensional models, the coefficients were not consistent in sign, and spurious unphysical singularities masked the true critical behaviour. G. A. Baker's application of the Padé approximant (1961), a piece of mathematics which had lain dormant since the end of the nineteenth century, led to remarkable progress, and sparked off similar applications in many other fields concerned with perturbation expansions (see e.g. Baker and Gammel 1970).

Gradually a body of reliable information was assembled on the critical properties of different theoretical models. Their behaviour differed from classical theory, but the differences were much less marked in three dimensions than in two dimensions. Also the theoretical predictions of critical exponents were much closer to experimental results than those of classical theory (for a review see Fisher (1967)).

These theoretical developments now stimulated a new wave of precise experimental measurements, making use of new techniques and new magnetic materials to attain a much greater accuracy than had been available previously (for a general review see Heller (1967)). In 1957 Fairbank, Buckingham and Kellers published the result of a new investigation into the nature of the λ-point of liquid helium-4. Improved low-temperature techniques enabled them to get very close to the λ-point, and domains and grain boundaries which limit the growth of specific heats of crystals are absent in liquid helium. They concluded that the specific heat is logarithmically infinite on both sides of the λ-point (Figure 1.8) and is in many respects similar to the Onsager specific heat depicted in Figure 1.4. Of course they did not suggest that there was any basic reason for the similarity between a three-dimensional quantum fluid and a two-dimensional Ising ferromagnet, but at least they demonstrated that logarithmically infinite specific heats are a practical reality.

Conventional ferromagnets like iron, nickel and cobalt have high Curie temperatures and accurate measurements of critical behaviour are difficult. However, new ferromagnets were discovered like europium sulfide ($T_c \sim 16.50$

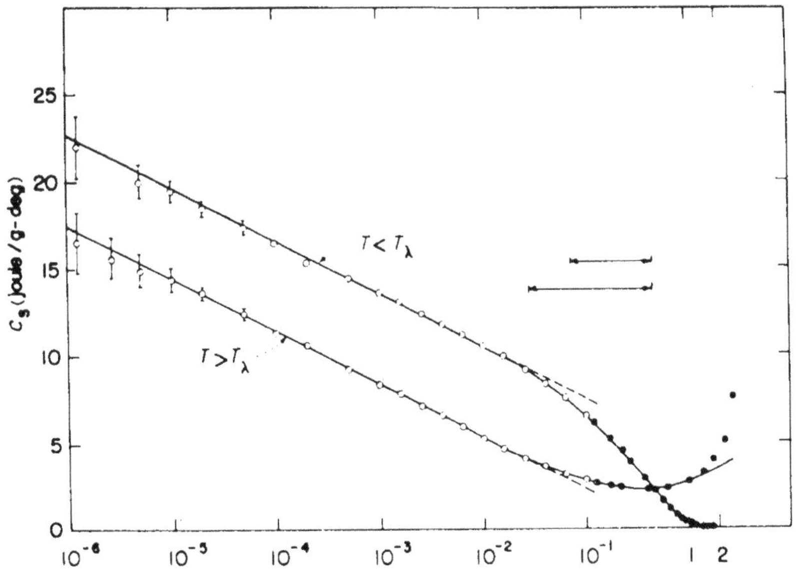

Figure 1.8 Accurate measurements of the specific heat of liquid helium as a function of $\log|T - T_\lambda|$. The solid lines represent the empirical equation $C_s = 4.55 - 3.00 \log_{10}|T - T_\lambda| - 5.20\Delta$ ($\Delta = 0$, $T < T_\lambda$; $\Delta = 1$, $T > T_\lambda$) (from Buckingham and Fairbank 1961).

K) which are much more amenable to accurate measurements, and nuclear magnetic resonance can be used to measure the fraction of spins oriented in a particular direction at a given temperature, and hence to deduce the spontaneous magnetization. Figure 1.9 is an example of such a measurement by Heller and Benedek in 1965 from which they deduced that the cube of the spontaneous magnetization is very nearly linear. (They had previously (1962) obtained a similar result for the sublattice magnetization of the antiferromagnet manganese fluoride, MnF_2.)

It was gratifying to find the new experimental results fitting reasonably well to the theoretical calculations, even though it was well understood that the theoretical models were simple and crude, and did not adequately map the true physical interactions. (For a review of experiments on model systems see Domb and Miedema (1964), and de Jongh and Miedema (1974).) It seemed as if critical behaviour was insensitive to the details of the interaction mechanism.

As the available theoretical data increased certain regularities were noted empirically. In a review article in 1960 Domb noted that critical behaviour depended very significantly on dimension, but rather little on lattice structure in a given dimension. This was borne out by further exact calculations in two dimensions. Extensions of Onsager's work to other lattices gave identical results for critical exponents, and small differences in critical amplitudes. Domb and Sykes (1962) noted that critical exponents varied from the Ising to the Heisenberg model but seemed to be independent of spin value for a given model. (Later Jasnow and Wortis (1968) suggested more generally that critical exponents might be determined by the symmetry of the ordered state.)

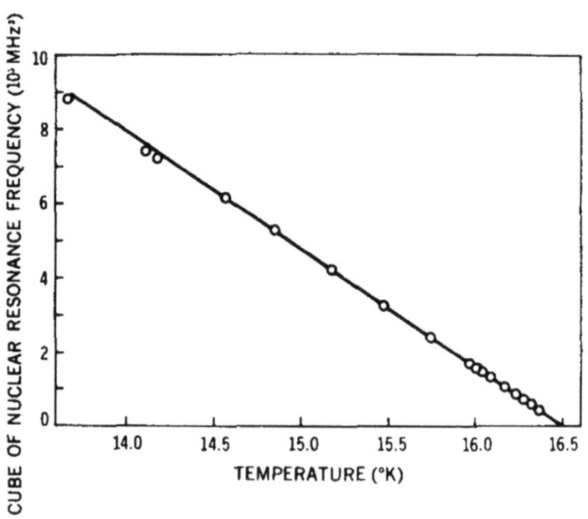

Figure 1.9 Temperature dependence of the cube of ^{153}Eu in EuS nuclear resonance frequency between 13.60 and 16.33 K. In this range the law $M_0 \sim D(1-t)^{1/3}$ is accurately obeyed (Heller and Benedek 1965).

An important question raised in the early 1960s was the relationship between the exponents characterizing critical behaviour for different thermodynamic properties. Were they independent or did they satisfy any relations or restrictive conditions? In studying the properties of a particular model introduced by Fisher, the droplet model, Essam and Fisher (1963) observed that the critical exponent α' characterizing the low-temperature specific heat ($C_H \sim (-\tau)^{-\alpha'}$), β the spontaneous magnetization ($M_0 \sim (-\tau)^{\beta}$) and γ the initial high-temperature susceptibility ($\chi_0 \sim \tau^{-\gamma}$), satisfied the exact relation

$$\alpha' + 2\beta + \gamma = 2. \tag{1.43}$$

They noted that this relation was also satisfied by the exponents of the two-dimensional Ising model.

This work stimulated Rushbrooke (1963) to investigate whether anything could be said purely from thermodynamics (which would then apply to all models). He found that there was indeed a thermodynamic relation

$$\alpha' + 2\beta + \gamma \geq 2. \tag{1.44}$$

Several other thermodynamic inequalities were discovered by other investigators (for more details see e.g. Stanley (1971)).

Two other exact calculations of the partition function for particular model systems became available during this period. The first considered the effect of weak long-range interactions, conveniently characterized in dimension d by

$$\operatorname*{Lt}_{\lambda \to 0} J\lambda^d \exp(-\lambda r). \tag{1.45}$$

The strength of the total interaction in finite as $\lambda \to 0$. For a continuum fluid model Kac, Uhlenbeck and Hemmer (1963) found that they reproduced exactly in one dimension* the results of van der Waals *with the associated Maxwell construction.* Lattice models considered by Kac, Hemmer, Baker and Siegert (for details see review by Hemmer and Lebowitz (1976)) reproduced exactly the results of Weiss for ferromagnets and of Bragg and Williams for alloys. Classical theory was thus partially reinstated as much more than an arbitrary empirical approximation. It is a valid theory for very long-range forces, but does not accord with experiment because intermolecular forces in nature are usually of shorter range.

The second was a mathematical exercise by Berlin and Kac in 1952 who considered an Ising model in which the spins could have any value subject to the condition that the sum of their squares remained equal to N. They were able to provide an exact solution for this, the 'spherical model', but the interaction seemed to bear no relation to physical reality. Nevertheless we shall find that this model, in common with other models with no possibility of

* The $(v - b)$ excluded volume term is correct only in one dimension. In two and three dimensions the appropriate hard-sphere partition function must be used (see §2.3).

physical realization, has played a significant part in achieving a proper understanding of critical behaviour.

1.6 Reconciliation

The first international conference devoted specifically to critical phenomena took place in April 1965 in Washington at the National Bureau of Standards. It could reasonably be termed the founding conference of critical phenomena, since for the first time all the different strands of the subject were woven into a coherent fabric, and the major obstacles to progress were clearly delineated. The host to the conference, and major driving force behind its organization, was Mel Green, and together with Jan Sengers he edited the proceedings. NBS Miscellaneous Publication 273, which cost (and perhaps still costs) only two dollars a bound copy, is still worthy of study by anyone seriously interested in the development of the field. The conference set the pattern for other 'get togethers' on critical phenomena which served as important catalysts to progress.

The keynote address was given by G. E. Uhlenbeck, and it is particularly fascinating now to quote the actual words of one of his concluding paragraphs:

> If there is such an universal, but not-classical behaviour, then there must be a universal explanation which means that it should be largely independent of the nature of the forces. The only corner where this can come from is I think the fact that the forces are not long range. The Onsager solution gives I think a strong hint. It may well be so that away from the critical points the classical theories give a good enough description but that they fail close to the critical point where the substance remembers so to say Onsager. I think that to show something like this is the central theoretical problem. One can call it the reconciliation of Onsager with van der Waals.

Uhlenbeck was looking for a new type of universality for short-range forces which would replace the universality of the classical theories which are valid only for very long-range forces. The description of this new universality is one of the aims of this monograph, and will be undertaken in detail in later chapters. For the rest of this chapter we summarize briefly the major items of progress contributing to the establishment of the modern theory.

The first requirement in effecting the reconciliation was a coherent description of the 'non-classical universality' analogous to the van der Waals description of classical behaviour. Such a description in the form of a non-classical equation of state valid in the critical region emerged within a few months of the above conference. It was suggested independently by three different groups; Widom (1965) in the USA who searched for a generalization of van der Waals' equation which could accommodate non-classical exponents;

Domb and Hunter (1965) in the UK as a result of an analysis of the behaviour of series expansions of higher derivatives with respect to magnetic field at the critical point; and Patashinskii and Pokrovskii (1966) in the USSR who considered the behaviour of multiple correlations near the critical point. The results were tied together neatly by Griffiths (1967) in the form*

$$H = M^{\delta} h(\tau M^{-1/\beta}) \tag{1.46}$$

as the equation of state of a ferromagnet. (For a fluid H was replaced by $P - P_c$ or $\mu - \mu_c$ where μ is the chemical potential, and M by $v - v_c$ or $\rho - \rho_c$.) Here β and δ are two parameters which determine all the exponents, and $h(x)$ is an analytic function. Classical theory corresponds to $\delta = 3$, $\beta = \frac{1}{2}$, $h(x)$ linear. Non-classical results could be accommodated with different values of δ, β and a different function $h(x)$.

The two characteristic features of equation (1.46) are first that critical relations between exponents like (1.43) suggested by Essam and Fisher are indeed satisfied exactly; second that critical data satisfy a 'scaling' relation, i.e. if $HM^{-\delta}$ is plotted against $tM^{-1/\beta}$, the two-dimensional data will all fall on a single curve $h(x)$. These two predictions were subjected to experimental test for a large number of systems, both fluids and magnets, and were found to be well satisfied by experimental data (e.g. Figure 1.10).

But equation (1.46) says nothing about possible values of β, δ and $h(x)$. We have already mentioned empirical information accumulating about regularities in the pattern of critical exponents. These were collected together in the *hypothesis of universality* put forward by Kadanoff in 1970 at a summer school on critical phenomena (Kadanoff 1971); an analogous *smoothness postulate* was advanced independently by Griffiths (1970). This non-classical universality is more sophisticated than had been envisaged by Uhlenbeck: different classes are defined by the space dimension d, and the spin dimension† n, and within a given class critical behaviour is universal. If a third parameter σ is introduced to take account of the range of the intermolecular forces, both classical and non-classical behaviour can be taken into account in a wider pattern of universality. Once d, n and σ have been specified the exponents δ, β and the function $h(x)$ are determined, and the behaviour is smooth and universal within the class. Changes in critical exponents with accompanying discontinuities occur in the *cross-over* between one universality class and another (e.g. in the transition from a two-dimensional to a three-dimensional system by the introduction of a new interaction).

Two highly theoretical ideas were now pursued which, although remote from physical reality, were subsequently to make an important contribution. Stanley (1968) investigated the behaviour of Ising models as a function of spin

* This form is valid only for positive M. More precisely the function on the RHS of (1.46) should be written $M\,|\,M\,|^{\delta-1}h(t\,|\,M\,|^{-1/\beta})$ to ensure the correct $(H\text{–}M)$ symmetry for negative M.

† The letter D was introduced originally to denote spin dimension by Stanley (1968). But Wilson subsequently used n, and this has now become widely accepted.

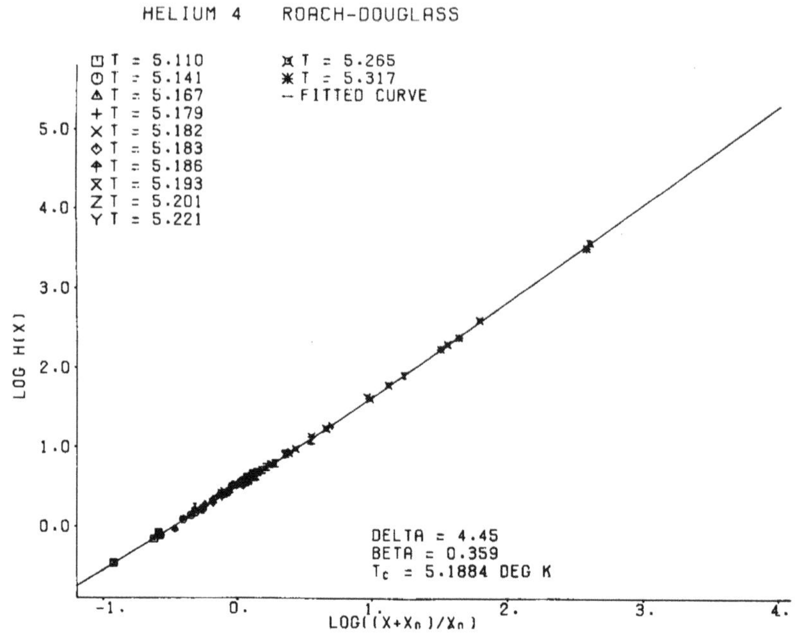

Figure 1.10 Scaling relation for the fluid helium-4 (Vincentini-Missoni 1971). Experimental data from Roach (1968).

dimension n. An increase of n gives greater freedom to the spin and corresponds to a decrease of 'cooperative strength'. He found that as $n \to \infty$ the spherical model solution (Berlin and Kac 1952) is retrieved. Hence this model has a place in the general framework.

Joyce (1966) had already examined the behaviour of the spherical model with varying dimensions and ranges of force. He found that classical behaviour results with long-range forces in low space dimensions d, but also with short-range forces if $d \geq 4$ (with possible logarithmic correction terms). Hence one could reasonably conclude that the same result will also hold for finite n since the cooperation is then stronger. This conclusion was demonstrated convincingly by diagrammatic methods (Larkin and Khmelnitskii 1969) which give precise results when $d \geq 4$.

We have noted that the behaviour of critical correlations as determined from model calculations differs from the classical results of Ornstein and Zernike (Fisher and Burford 1967). But the non-classical correlation exponents and functions fit in with the universality class picture above. Are the correlation exponents determined by δ and β of equation (1.46)? In regard to this question the treatment of Patashinskii and Pokrovskii (1966) went further than the others mentioned above and suggested the relation

$$d\nu = \beta(\delta + 1) \tag{1.47}$$

where ν is the non-classical exponent which characterizes the range of correlation ($\kappa \sim \tau^{\nu}$) in (1.37). But the argument was put more cogently by Kadanoff (1966) in a key paper in which he tried to find a theoretical basis for the scaling properties which had been discovered in critical behaviour.

Kadanoff noted that near T_c the coherence length ξ becomes very large, and hence it is possible to find a length L large compared with the lattice spacing but small compared with ξ. He then considered replacing the interaction of individual spins by the interaction of blocks of L^d spins. One might perhaps expect that in each block the spins would be nearly all up or nearly all down, and the original Ising model of spin σ_i and interaction J could then be replaced by a new model with block spin $\tilde{\sigma}_i$ and interaction $\tilde{J}$. If the new block spin model was effectively the same as the original model, the free energies would be related by

$$f(h, \tau) = L^{-d} f(\tilde{h}, \tilde{\tau}) \quad (h = \beta H). \tag{1.48}$$

But how would $\tilde{h}$ and $\tilde{\tau}$ be related to h and τ? Kadanoff suggested that

$$\tilde{h} = L^x h, \quad \tilde{\tau} = L^y \tau. \tag{1.49}$$

It is then easy to show that all the exponents (including ν) can be expressed in terms of x and y, and that both (1.46) and (1.47) result.

Relation (1.47) is satisfied by the Ising model in two dimensions, yet numerical calculations for the three-dimensional model showed a small but persistent discrepancy. Also there are several features of Kadanoff's argument which do not stand up to critical examination. But his ideas stimulated the next major theoretical advance which completed the reconciliation sought by Uhlenbeck.

1.7 Renormalization Group: Respectability

In a talk on 'The Curie Point' given in honour of Uhlenbeck's 70th birthday in October 1969 (Domb 1971), I summarized the current situation as follows:

> From the point of view of practical calculations of the behaviour near the Curie point we are well on the way to satisfying the needs of the experimentalist for most models of interest. Unifying features have been discovered which suggest that the critical behaviour of a larger variety of theoretical models can be described by a simple type of equation of state. But the rigorous mathematical theory needed to make the above developments 'respectable' is still lacking.

The application of the renormalization group (RG) which K. G. Wilson introduced so effectively a year or two later was not rigorous by mathematical standards, and the search for a proper mathematical description of his ideas still continues. But he was able to account convincingly for the striking empirical discoveries of the previous section, and the 'respectability' which his work achieved is adequate for most theoreticians in the field.

A suggestion that the RG could be relevant to critical phenomena was made at the summer school which Mel Green organized in 1970 in Varenna (de Pasquale, di Castro and Jona-Lasinio 1971). However, no precise indication was forthcoming as to how it should be used.

The RG had been developed nearly 20 years earlier (Stueckelberg and Peterman 1953, Gell-Mann and Low 1954), in connection with field theory, but it had no apparent practical consequences, and had not been taken very seriously. Wilson saw that the theory was capable of providing an understanding of universality, and a framework for the detailed calculation of critical behaviour. He acknowledged his debt to Kadanoff for setting his thoughts in the right direction, but he converted Kadanoff's vague block spin mappings into a precise tool for calculation.

Wilson's work sparked off a flood of new ideas and projects (for more details see the reviews and texts by Fisher (1974), Wilson (1975), Domb and Green (1976), Ma (1976), Pfeuty and Toulouse (1977)). Anyone interested in the development of scientific concepts is particularly recommended to read Wilson's first two papers (1971) and his first definitive review (Wilson and Kogut 1974) where he describes the evolution of his own ideas, explains the philosophy of the RG approach, and shows with simple illustrated examples how it can be applied exactly to simple artificial models and approximately to real systems.

The RG is a transformation of the original Hamiltonian with N degrees of freedom into a new Hamiltonian $\mathscr{H}'$

$$\mathscr{H}' = \mathbf{R}[\mathscr{H}] \tag{1.50}$$

with a reduced number of degrees of freedom N'. The transformation is chosen so that the partition function is preserved,

$$Z_{N'}(\mathscr{H}') = Z_N(\mathscr{H}). \tag{1.51}$$

A wide choice of operators $\mathbf{R}$ is possible satisfying these conditons. The block spin device of Kadanoff can be used by eliminating the internal interactions in the block in some suitable manner, or half the spins can be eliminated by partial summation in a *decimation* procedure. Choices of this type are usually referred to as 'real-space renormalization'. Alternatively, the transformation can be carried out in momentum space, and high-momentum variables corresponding to short-range fluctuations can be integrated out.

The most important feature of the transformation is that it can be iterated,

$$\mathscr{H}' = \mathbf{R}[\mathscr{H}], \quad \mathscr{H}'' = \mathbf{R}[\mathscr{H}'], \quad \ldots \tag{1.52}$$

and the universality properties follow from the limiting behaviour of such iterative processes. Elementary iterative processes of the form

$$x_{n+1} = f(x_n) \tag{1.53}$$

are well known in numerical analysis (see e.g. Hartree 1958) as a useful method of approximating to a root of the equation

$$x = f(x). \tag{1.54}$$

They have the advantage that the final solution is within wide limits independent of the starting point. A root of (1.54) is called a *fixed point*, x^* of the transformation (1.53). If equation (1.54) has a number of different roots, each of these fixed points will have a characteristic 'range of attraction', i.e. a range of starting points within which convergence will occur to the particular root.

Equation (1.52) involves the generalization of the above process (1.53) to a multidimensional space

$$\mathbf{K}_{n+1} = f(\mathbf{K}_n) \tag{1.55}$$

corresponding to a number of parameters in the Hamiltonian, but many of the general characteristics remain the same, including the possibility of convergence to fixed points $\mathbf{K}^*$.

We can now see qualitatively how the above properties provide a basis for the explanation of universality classes. Each fixed point $\mathbf{K}$ corresponds to one universality class, and Hamiltonians with a wide variety of parameters all converge to the same fixed point.†

Critical behaviour is determined by the behaviour of (1.55) near $\mathbf{K}^*$, and if a linear expansion is undertaken near this point, a linear operator can be derived whose eigenvalues are related to critical exponents. A scaling equation of state equivalent to (1.46) can readily be deduced. The non-analytic critical behaviour arises from analytic functions $f(\mathbf{K})$ as a result of the limiting procedure as the number of iterations tends to infinity.

Although Kadanoff's idea led naturally to real-space renormalization, the major progress in actual calculations was achieved by applying perturbation methods which had been developed in quantum field theory to momentum-space renormalization. To make this application possible Wilson went back to Landau's ideas, not in the original macroscopic form of equation (1.39), but in the later microscopic Ginzburg–Landau formulation (1950). This corresponds to a continuous spin with local Hamiltonian

$$\mathscr{H} = (\nabla s)^2 + Rs^2 + Us^4 - Hs \tag{1.56}$$

where R and U are analytic functions of T. Few of the older workers in the field (remembering the failure of the macroscopic Landau theory) would have been prepared to believe that an expansion of the form (1.56) was sufficient to lead to non-classical results of Ising model type. But Wilson was able to indicate that any higher-order terms in s^6, s^8, etc., make no significant contribution to critical behaviour (the coefficients are *irrelevant parameters*).

† The difference between stable, unstable and partially unstable fixed points will be discussed in detail in Chapter 7.

The parameter chosen (Wilson and Fisher 1972) for the expansion of critical exponents and scaling functions was

$$\epsilon = 4 - d \tag{1.57}$$

where d is the space dimension, following earlier ideas by Fisher and Gaunt (1964) which had seemed remote from reality when they were published. $\epsilon = 0$ corresponds to classical theory.

Only a few terms of the ϵ expansion were available; it turns out to be an asymptotic expansion, and the value $\epsilon = 1$ corresponding to three-dimensional systems is not small. But the results were in satisfactory agreement with those of series expansions, and later work has derived more terms and devised improved methods of summation (Le Guillou and Zinn-Justin 1977). As mentioned above, there were some small discrepancies with the results of series expansions, but these have now been ironed out with moderate success (Nickel (1981) and § 7.8).

Real-space renormalization methods worked well in two dimensions, and were able to reproduce the Onsager solution to a high degree of accuracy (see e.g. Niemeijer and van Leeuwen 1976). More recently methods have been devised by which they can be applied in three dimensions, and they have contributed accurate numerical information.

Wilson has described his own work as a second stage of the Landau theory, and has shown precisely where the original Landau theory breaks down (Wilson 1974). The microscopic Hamiltonian (1.56) is correct. But the macroscopic form of the free energy obtained by averaging over a region

$$F = \int d^3\mathbf{x}\{[\nabla M(x)]^2 + RM^2(x) + UM^4(x) - B(x)M(x)\} \tag{1.58}$$

is incorrect since it ignores the variation of R and U with L, the size of the region, which is non-analytic (as had already been suggested previously by Kadanoff (1966)).

The RG theory did not stop at explaining the empirical results of the previous section. Like all good scientific theories it suggested new avenues of exploration, and was able to deal with problems for which previous methods had not been successful. From the point of view of this monograph the most notable are the corrections to the equation of state (1.46) (Wegner 1972, Le Guillou and Zinn-Justin 1977, 1980 and §7.5.7) which incidentally enabled critical behaviour as derived from series expansions to be interpreted more precisely; and systems with long-range dipolar interactions (see Aharony (1976) and §7.5.6) for which the derivation of series expansions is an extremely difficult task.

The RG method has been applied with success to a number of other problems in condensed matter physics. The award of the 1982 Nobel prize in physics to Kenneth Wilson is an appropriate recognition by the wider physics community of the significance of his ideas. His own account of their development is given in his Nobel address (Wilson 1983).

General Notation (in Chapter 1 and throughout the book)

Thermodynamics and Statistical Mechanics

P pressure, T temperature, V volume, ρ density.
N number of molecules in assembly, $V/N = v = 1/\rho$.
U internal energy, S entropy, $U/N = u$, $S/N = s$.
$F(T,V,N)$ free energy of Helmholtz, $F/N = f$.
$G(T,P,N)$ free energy of Gibbs, $G/N = g$.
$\psi = -F/T$ (Planck function), $\Phi = PV/T$.
C_P specific heat at constant pressure $= T(\partial S/\partial T)_P$, C_V specific heat at constant volume $= T(\partial S/\partial T)_V$, K_T isothermal compressibility $= -1/V(\partial V/\partial P)_T = 1/\rho(\partial \rho/\partial P)_T$.
ε refractive index, λ wavelength, ν frequency.
ν_i number of molecules of species i, μ_i chemical potential of species i.
$\mathscr{H}$ Hamiltonian, $\mathbf{x}_i$ position of ith molecule, $\mathbf{p}_i$ momentum of ith molecule.
$\phi(\mathbf{r}_{ij}) = \phi(\mathbf{x}_i - \mathbf{x}_j)$ interaction potential between molecules at $\mathbf{x}_i$ and $\mathbf{x}_j$.
ϵ_r typical molecular energy level, E total energy of assembly.
$Z_N = Z(T,V,N)$ partition function (p.f.).
$\zeta = \zeta(T,V,\mu_i)$ grand partition function (g.p.f.).
h Planck's constant, k Boltzmann's constant, $\beta = 1/kT$.
$\langle\ \rangle$ average over a probability distribution, in statistical mechanics over a Boltzmann distribution.
W probability of a macrostate, Ω number of complexions in a macrostate.

Magnetism

H magnetic field, $h = \beta H$, M magnetization, $M/N = m$.
$A(T,M) = U - TS$, $A/N = a$, $F(T,H) = A - HM$.
χ_T isothermal susceptibility $= (\partial M/\partial H)_T$.
χ_0 initial isothermal susceptibility at $H = 0$.
C_H specific heat at constant field $= T(\partial S/\partial T)_H$.
C_M specific heat at constant magnetization $= T(\partial S/\partial T)_M$.

Critical Phenomena

T_c critical temperature, $t = T/T_c$,

$$\tau = T/T_c - 1 \quad (T > T_c)$$

$$\tau' = 1 - T/T_c \quad (T < T_c).$$

σ_i a variable representing spin in Ising model of spin $\frac{1}{2}$ taking values ± 1.

s_i spin variable in general Ising model of spin s, taking values $-s$, $-(s-1)$, $\ldots, s$.
$s(x)$ spin variable with continuum values.
J interaction energy between neighbouring spins in Ising model.
Notation for critical exponents is given in Table 5.4.
d space dimension, n spin dimension.
S long-distance order in Bragg–Williams theory.
η order parameter in Landau theory.
ξ coherence length $= \kappa^{-1}$.
$h(x)$ analytic function arising in scaling form of equation of state (1.46).

References

AHARONY, A. (1976) Ch. 6 in *Phase Transitions and Critical Phenomena*, Vol. 6, ed. C. DOMB and M. S. GREEN, London and New York: Academic Press.
ANDREWS, T. (1869) *Philos. Trans. R. Soc.* **159**, 575.
BAIN, E. C. (1923) *Trans. Am. Inst. Min. (Metall.) Eng.* **68**, 625.
BAKER, G. A. (1961) *Phys. Rev.* **124**, 768.
BAKER, G. A. and GAMMEL, J. L. (1970) *The Padé Approximant in Theoretical Physics*, New York and London: Academic Press.
BARRETT, W. F. (1874) *Philos. Mag.* **47**, 51.
BAUER, C. (1880) *Wiedemann Ann. Phys. Chem.* **11**, 394.
BERLIN, T. H. and KAC, M. (1952) *Phys. Rev.* **86**, 821.
BETHE, H. A. (1935) *Proc. R. Soc. A* **216**, 45.
BOLTZMANN, L. (1868) *Sitz. K. Akad. Wiss. Wien, Math., Natur Klasse* **58**, 517.
BOLTZMANN, L. (1871) *Wien. Ber.* **63**, 679.
BOLTZMANN, L. (1872) *Wien. Ber.* **66**, 275 (English translation, S. G. BRUSH (1965) *Kinetic Theory*, Vol. 2, Oxford: Pergamon Press, p. 88).
BOLTZMANN, L. (1877) *Wien. Ber.* **76**, 373.
BOLTZMANN, L. (1884) *Wien. Ber.* **89**, 714.
BORELIUS, G. (1934) *Ann. Phys., Leipzig* **20**, 57.
BRAGG, W. L. and WILLIAMS, E. J. (1934) *Proc. R. Soc. A* **145**, 699.
BRAGG, W. L. and WILLIAMS, E. J. (1935) *Proc. R. Soc. A* **151**, 540.
BRUSH, S. G. (1965) *Kinetic Theory*. Oxford: Pergamon Press.
BRUSH, S. G. (1983) *Statistical Physics and the Atomic Theory of Matter*, Princeton, NJ: Princeton University Press.
BUCKINGHAM, M. J. and FAIRBANK, W. M. (1961) *Prog. Low Temp. Phys.* **3**, 80.
BUYS-BALLOT, C. H. D. (1858) *Ann. Phys., Leipzig* **103**, 240.
CERNUSCHI, F. and EYRING, H. (1939) *J. Chem. Phys.* **7**, 547.
CLAUSIUS, R. (1857) *Ann. Phys., Leipzig* (Ser. 2) **100**, 353; (1858) **105**, 239.
CLAUSIUS, R. (1870) *Ann Phys., Leipzig* **141**, 124 (English translation *Philos. Mag.*, 1870, **40**, 122).
CURIE, P. (1895) *Ann. Chim. Phys.* **5**, 289.
DEHLINGER, U. (1930) *Z. Phys.* **64**, 359.
DEHLINGER, U. (1932) *Z. Phys.* **74**, 267.
DEHLINGER, U. (1933) *Z. Phys.* **83**, 832.
DE JONGH, L. J. and MIEDEMA, A. R. (1974) *Adv. Phys.* **23**, 1.

DE PASQUALE, F., DI CASTRO, C. and JONA-LASINIO, G. (1971) *Proc. Enrico Fermi School on Critical Phenomena*, ed. M. S. GREEN, New York and London: Academic Press, p. 123.
DOMB, C. (1960) *Adv. Phys.* **9**, 149, 245.
DOMB, C. (1971) *Statistical Mechanics at the Turn of the Decade*, ed. E. G. D. COHEN, New York: Marcel Dekker.
DOMB, C. and GREEN, M. S. (1974) *Phase Transitions and Critical Phenomena*,* Vol. 3 Academic Press, and (1976), Vol. 6, London and New York: Academic Press.
DOMB, C. and HUNTER, D. L. (1965) *Proc. Phys. Soc.* **86**, 1147.
DOMB, C. and MIEDEMA, A. R. (1964) *Progress in Low Temperature Physics*, ed. C. J. GORTER, Ch. 4, Amsterdam: North-Holland.
DOMB, C. and SYKES, M. F. (1962) *Phys. Rev.* **128**, 168.
EHRENFEST, P. (1933) *Proc. Kon. Akad. Wetenschap, Amsterdam* **36**, 147.
EINSTEIN, A. (1910) *Ann. Phys., Leipzig* **33**, 1276.
ESSAM, J. W. and FISHER, M. E. (1963) *J. Chem. Phys.* **38**, 147.
FAIRBANK, W. M., BUCKINGHAM, M. J. and KELLERS, C. F. (1957) *Proc. 5th Int. Conf. on Low Temperature Physics*, University of Wisconsin Press.
FARADAY, M. (1839–55) *Experimental Researches in Electricity*, (Vol. 3, pp. 54–6) 2343–7.
FISHER, M. E. (1964) *J. Math. Phys.* **5**, 944.
FISHER, M. E. (1967) *Rep. Prog. Phys.* **30**, 615.
FISHER, M. E. (1974) *Rev. Mod. Phys.* **46**, 597.
FISHER, M. E. and BURFORD, R. J. (1967) *Phys. Rev.* **156**, 583.
FISHER, M. E. and GAUNT, D. S. (1964) *Phys. Rev.* **133A**, 224.
GELL-MANN, M. and LOW, F. E. (1954) *Phys. Rev.* **95**, 1300.
GIBBS, J. W. (1889) *Proc. Am. Acad.* **16**, 458 (Collected Works, Longmans Green, New York, 1928, Vol. 2, p. 261, reprinted Dover, New York, 1961).
GIBBS, J. W. (1902) *Elementary Principles in Statistical Mechanics*, New York: Scribner's, and London: Arnold; reprinted Dover, New York, 1960.
GILBERT, W. (1600) *De Magnete* (translated by P. F. MOTTELAY, New York, 1893, and for the Gilbert Club, London, 1900), p. 66.
GINZBURG, V. L. and LANDAU, L. D. (1950) *JETP* **20**, 1064.
GORSKY, W. (1928) *Z. Phys.* **50**, 84.
GRIFFITHS, R. B. (1967) *Phys. Rev.* **158**, 176.
GRIFFITHS, R. B. (1970) *Phys. Rev. Lett.* **24**, 1479.
GUGGENHEIM, E. A. (1945) *J. Chem. Phys.* **13**, 253.
HABGOOD, H. W. and SCHNEIDER, W. G. (1954) *Can. J. Chem.* **32**, 98.
HARTREE, D. R. (1958) *Numerical Analysis*, Oxford p. 211.
HEISENBERG, W. (1928) *Z. Phys.* **49**, 619.
HELLER, P. (1967) *Rep. Prog. Phys.* **30**, 731.
HELLER, P. and BENEDEK, G. (1962) *Phys. Rev. Lett.* **8**, 428.
HELLER, P. and BENEDEK, G. (1965) *Phys. Rev. Lett.* **14**, 71.
HEMMER, P. C. and LEBOWITZ, J. L. (1976) DG **5b**, Ch. 2.
HOPKINSON, J. (1889) *Philos. Trans. R. Soc. A* **180**, 443.
ISING, E. (1925) *Z. Phys.* **31**, 253.

* Hereinafter abbreviated to DG.

Jasnow, D. and Wortis, M. (1968) *Phys. Rev.* **176**, 739.
Johansson, C. H. and Linde, J. O. (1925) *Ann. Phys., Leipzig* **78**, 439.
Joyce, G. S. (1966) *Phys. Rev.* **146**, 349.
Kac, M., Uhlenbeck, G. E. and Hemmer, P. C. (1963) *J. Math. Phys.* **4**, 216.
Kadanoff, L. P. (1966) *Physica* **2**, 263.
Kadanoff, L. P. (1971) *Proc. Enrico Fermi School on Critical Phenomena*, ed. M. S. Green, New York and London: Academic Press, p. 100.
Kaufman, B. and Onsager, L. (1949) *Phys. Rev.* **76**, 1244.
Kramers, H. A. and Wannier, G. H. (1941) *Phys. Rev.* **60**, 252, 263.
Landau, L. D. (1937) *Phys. Z. Sow.* **11**, 26, 545 (English translation (1965) *Collected Papers of L. D. Landau*, ed. D. ter Haar, Oxford: Pergamon Press, pp. 193–216) (see also Landau, L. D. and Lifshitz, E. M. (1959) *Statistical Physics*, Oxford: Pergamon Press).
Langevin, P. (1905) *J. Phys.* **4**, 678; *Ann. Chim. Phys.* **5**, 70.
Larkin, A. I. and Khmelnitskii, D. E. (1969) *Sov. Phys. JETP* **29**, 1123.
Le Guillou, J. C. and Zinn-Justin, J. (1977) *Phys. Rev. Lett.* **39**, 95.
Le Guillou, J. C. and Zinn-Justin, J. (1980) *Phys. Rev. B* **21**, 3976.
Levelt-Sengers, J. M. H. (1974) *Proc. Van der Waals Centennial Conf. on Statistical Mechanics*, Amsterdam: North Holland, p. 73.
Ma, S. K. (1976) *Modern Theory of Critical Phenomena*, Reading, PA: Benjamin.
Maxwell, J. C. (1859) *Report of the 29th Meeting of the British Association*; reproduced in *The Letters and Papers of James Clerk Maxwell*, Vol. I, ed. P. M. Harman, Cambridge: Cambridge University Press, 1990, p. 615.
Maxwell, J. C. (1871) *Theory of Heat*, London and New York: Longmans Green.
Maxwell, J. C. (1874, 1875) *Nature*, Vols 10, 11 (see *The Scientific Papers of J. C. Maxwell*, ed. W. D. Niven, Vol. 2, Cambridge, 1890, p. 407, 418; reprinted Dover, New York, 1965).
Nickel, B. (1981) *Proc. 14th Int. Conf. on Statistical Mechanics, Edmonton, Canada, Physica* **106A**, 48.
Niemeijer, Th. and van Leeuwen, J. M. J. (1976) DG **6**.
Onsager, L. (1944) *Phys. Rev.* **65**, 117.
Onsager, L. (1949) *Proc. Florence Conference on Statistical Mechanics, Il Nuovo Cimento* **6**, 261.
Ornstein, L. S. and Zernike, F. (1914) *Proc. Akad. Sci. Amsterdam* **17**, 793.
Ornstein, L. S. and Zernike, F. (1916) *Proc. Akad. Sci. Amsterdam* **18**, 1520 (reproduced in *The Equilibrium Theory of Classical Fluids*, ed. A. L. Frisch and L. J. Lebowitz, Reading, PA: Benjamin, 1964).
Patashinskii, A. Z. and Pokrovskii, V. L. (1966) *Zh. Eksp. Teor. Fiz.* **50**, 439 (*Sov. Phys. JETP* **23**, 292).
Peierls, R. E. (1936) *Proc. Cambridge Philos. Soc.* **32**, 477.
Pfeuty, P. and Toulouse, G. (1977) *Introduction to the Renormalization Group and to Critical Phenomena*, London: John Wiley.
Pippard, A. B. (1957) *Classical Thermodynamics*, Ch. 9, Cambridge: Cambridge University Press.
Roach, P. R. (1968) *Phys. Rev.* **170**, 213.
Rowlinson, J. S. (1969) *Nature* **224**, 541.
Rushbrooke, G. S. (1949) *Statistical Mechanics*, Oxford.
Rushbrooke, G. S. (1963) *J. Chem. Phys.* **39**, 842.

SMOLUCHOWSKI, M. (1908) *Ann. Phys., Leipzig* **25**, 205; (1912) *Philos. Mag.* **23**, 165.
STANLEY, H. E. (1968) *Phys. Rev.* **176**, 718.
STANLEY, H. E. (1971) *Introduction to Phase Transitions and Critical Phenomena*, Ch. 4, Oxford: Oxford University Press.
STUECKELBERG, E. C. G. and PETERMAN, A. (1953) *Helv. Phys. Acta* **26**, 499.
TAMMANN, G. (1919) *Z. Anorg. Chem.* **107**, 1.
UHLENBECK, G. E. (1965) *Critical Phenomena*, NBS Misc. Publ. 273, ed. M. S. GREEN and J. V. SENGERS, Washington, DC: NBS, p. 3.
VAN DER WAALS, J. H. (1873) *Over de continuiteit van der gas en vloeisoftoestand*, Thesis, Leiden.
VERSCHAEFFELT, J. E. (1900) *Proc. Kon. Akad. Sci. Amsterdam* **2**, 588.
VINCENTINI-MISSONI, M. (1971) DG **2**, Ch. 2.
WEGNER, F. J. (1972) *Phys. Rev. B* **5**, 4529.
WEISS, P. (1907) *J. Phys.* **6**, 661.
WEISS, P. and KAMERLINGH ONNES, H. (1910) *J. Phys.* **9**, 555.
WIDOM, B. (1965) *J. Chem. Phys.* **43**, 3898.
WILSON, K. G. (1971) *Phys. Rev. B* **4**, 3174, 3184.
WILSON, K. G. (1974) *Physica* **73**, 119.
WILSON, K. G. (1975) *Rev. Mod. Phys.* **47**, 773.
WILSON, K. G. (1983) *Rev. Mod. Phys.* **55**, 583.
WILSON, K. G. and FISHER, M. E. (1972) *Phys. Rev. Lett.* **28**, 240.
WILSON, K. G. and KOGUT, J. (1974) *Phys. Rep.* **12C**, 77.
YANG, C. N. (1952) *Phys. Rev.* **85**, 808.

2

Fluids: Classical Theory

2.1 Thermodynamic Background

Thermodynamics is 'a science with secure foundations, clear definitions, and distinct boundaries'. Thus wrote Maxwell in 1878, and his words were echoed by Gibbs in the obituary tribute to Clausius to which we have referred in the previous chapter. But Gibbs traced the origin of these foundations, definitions and boundaries to the first paper which Clausius published on thermodynamics in 1850 entitled 'Ueber die bewegende Kraft der Wärme, und die Gesetze, welche sich daraus für die Wärmelehre selbst ableiten lassen' ('On the Motive Power of Heat and on the Laws which can be deduced from it for the Theory of Heat').

It is interesting that only seven years elapsed between this paper and the paper quoted previously which laid the foundations of statistical mechanics.

The logical structure and self-contained character of classical thermodynamics are well described in the texts by Pippard (1957) and Callen (1960) and it is on these that we shall base our own discussion. There are two separate parts of the second law of thermodynamics. The reversible aspect states that in any reversible cycle

$$\oint \frac{dQ}{T} = 0 \tag{2.1}$$

(where dQ is the heat absorbed and T the absolute temperature) so that in any reversible change from state A to state B

$$\int_A^B \frac{dQ}{T} \tag{2.2}$$

does not depend on the path from A to B. The integral (2.2) can then be used to define a function of state S, the entropy, satisfying

$$dQ = T\,dS = dU + P\,dV. \tag{2.3}$$

$S(U,V)$ is a fundamental thermodynamic relation from which all the equilibrium thermodynamic properties of the system can be derived. Equation (2.3) is the starting point for the definition of other functions of state which give rise to the basic relations of equilibrium thermodynamics.

The irreversible aspect states that in an irreversible cycle

$$\oint \frac{dQ}{T} < 0. \tag{2.4}$$

From this we deduce that in an irreversible change from A to B

$$\int_A^B \frac{dQ}{T} < S_B - S_A \tag{2.5}$$

and hence that in a thermally isolated system the entropy must increase. This leads to the condition of equilibrium that the entropy must be a maximum if U and V are kept constant. This second aspect is of great importance in phase transitions; it enables conditions of stability to be derived for a particular phase, and conditions of coexistence between phases to be formulated.

Let us first consider the various thermodynamic functions usually introduced. It is important to differentiate between *extensive* variables like volume V, internal energy U and entropy S whose value is proportional to the number of moles in the system, and *intensive* variables like temperature T, pressure P and chemical potential μ whose value is independent of the number of moles present. The term *fields* has been used by Griffiths and Wheeler (1970) to describe the latter variables. It is often useful to convert extensive variables into *densities* by dividing by the number n of moles, e.g. V/n, U/n, S/n or by the number of molecules N, $V/N = v$, $U/N = u$, $S/N = s$. The former is more usual in thermodynamics, the latter in statistical mechanics, and since our concern will be largely with statistical mechanics we shall use the latter. Like fields v, u, s do not depend on the number of molecules in the system. However, fields should be differentiated from densities by the property that they take on identical values for two phases in equilibrium.

The first fundamental relation $S(U,V)$ with differential relations (2.3) uses extensive variables. The equilibrium condition of maximum entropy can be put in the form (combining the reversible and irreversible portions)

$$(\Delta S)_{U,V} \leq 0 \tag{2.6}$$

where the subscripts mean that the variables are kept constant. An alternative fundamental relation with extensive variables is $U(S,V)$

$$dU = T\,dS - P\,dV \tag{2.7}$$

with the equilibrium condition that U must be a minimum (Callen 1960, Ch. 5)

$$(\Delta U)_{S,V} \geq 0. \tag{2.8}$$

It is often convenient in thermodynamics to transform to intensive variables, first replacing the entropy S by temperature T, and then the volume V by pressure P. This is done by means of a Legendre transformation and is represented geometrically in Figure 2.1. The transformation is normally one to one and reversible, and is of the form

$$y = F(x), \quad p = F'(x), \quad G(p) = y - xp. \tag{2.9}$$

Starting with S,V as independent variables and a thermodynamic function $U(S,V)$, we can change from S to T by means of

$$T = \left(\frac{\partial U}{\partial S}\right)_V, \quad F = U - TS = F(T,V) \tag{2.10}$$

$$\mathrm{d}F = -S\,\mathrm{d}T - P\,\mathrm{d}V. \tag{2.11}$$

$F(T,V)$ is the free energy of Helmholtz. A new condition of equilibrium can be derived for a system at constant volume and temperature (i.e. in contact with a heat bath), that $F(T,V)$ must be a minimum,

$$(\Delta F)_{T,V} \geq 0. \tag{2.12}$$

Finally we can change from V to P by means of another Legendre transformation,

$$-P = \left(\frac{\partial F}{\partial V}\right)_T, \quad G = F + PV = G(T,P) \tag{2.13}$$

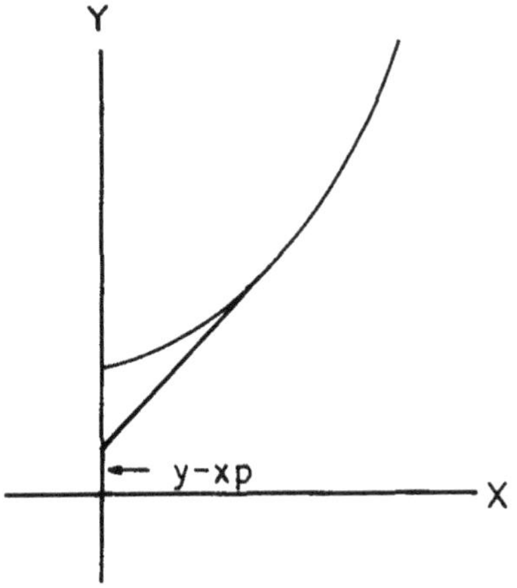

Figure 2.1 Legendre transformation.

$$dG = -S\,dT + V\,dP. \tag{2.14}$$

$G(T,P)$ is the free energy of Gibbs. The condition of equilibrium for a system at constant temperature and pressure can then be derived, that $G(T,P)$ must be a minimum,

$$(\Delta G)_{T,P} \geq 0. \tag{2.15}$$

It is sometimes useful to have as an independent variable the density ρ $(=N/V = 1/v)$ instead of v, and to define other densities per unit of volume instead of per molecule. This is particularly appropriate in pursuing the analogy between fluids and magnets (Griffiths 1967). We define

$$a(T,\rho) = \frac{F(T,V)}{V} = \frac{Nf(T,v)}{V} = \rho f(T,v). \tag{2.16}$$

Then

$$\frac{\partial a}{\partial T} = -\rho s, \quad \frac{\partial a}{\partial \rho} = f - \frac{1}{\rho}\left(\frac{\partial f}{\partial v}\right)_T = f + \frac{P}{\rho} = \mu. \tag{2.17}$$

The condition of equilibrium (2.8) enables us to derive the condition for a phase to be stable. This can be summarized by saying that the function $u(s,v)$ must be convex upwards, i.e. must lie above its tangent. If the condition is not satisfied it will be possible to split into two phases having a lower value of u. We now consider this in more detail.

2.2 Stability of a Phase: First-order Transitions

It was Gibbs who in a memoir published by the Connecticut Academy in 1873 first drew attention to the geometrical properties of the $u(s,v)$ surface. He first defined a *primitive thermodynamic surface* which describes the properties of a homogeneous body in a uniform state throughout, and then proceeded to examine where points on this surface are unstable and will break up into two separate phases. The extensive variables s,v have the advantage that mixtures of two phases in equilibrium in different proportions are represented by different points. In fact if Q,R are two different points on the primitive surface, any mixture of the phases Q and R is represented by a point on the line QR. This does not apply to intensive variables, e.g. to $g(P,T)$.

Gibbs first considered the condition for local stability. If at any point O of the primitive surface a portion of the surface lies below the tangent plane, it will clearly be possible to find a second point Q on this surface near O for which the line OQ lies below the surface. This portion representing a mixture of two phases has a lower value of u than the single phase at O and will therefore be more stable. Thus, if O represents a stable point the primitive surface must lie above the tangent plane at O.

From this property two conditions for the local stability of a phase can be deduced: the specific heat at constant volume* and the isothermal compressibility must be positive,

$$C_v = \left(\frac{\partial u}{\partial T}\right)_v > 0 \tag{2.18}$$

$$K_T = -\frac{1}{v}\left(\frac{\partial v}{\partial P}\right)_T > 0. \tag{2.19}$$

Gibbs next passed to global stability. Even if the tangent plane at O lies completely below the primitive surface near O, it may cut this surface at a point Q removed from O. The line OQ representing a mixture of two phases will again have a lower value of u than at O and will be more stable. Hence the point O is locally stable but not globally stable – this is usually termed *metastable*. To find the mixture of phases with the lowest value of u we must draw the tangent planes at all points of the primitive surface and select lines on them which are tangent at two points O′, O″ of the primitive surface. Such lines generate a ruled surface, and this, together with the original primitive surface, will be termed the *derived thermodynamic surface* (often called the convex envelope of the primitive surface). The ruled portion will be called the *coexistence region*.

Since the tangent planes at O′, O″ of the primitive surface are identical we must have (equating the slopes at O′, O″)

$$T' = T'', \quad P' = P'' \tag{2.20}$$

and equating the points at which they cut the u axis,

$$\mu' = u' - T's' + P'v' = u'' - T's'' + P''v'' = \mu''. \tag{2.21}$$

Equations (2.20) and (2.21) represent the conditions to be satisfied by two phases in equilibrium. From (2.21) by equating $\mu' + d\mu'$ and $\mu'' + d\mu''$ and expressing μ as a function of (P,T) it is easy to derive the Clausius–Clapeyron equation relating the equilibrium at (P,T) with that at $(P + dP, T + dT)$:

$$\frac{dP}{dT} = \frac{\Delta s}{\Delta v}. \tag{2.22}$$

If we now use a Legendre transformation to pass from s to T as independent variable, this transformation will be degenerate in the region representing two phases in equilibrium. A cross-section of the derived thermodynamic $u(s, v)$ surface by a plane v = constant will have a linear portion corresponding to the first-order transition, and all the different points R on this line will transform into a single point R' (Figure 2.2). The function $f(T,v)$ which results from the transformation has one extensive variable v, and the condition (2.12)

* We shall use the capital letter C for specific heat for convenience of notation; it will always refer to the specific heat per molecule.

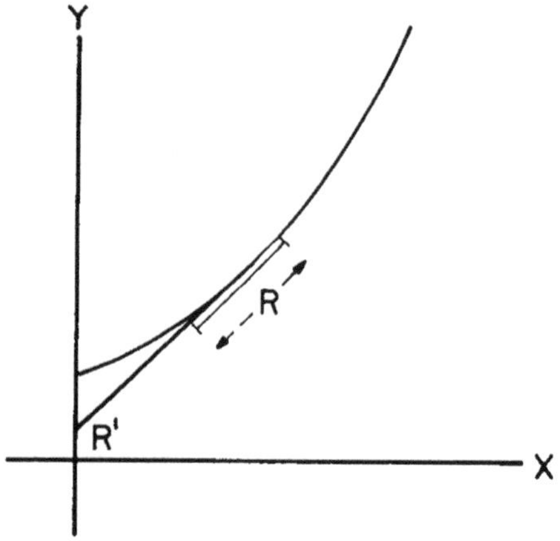

Figure 2.2 Degeneracy of transformation.

enables us to pursue a discussion similar to that for $u(s,v)$. For local stability at any point the curve $f(T,v)$ must always lie above its tangents, and for global stability this tangent must not cut the primitive $f(T,v)$ curve at any other point. If it does, the substance will split into a mixture of two phases with values of volume v_1, v_2 corresponding to the two points of contact of the tangent (Figure 2.3). The double-contact tangent corresponds to the coexistence of two phases in equilibrium, and is equivalent to the ruled surface in the u,s,v space. For different values of temperature we will obtain different tangents, and if we represent their limits of volume v_1, v_2 in the (v,T) plane we will obtain a *coexistence region* in this plane (Figure 2.4). The boundary of this

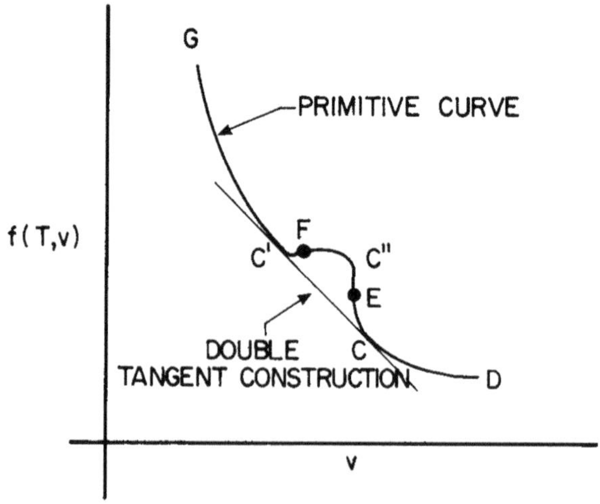

Figure 2.3 First-order transitions: Helmholtz free energy (after Gibbs).

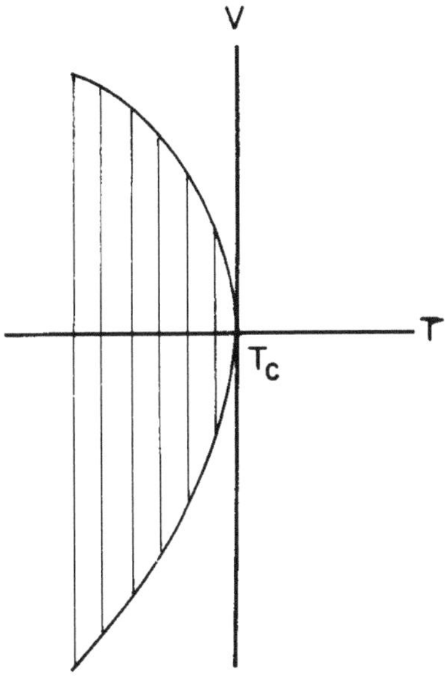

Figure 2.4 Shaded portion represents coexistence region.

region is usually called the *phase boundary curve* or *coexistence boundary curve*.

Finally, if we use a second Legendre transformation to pass from v to P as independent variable, we will again encounter a degeneracy in the region representing a first-order transition. All of the points on the double-contact tangent will transform into a single point having the pressure corresponding to the tangent ($P = -(\partial f/\partial v)_T$). Thus the region of coexistence represented by a ruled surface in the (s,v) space, and by a line in the (T,v) space, corresponds to a single point in the (T,P) space. When we vary the temperature we obtain a *coexistence curve* of phase equilibrium in (T,P) space (Figure 2.5).

The function $g(T,P)$ $(=\mu(T,P))$ is particularly convenient for the representation of first-order phase transitions. For equilibrium at a given temperature and pressure $g(T,P)$ must be a minimum (equation (2.15)). The second derivatives are given by

$$\left(\frac{\partial^2 g}{\partial T^2}\right)_P = -\left(\frac{\partial s}{\partial T}\right)_P = -\frac{C_p}{T} < 0 \tag{2.23}$$

$$\left(\frac{\partial^2 g}{\partial P^2}\right)_T = \left(\frac{\partial v}{\partial P}\right)_T = -vK_T < 0 \tag{2.24}$$

so that the $g(T,P)$ curve is convex downwards both in the T = constant and P = constant cross-sections for a stable phase. A typical $g(T,P)$ curve (P = constant) for a first-order phase transition is shown in Figure 2.6.

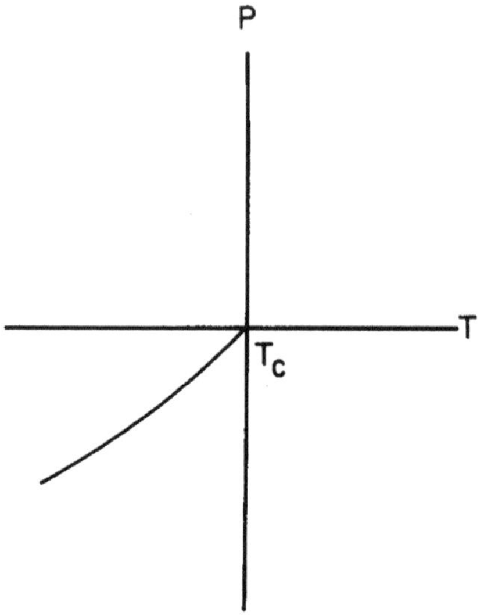

Figure 2.5 Coexistence curve.

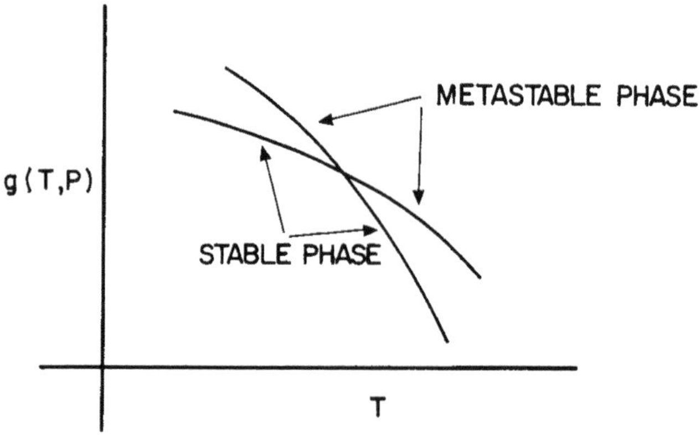

Figure 2.6 First-order transitions: Gibbs free energy.

2.3 Theory of van der Waals

Let us recall from §1.4.1 the experimental facts about fluids discovered by Andrews. There exists a critical temperature T_c below which a gas can be liquefied, and above which it cannot be liquefied by increase of pressure. For temperatures below T_c there is a first-order transition between liquid and gaseous phases, and these phases merge symmetrically at the critical point. For temperatures above T_c there are no discontinuities.

Van der Waals endeavoured to account for these experimental facts by taking account of intermolecular forces, $\phi(r)$, which he assumed to consist of a weak long-range attraction and a hard-core repulsion,

$$\phi(r) = \phi_{\mathrm{attr}} + \phi_{\mathrm{hard\ core}}. \tag{2.25}$$

Using the Gibbs formulation §1.2, and the canonical ensemble, the following partition function is derived:

$$Z(T,V,N) = \frac{1}{N!\Lambda^{3N}} \int \cdots \int \mathrm{d}\mathbf{r}_1 \cdots \mathrm{d}\mathbf{r}_N \exp - \left(\beta \sum_{i<j} (\phi \,|\mathbf{r}_i - \mathbf{r}_j|) \right) \tag{2.26}$$

$$(\Lambda = h/(2\pi mkT)^{1/2})$$

(The factor $1/N!$ is introduced into equation (1.17) to take account of quantum statistics; see e.g. Mandl (1970, Ch. 7).) The integral in (2.26) is usually called the *configurational integral.* We shall derive the van der Waals equation using the method which Ornstein introduced in his thesis in 1908 (see van Kampen 1964, Uhlenbeck 1968) instead of the virial theorem (1.23) used by van der Waals. Also we shall assume ϕ_{attr} to be very long range as in (1.45),

$$\phi_{\mathrm{attr}} = -cJ\lambda^d \exp(-\lambda r) \tag{2.27}$$

where λ will be allowed to tend to zero, and the normalizing constant c is chosen so that

$$-\int \phi_{\mathrm{attr}}(r)\, \mathrm{d}\mathbf{r} = J. \tag{2.28}$$

In one dimension $c = 1$, in two dimensions $c = \frac{1}{2}\pi$, in three dimensions $c = \frac{1}{8}\pi$.

The basic assumption of *mean field theory* is that the fluctuating field seen by an individual molecule can be replaced by its average. This is a reasonable assumption if a large number of molecules contribute to the fluctuating field, and it is rigorously correct for the potential (2.27) as $\lambda \to 0$ (Hemmer and Lebowitz 1976). For shorter-range potentials it provides an approximation which gets progressively worse as the number of molecules contributing to the field decreases.

Making this assumption we replace the sum of the attractive fields seen by the ith molecule $\sum_j \phi_{\mathrm{attr}}(|\mathbf{r}_i - \mathbf{r}_j|)$ by its average represented by the integral

$$\frac{N}{V}\int \phi_{\mathrm{attr}}(r)\, \mathrm{d}\mathbf{r} = -\frac{JN}{V}. \tag{2.29}$$

Such a replacement of a sum by an integral is reasonable if there is no significant change in the intermolecular force over the mean distance between molecules. Note that (2.29) is independent of $\mathbf{r}_i$. We now sum over all $i < j$, and ignoring surface contributions we obtain

$$\left\langle \sum_{i<j} \phi_{\text{attr}}(|\mathbf{r}_i - \mathbf{r}_j|) \right\rangle = -JN^2/2V \tag{2.30}$$

the factor 2 entering because of the $i < j$ condition. Thus, the attractive forces give a net factor $\exp(\beta JN^2/2V)$ in (2.26) which can be taken outside the integral. If the radius of the hard core is σ,

$$\exp(-\beta\phi_{\text{hard core}}(r)) = S(r) = \begin{cases} 0 & (r < \sigma) \\ 1 & (r > \sigma). \end{cases} \tag{2.31}$$

Hence we find that

$$Z(T,V,N) = \frac{1}{N!\Lambda^{3N}} \exp\left(\frac{\beta JN^2}{2V}\right) \int \cdots \int d\mathbf{r}_1 \cdots d\mathbf{r}_N \prod_{i<j} S(|\mathbf{r}_i - \mathbf{r}_j|) \tag{2.32}$$

where $\prod$ represents the product. This integral is the well-known hard-sphere integral which can be evaluated exactly only in one dimension. In two and three dimensions a number of terms of the virial series have been evaluated, and estimates have been obtained from model computer calculations. Let us denote the value of the integral by $Z_{\text{h.c.}}(V,N)$. Then the free energy is given by

$$\begin{aligned} F &= -kT \ln Z(T,V,N) \\ &= -JN^2/2V + NkT(3 \ln \Lambda + \ln N - 1) - kT \ln Z_{\text{h.c.}}(V,N). \end{aligned} \tag{2.33}$$

We can obtain the pressure by differentiation,

$$\begin{aligned} P &= -\left(\frac{\partial F}{\partial V}\right)_T = P_{\text{h.c.}} - \frac{JN^2}{2V^2} = P_{\text{h.c.}} - \frac{a}{v^2} \\ P &+ \frac{a}{v^2} = P_{\text{h.c.}}. \end{aligned} \tag{2.34}$$

Van der Waals assumed a constant *excluded volume*, i.e.

$$Z_{\text{h.c.}} = (V - Nb)^N. \tag{2.35}$$

This is exact in one dimension since the excluded volumes are simply additive and it is possible to establish a one-to-one correspondence between configurations of N molecules of size b in a space of length V, and configurations of N molecules of zero size in a space of length $(V - Nb)$ (first noted by Rayleigh in 1891). In two or three dimensions, the geometrical problems involved in fitting circles or spheres in a given space are complex; (2.35) can be regarded as a reasonable approximation only at low densities, as noted by Maxwell and

Boltzmann. Taking the form (2.35) for $Z_{\text{h.c.}}$, we find that

$$P_{\text{h.c.}} = kT/(v - b)$$
$$P = -\frac{a}{v^2} + \frac{kT}{v - b} \tag{2.36}$$

the standard van der Waals equation.

Quite remarkably this simple equation can be made to account qualitatively for the basic experimental results of Andrews. At sufficiently low temperatures there are three values of P for a given v, at sufficiently high temperatures there is only one, and the critical temperature T_c divides these

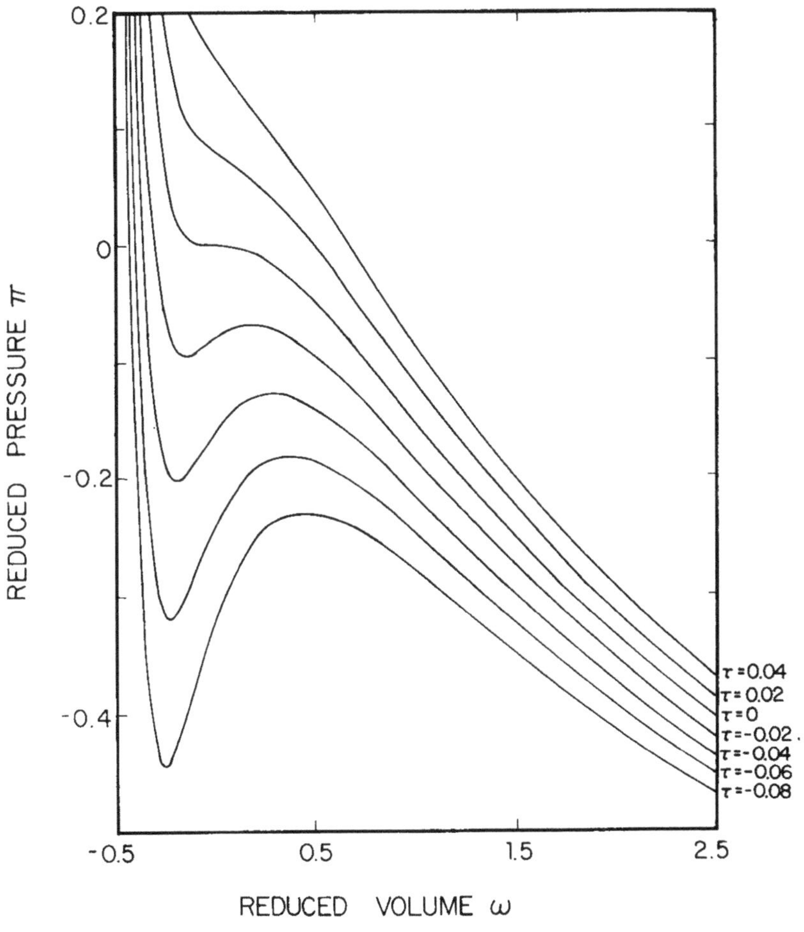

Figure 2.7 Van der Waals' isotherms.

two regions. The well-known isotherms corresponding to (2.36) are reproduced in Figure 2.7. To take proper account of liquid–vapour equilibrium a horizontal line must be drawn at an appropriate place to represent stable equilibrium and the portion of the original van der Waals curve which this line replaces should be interpreted as representing metastable and unstable states. Such a region had already been envisaged by a colleague of Andrews, James Thomson (the elder brother of William, Lord Kelvin), in 1871 to account for metastable supercooled gases and superheated liquids.

2.3.1 Thermodynamic Functions for a van der Waals Fluid

The van der Waals equation can conveniently be used to illustrate the thermodynamic discussion of §2.2. By integrating the equation

$$P = -\left(\frac{\partial f}{\partial v}\right)_T = -\frac{a}{v^2} + \frac{kT}{v-b}$$

we obtain

$$f = -\frac{a}{v} - kT\ln(v-b) + \Phi(T) \tag{2.37}$$

where the function $\Phi(T)$ can readily be determined from the consideration of an ideal gas ($a = 0$, $b = 0$). We find (e.g. Rushbrooke 1949)

$$\Phi(T) = -kT\left(\ln\frac{(2\pi mkT)^{3/2}}{h^3} + 1\right). \tag{2.38}$$

If we wish to use the function $a(T,\rho)$ of (2.16) we find that

$$a(T,\rho) = -a\rho^2 - kT\rho\ln\frac{1-\rho b}{\rho} + \rho\Phi(T). \tag{2.39}$$

For $T > T_c$ (2.37) gives rise to convex upwards curves which are locally and globally stable. But when $T < T_c$ this is no longer the case (Figure 2.3), and there are regions of metastability (C′F, EC) and instability (FC″E).

The internal energy is determined by standard thermodynamics

$$u = -T^2\frac{\partial}{\partial T}\left(\frac{f}{T}\right)_v = -\frac{a}{v} + \frac{3}{2}kT. \tag{2.40}$$

The internal pressure, which is defined as $(\partial u/\partial v)_T$, is equal to a/v^2, and this is consistent with the original van der Waals picture (1.24).

The entropy can similarly be determined by standard thermodynamics,

$$s = -\left(\frac{\partial f}{\partial T}\right)_v = k\ln(v-b) + k\left(\ln\frac{(2\pi mkT)^{3/2}}{h^3} + \frac{5}{2}\right). \tag{2.41}$$

The thermodynamic surface $u(s,v)$ is obtained by eliminating T between (2.40) and (2.41), to give

$$s - s_0 - k \ln(v - b) = \frac{3}{2} k \ln T = \frac{3}{2} k \ln \frac{2(u + a/v)}{3k}$$
$$s_0 = k\left(\ln \frac{(2\pi mk)^{3/2}}{h^3} + \frac{5}{2}\right) \tag{2.42}$$

or

$$u = -\frac{a}{v} + \frac{3}{2}(v - b)^{-2/3} \exp[2(s - s_0)/3k]. \tag{2.43}$$

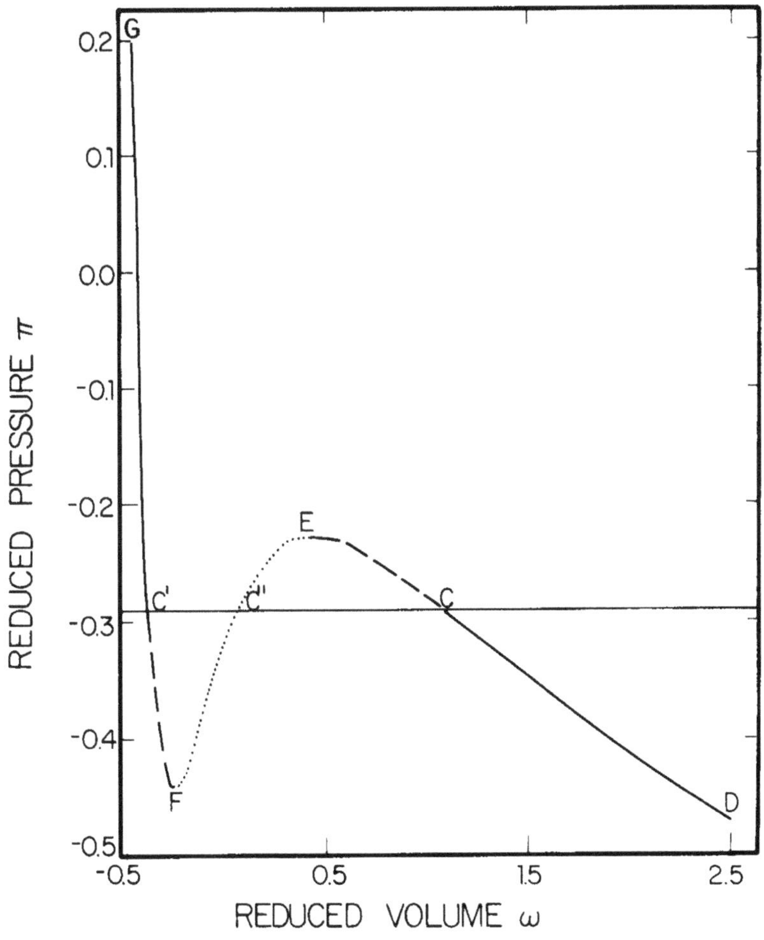

Figure 2.8 Maxwell construction for a van der Waals' fluid.

Finally the Gibbs function is obtained from (2.36) and (2.37):

$$g(T,P) = f + Pv = -\frac{2a}{v} + \frac{kTv}{(v-b)} - kT\ \ln(v-b) + \Phi(T). \tag{2.44}$$

To see the qualitative behaviour of $g(T,P)$ as a function of P we use the method of Pippard (1957) integrating the equation $v = (\partial g/\partial P)_T$,

$$g(T,P) = g(T,P_0) + \int_{P_0}^{P} v\ \mathrm{d}P. \tag{2.45}$$

On the $g(T,P)$ curve as a function of P, v represents the slope, and $(\partial v/\partial P)_T$ the curvature. We take the van der Waals isotherm DEFG in Figure 2.8 and show the corresponding $g(T,P)$ curve in Figure 2.9 as P increases, $\partial v/\partial P$ being negative. Near E

$$(P - P_E) \sim -(v - v_E)^2, \quad (v - v_E) \sim (P_E - P)^{1/2}, \quad (\partial v/\partial P) \sim -(P_E - P)^{-1/2} \tag{2.46}$$

so that $(\partial v/\partial P)$ becomes infinite at E and the $g(T,P)$ curve has a cusp. From E to F, v decreases as P decreases, $\partial v/\partial P$ being positive (this concave portion is

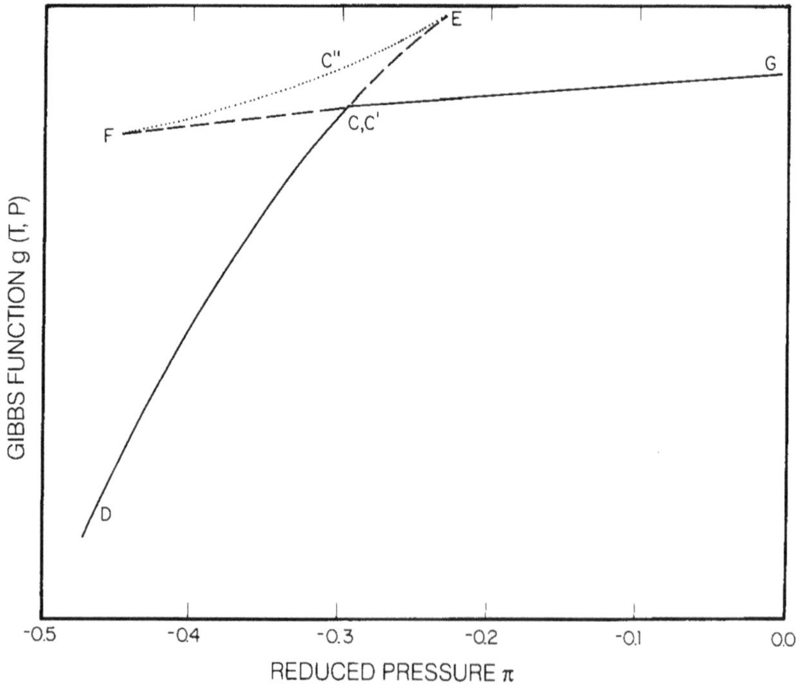

Figure 2.9 Gibbs function for a van der Waals' fluid. The dashed curve is metastable, the dotted curve is unstable.

unstable). Near F

$$(P - P_F) \sim (v - v_F)^2, \quad (v - v_F) \sim (P - P_F)^{1/2}, \quad (\partial v/\partial P) \sim (P - P_F)^{-1/2} \tag{2.47}$$

and there is another cusp. From F to G, v decreases as P increases and $\partial v/\partial P$ is negative but becomes small as G is approached. The dashed curves in Figure 2.9 correspond to metastable regions, the dotted curve to unstable regions.

2.3.2 Maxwell Construction

To determine where the horizontal line must be drawn in the van der Waals isotherms Maxwell (1875) suggested the use of a thermodynamic cycle CEC″FC′C″C (Figure 2.8). The cycle would be carried out at constant temperature, and hence by the second law there could be no net external work. External work is represented by $\int P\, dv$. Hence the areas CEC″C and C″FC′C″ must be equal.

The objection to this argument is that whilst the substance can exist in the metastable states CE, C′F, it cannot exist in the unstable region FC″E. Hence alternative thermodynamic arguments are introduced that the stable state is determined by the convex envelope construction for $f(T,V)$ in Figure 2.3, or by the intersection of the branches of $g(T,P)$ in Figure 2.9; these lead to the same equal-area result. However, Griffiths (1967) has pointed out that such arguments are equally invalid since they make use of integration through the unstable state in (2.36) to define $f(T, v)$ and $g(T,P)$.

The justification of the Maxwell construction was given by Kac, Uhlenbeck and Hemmer (1963) in their rigorous treatment of the potential (2.27), and the arguments are clearly presented in the review by Hemmer and Lebowitz (1976). It is emphasized there that statistical mechanics carried out properly must lead to equilibrium states – metastable states will only result if some restriction has been imposed.

The nature of the restriction imposed in standard van der Waals' theory has been pointed out by van Kampen (1964); the use of the mean field theory in the treatment of §2.3 imposes a restriction which requires the system to be in a single phase. By dividing the system into a large number of cells, each small compared with the range of force but large enough to contain many particles, and allowing a non-uniform distribution among the cells, and then minimizing the free energy, van Kampen was able to derive the true equilibrium as given by the Maxwell construction.

Griffiths (1967) has suggested an alternative *hypothesis of analyticity* for deriving the true equilibrium: that the free energy $f(v,T)$ is analytic everywhere except on the phase boundary. Only the equal-areas construction satisfies this condition; an attempt to place the horizontal part of the isotherm in any other position inevitably leads to some sort of phase transition in the one-phase region.

2.3.3 Corresponding States

It is well known that the van der Waals equation leads to a law of corresponding states, i.e. if we use P/P_c, v/v_c, T/T_c as reduced variables we obtain a universal equation of state with no free parameters.

Critical parameters are given by

$$\left(\frac{\partial P}{\partial v}\right)_T = 0 = \left(\frac{\partial^2 P}{\partial v^2}\right)_T \tag{2.48}$$

$$\frac{2a}{v^3} - \frac{kT}{(v-b)^2} = 0, \quad -\frac{6a}{v^4} + \frac{2kT}{(v-b)^3} = 0 \tag{2.49}$$

$$v_c = 3b, \quad kT_c = 8a/27b, \quad P_c = a/27b^2 \tag{2.50}$$

$$P_c v_c / kT_c = 3/8.$$

To examine critical behaviour we write

$$\frac{P}{P_c} = 1 + \pi, \quad \frac{v}{v_c} = 1 + \omega, \quad \frac{T}{T_c} = 1 + \tau, \quad \frac{\rho}{\rho_c} = 1 + r \tag{2.51}$$

where π, ω, τ, r can be regarded as small. We obtain the universal equation

$$(1 + \pi) + \frac{3}{(1+\omega)^2} = \frac{8(1+\tau)}{2 + 3\omega} \tag{2.52}$$

or expanding,

$$\pi = 4\tau - 6\tau\omega + 9\tau\omega^2 - \frac{3}{2}\,\omega^3 - \frac{27}{2}\,\tau\omega^3 + \frac{21}{4}\,\omega^4 + \frac{81}{5}\,\omega^4\tau - \frac{99}{8}\,\omega^5 \cdots. \tag{2.53}$$

The last equation, (2.53), is the key to the analysis of behaviour in the critical region.

In the next chapter we will need to consider the analogy between fluids and magnets, and for this purpose it is often more convenient to use the chemical potential μ instead of the pressure P, and to use ρ as independent variable instead of v. For a system with one component μ is equal to the Gibbs function g, and we can write (2.44) in the form

$$\mu - \Phi(T) = -2a\rho - kT \ln[(1 - b\rho)/\rho] + kT/(1 - b\rho).$$

In terms of reduced variables we find for the exact equation of state

$$\frac{\mu - \mu_0(T)}{P_c v_c} = -6r - \frac{8}{3}(1+\tau)\ln\frac{1 - r/2}{1 + r} + 4(1+\tau)[(1 - r/2)^{-1} - 1] \tag{2.52'}$$

where $\mu_0(T) = \Phi(T) + 6P_c v_c + kT(\frac{3}{2} + \ln\frac{3}{2} - \ln v_c)$.

Expanding for the critical region,

$$\frac{\mu - \mu_0(T)}{P_c v_c} = 6r\tau + \frac{3}{2} r^3(1+\tau) - \frac{3}{8} r^4(1+\tau). \tag{2.53'}$$

Comparing (2.53) with (2.53′) we see that the square term has disappeared.

2.4 Critical Behaviour of a van der Waals Fluid ($T > T_c$)

2.4.1 *Critical Isotherm* ($\tau = 0$)

From (2.53)

$$\pi \simeq -\frac{3}{2}\omega^3 + \frac{21}{4}\omega^4, \tag{2.54}$$

i.e. a cubic curve.

2.4.2 *Isothermal Compressibility on Critical Isochore* ($\omega = 0$)

$$K_T = -\frac{1}{v}\left(\frac{\partial v}{\partial P}\right)_T \simeq -\frac{1}{P_c}\left(\frac{\partial \omega}{\partial \pi}\right)_\tau = \frac{1}{6P_c\tau} = \frac{2}{3P_c\pi} \tag{2.55}$$

using (2.53). This is strongly divergent. Note that

$$\left(\frac{\partial P}{\partial T}\right)_v = \frac{P_c}{T_c}\left(\frac{\partial \pi}{\partial \tau}\right)_\omega = \frac{4P_c}{T_c} \tag{2.56}$$

which is finite.

2.4.3 *Specific Heat at Constant Volume, C_v*

On the critical isochore $v = v_c$, $\omega = 0$. Hence from (2.40)

$$C_v = \left(\frac{\partial u}{\partial T}\right)_v = \frac{3}{2}k. \tag{2.57}$$

This is non-singular. When we calculate C_v for $T < T_c$ we shall find that it has a discontinuity.

2.4.4 *Specific Heat at Constant Pressure, C_P*

From (2.42) we find that

$$C_P = T\left(\frac{\partial s}{\partial T}\right)_P = \frac{3}{2}k + \frac{kT}{v-b}\left(\frac{\partial v}{\partial T}\right)_P \simeq \frac{3}{2}k + \frac{kv_c}{v_c - b}\left(\frac{\partial \omega}{\partial \tau}\right)_\pi. \tag{2.58}$$

To determine $(\partial\omega/\partial\tau)_\pi$ on the critical isobar we must solve (2.53) for ω as a function of τ when $\pi = 0$. It is easy to see that the dominant behaviour results from the τ and ω^3 terms, the rest giving rise to terms of higher order. Hence

$$\omega \sim (8\tau/3)^{1/3} \tag{2.59}$$

and

$$C_P \sim \frac{3}{2}k(1 + 2.3^{-4/3}\tau^{-2/3}). \tag{2.60}$$

This is strongly divergent.

2.4.5 Thermal Expansion

The thermal expansion coefficient, α, is defined by

$$\alpha = \frac{1}{v}\left(\frac{\partial v}{\partial T}\right)_P \tag{2.61}$$

and is readily shown by elementary thermodynamics to be equal to

$$-\frac{1}{v}\left(\frac{\partial P}{\partial T}\right)_v \Bigg/ \left(\frac{\partial P}{\partial v}\right)_T. \tag{2.62}$$

Hence from (2.55) and (2.56) the critical behaviour of α is

$$\alpha = K_T(\partial P/\partial T)_v \simeq \frac{2}{3T_c\tau} \tag{2.63}$$

so that it is strongly divergent like K_T.

2.4.6 Adiabatic Behaviour

The adiabatic behaviour can readily be derived by putting $s = \text{constant}$ in (2.42). Thus, the equation of the adiabatic through T_c is

$$\ln(v - b) + \frac{3}{2}\ln T = \ln(v_c - b) + \frac{3}{2}\ln T_c. \tag{2.64}$$

Near the critical point (2.64) gives

$$\omega \simeq -\tau. \tag{2.65}$$

The adiabatic compressibility is defined by

$$K_s = -\frac{1}{v}\left(\frac{\partial v}{\partial P}\right)_s \tag{2.66}$$

and since from (2.7)

$$P = -(\partial u/\partial v)_s \tag{2.67}$$

we find from (2.43) that

$$K_s^{-1} = v\left(\frac{\partial^2 u}{\partial v^2}\right)_s = -\frac{2a}{v^2} + \frac{10}{9}\frac{v}{(v-b)^2}\left(u + \frac{a}{v}\right). \tag{2.68}$$

From (2.40) and (2.50) we calculate that K_s^{-1} is non-zero at the critical point and equal to $4P_c$.

The coefficient of adiabatic expansion, α_s, defined by

$$\alpha_s = \frac{1}{v}\left(\frac{\partial v}{\partial T}\right)_s \tag{2.69}$$

can also be obtained from (2.41):

$$\alpha_s = -\frac{3}{2}\frac{v-b}{vT}. \tag{2.70}$$

The critical value is finite and equal to $-1/T_c$.

2.5 Critical Behaviour of a van der Waals Fluid ($T < T_c$)

Below T_c the situation is considerably more complicated because of the splitting of the system into two phases and the associated cubic equation to be solved. Since both π and τ are negative we write for convenience $\pi' = -\pi$ and $\tau' = -\tau$. Equation (2.53) in the new notation becomes

$$\begin{aligned}\pi' - 4\tau' = {} & -6\tau'\omega + 9\tau'\omega^2 + \frac{3}{2}\omega^3 - \frac{27}{2}\tau'\omega^3 - \frac{21}{4}\omega^4 \\ & + \frac{81}{5}\tau'\omega^4 + \frac{99}{8}\omega^5 + \cdots.\end{aligned} \tag{2.71}$$

We shall denote by π' the equilibrium value corresponding to the Maxwell construction (Figure 2.10)

$$\int_{\omega_1}^{\omega_3} (\pi' - \pi_0')\, d\omega = 0. \tag{2.72}$$

Relation (2.72) can be developed from (2.71) in the form

$$\begin{aligned}\pi_0'(\omega_3 - \omega_1) &= 4\tau'(\omega_3 - \omega_1) - 3\tau'(\omega_3^2 - \omega_1^2) \\ &\quad + 3\tau'(\omega_3^3 - \omega_1^3) + \tfrac{3}{8}(\omega_3^4 - \omega_1^4) \\ \pi_0' &= 4\tau' - 3\tau'(\omega_3 + \omega_1) + 3\tau'(\omega_3^2 + \omega_3\omega_1 + \omega_1^2) + \cdots.\end{aligned} \tag{2.73}$$

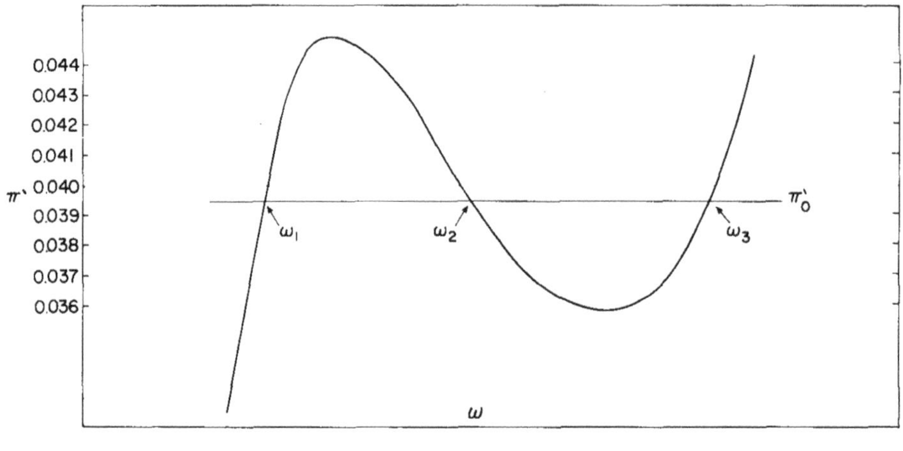

Figure 2.10 Maxwell construction for $\tau' = 0.01$. (The approximate formulae of §2.5 are valid only within this distance from the critical point. Symmetry disappears quickly with increasing τ'.)

To the lowest order in τ' it is not difficult to verify the solution

$$\pi_0' \simeq 4\tau', \quad \omega_1 \simeq -2\tau'^{1/2}, \quad \omega_2 \simeq 0, \quad \omega_3 \simeq 2\tau'^{1/2}. \tag{2.74}$$

The development of a series solution in terms of temperature is quite tricky. If we expand ω_1, ω_3 and π_0' as a power series in $\tau'^{1/2}$, equations (2.71) and (2.73) are sufficient to determine all the coefficients. Fortunately a pattern of symmetry is soon revealed, and the expansions are of the following form:

$$\begin{aligned} \omega_1 &= -2\tau'^{1/2} + a_1\tau' + a_2\tau'^{3/2} + a_3\tau'^2 + \cdots \\ \omega_3 &= 2\tau'^{1/2} + a_1\tau' - a_2\tau'^{3/2} + a_3\tau'^2 + \cdots \\ \pi_0' &= 4\tau' + A_2\tau'^2 + A_3\tau'^3 + \cdots \end{aligned} \tag{2.75}$$

$$a_1 = 18/5 \quad a_2 = -147/25 \quad A_2 = -24/5.$$

The middle root, ω_2, does not seem to enter any of the equations. But it is latent in the calculation of ω_1 and ω_3, and can readily be determined from the original cubic equation (2.36) in v. From this we deduce for the sum of the three roots v_1, v_2, v_3

$$v_1 + v_2 + v_3 = \frac{kT}{P} + b \tag{2.76}$$

from which one can derive the corresponding relation in reduced variables,

$$\begin{aligned} \tfrac{3}{8}(\omega_1 + \omega_2 + \omega_3) &= (1 + \tau)/(1 + \pi) - 1 \quad \text{(generally)} \\ &= (1 - \tau')/(1 - \pi_0') - 1 \quad \text{(equilibrium)}. \end{aligned} \tag{2.77}$$

It will be seen that, like π'_0, the expansion of ω_2 does not involve any square root terms.

2.5.1 *Coexistence Boundary Curve (Figure 2.4)*

From (2.75)

$$\omega_1 = -2\tau'^{1/2} + \frac{18}{5}\tau' - \frac{147}{25}\tau'^{3/2} + \cdots$$
$$\omega_3 = 2\tau'^{1/2} + \frac{18}{5}\tau' + \frac{147}{25}\tau'^{3/2} + \cdots. \tag{2.78}$$

Thus although the odd terms give a symmetric curve above and below the τ' axis, the even terms destroy the symmetry.

The *diameter* of the coexistence boundary curve is defined as

$$\tfrac{1}{2}(v_{\text{liquid}} + v_{\text{gas}}) = v_c[1 + \tfrac{1}{2}(\omega_1 + \omega_3)] = v_c\left(1 + \frac{18}{5}\tau' + a_3\tau'^2 + \cdots\right). \tag{2.79}$$

Near T_c it is linear, and the term *rectilinear diameter* is often used to describe this situation.

2.5.2 *Coexistence Curve (Figure 2.5)*

In terms of our original variables this is given from (2.75) by

$$\pi_0 = 4\tau - \frac{24}{5}\tau^2 + \cdots. \tag{2.80}$$

2.5.3 *Isothermal Compressibility along the Coexistence Boundary*

For any point of volume ω in the shaded region in Figure 2.4 the system exists in two phases, the proportions of liquid and gas $\theta:(1-\theta)$ being determined by the equation

$$\omega = \theta\omega_1 + (1-\theta)\omega_3. \tag{2.81}$$

Any attempt to compress the substance at constant temperature results in a decrease in volume with no change in pressure. Hence the isothermal compressibility is infinite except on the coexistence boundary ($\theta = 0$ or 1) when the substance becomes homogeneous. We use the general formula (2.55) for K_T, and find from (2.71) that

$$\left(\frac{\partial\pi'}{\partial\omega}\right)_{\tau'} = -6\tau' + 18\tau'\omega + \frac{9}{2}\omega^2 + \cdots. \tag{2.82}$$

Hence to lowest order on the coexistence boundary ($\omega^2 \sim 4\tau'$),

$$K_T = \frac{1}{P_c(1+\omega)}\left(\frac{\partial\omega}{\partial\pi'}\right)_{\pi'} \simeq \frac{1}{12P_c\tau'}. \tag{2.83}$$

Comparing with (2.55), this is exactly one-half of the value for $T > T_c$. The next order terms are no longer symmetric on the liquid and gaseous sides of the coexistence boundary.

2.5.4 Specific Heat C_v on the Critical Isochore

The critical isochore corresponds to $\omega = 0$ in (2.81). Hence the ratio of liquid and gas components is given by

$$\theta:(1-\theta) = -\omega_3:\omega_1. \tag{2.84}$$

The internal energy is given by

$$u = \theta u_l + (1-\theta)u_g \tag{2.85}$$

with u_l and u_g corresponding to (2.40).

Expanding as far as terms in τ'^2

$$\begin{aligned} u &= \frac{3}{2}kT - \frac{a}{v_c}\frac{1}{\omega_3-\omega_1}\left(\frac{\omega_3}{1+\omega_1} - \frac{\omega_1}{1+\omega_3}\right) \\ &= \frac{3}{2}kT - \frac{a}{v_c}[1 - \omega_1\omega_3 + \omega_1\omega_3(\omega_1+\omega_3) + \omega_1^2\omega_3^2] \\ &\simeq \frac{3}{2}kT - \frac{a}{v_c}\left(1 + 4\tau' - \frac{56}{25}\tau'^2\right). \end{aligned} \tag{2.86}$$

Differentiating and substituting for a, v_c, T_c

$$C_v - \frac{3}{2}k \simeq \frac{9}{2}k\left(1 - \frac{28}{25}\tau'\right). \tag{2.87}$$

The specific heat at constant volume therefore has the form shown in Figure 2.11 on passing through the critical point.

2.5.5 Specific Heat C_P

In the shaded region in Figure 2.4 the addition of heat at constant pressure does not change the temperature of the system, but merely the ratio $\theta:(1-\theta)$ of liquid to gas. Hence the specific heat C_P is infinite. However, we can calcu-

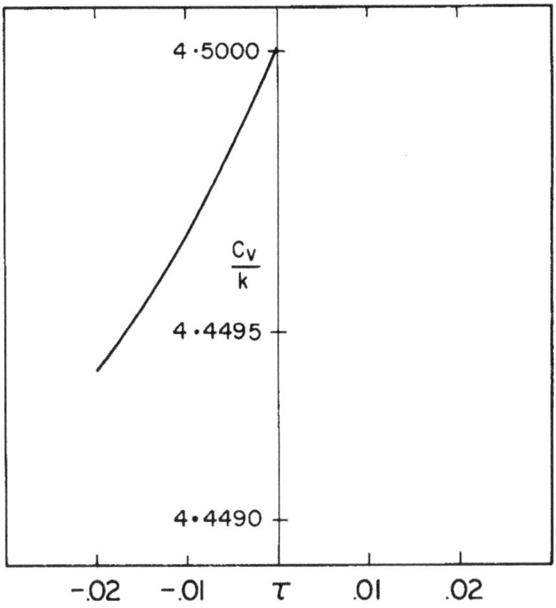

Figure 2.11 Configurational specific heat C_V of a van der Waals' fluid near T_c.

late C_P on the coexistence boundary curve. From (2.42)

$$C_P = T\left(\frac{\partial s}{\partial T}\right)_P = \frac{3}{2}k + \frac{kT}{v-b}\left(\frac{\partial v}{\partial T}\right)_P \simeq \frac{3}{2}k + \frac{kv_c}{v_c - b}\left(\frac{\partial \omega}{\partial \tau}\right)_\pi$$
$$= \frac{3}{2}k\left[1 - \left(\frac{\partial \omega}{\partial \tau'}\right)_{\pi'}\right]. \tag{2.88}$$

From (2.71)

$$\left(\frac{\partial \omega}{\partial \tau'}\right)_{\pi'} = \frac{-4 - 9\omega^2}{-6\tau' + 18\tau'\omega + \frac{9}{2}\omega^2} \simeq -\frac{1}{3\tau'} \tag{2.89}$$

on the coexistence boundary. Thus, C_P has a strong divergence.

2.5.6 *Adiabatic Behaviour* ($T < T_c$)

An adiabatic in the coexistence region has the equation

$$\theta s_l + (1 - \theta)s_g = \text{constant},$$

i.e. from (2.42)

$$\theta \ln[v_c(1+\omega_1)-b] + (1-\theta)\ln[v_c(1+\omega_3)-b] + \tfrac{3}{2}\ln T_c(1+\tau) = \text{constant} \tag{2.90}$$

$$\theta \ln(2+3\omega_1) + (1-\theta)\ln(2+3\omega_3) + \tfrac{3}{2}\ln(1+\tau) = \text{constant}.$$

Equations (2.90) determine the value of θ as a function of temperature for adiabatic compression. For the critical adiabatic, the constant in (2.90) has the value ln 2, and using (2.78) we find

$$\theta \simeq \tfrac{1}{2} - \tfrac{1}{10}\tau'^{1/2}, \quad 1-\theta \simeq \tfrac{1}{2} + \tfrac{1}{10}\tau'^{1/2}. \tag{2.91}$$

Along the adiabatic

$$v = \theta v_c(1+\omega_1) + (1-\theta)v_c(1+\omega_3) = v_c[1+\theta\omega_1+(1-\theta)\omega_3]$$

$$= v_c[1+4\tau' + O(\tau'^2)]. \tag{2.92}$$

Hence the adiabatic coefficient of thermal expansion,

$$\alpha_s = \frac{1}{v}\left(\frac{\partial v}{\partial T}\right)_s \simeq -\frac{4}{T_c}. \tag{2.93}$$

By comparison with (2.70) we find that α_s has a simple discontinuity at the critical point.

For the adiabatic compressibility, we note that in the coexistence region π_0 is a simple function of τ given by (2.80). Hence,

$$K_s^{-1} = -v\left(\frac{\partial P}{\partial v}\right) = -P_c\left(\frac{\partial \pi_0}{\partial \omega}\right)_s = -P_c\frac{d\pi_0}{d\tau}\frac{d\tau}{d\omega} = P_c. \tag{2.94}$$

By comparison with (2.68) we find that this also has a simple discontinuity at the critical point.

2.5.7 *Discontinuity in Volume and Entropy*

A first-order transition is accompanied by a discontinuity in volume Δv and in entropy Δs. In the critical region

$$\Delta v = v_c(\omega_3-\omega_1) \simeq v_c\left(4\tau'^{1/2} + 11\frac{19}{25}\tau'^{3/2}\right). \tag{2.95}$$

and Δs can best be determined from the Clausius–Clapeyron equation

$$\frac{dP}{dT} = \frac{P_c}{T_c}\frac{d\pi_0}{d\tau} = \frac{\Delta s}{\Delta v}. \tag{2.96}$$

Thus

$$\Delta s = \tfrac{3}{2}k\tau'^{1/2}(1+5.34\tau'). \tag{2.97}$$

2.6 Strong and Weak Divergences (Griffiths and Wheeler)

In the previous sections we have calculated the critical behaviour of a number of thermodynamic quantities and have found that they belong to two categories: those with strong singularities like C_P, K_T, α_T, and those with weak singularities like C_v, K_s, α_s. The strong singularities were either divergencies of the form τ^{-1} (equations (2.55), (2.63), (2.83)) or of the form $\tau^{-2/3}$ (equation (2.60)). The weak singularities were simple discontinuities. What determines whether a thermodynamic quantity has a strong or weak divergence, and whether this divergence is of the form τ^{-1} or $\tau^{-2/3}$? These questions were raised and answered satisfactorily by Griffiths and Wheeler (1970) and since their treatment applies to the general scaling form of the equation of state referred to in the previous chapter (equation (1.46)), we shall use this form with general parameters and then particularize to van der Waals. (A detailed analysis of the scaling form of the equation of state will be given later in Chapter 6.)

For a fluid the analogue of equation (1.46) can be written as

$$a\tau - \pi = \omega|\omega|^{\delta-1}h(\tau|\omega|^{-1/\beta}). \tag{2.98}$$

This corresponds to the van der Waals critical equation of state

$$4\tau - \pi = \tfrac{3}{2}\omega^3(1 + 4\tau\omega^{-2}) \tag{2.99}$$

derived from equation (2.53). The other terms in equation (2.53) are of higher order, and give rise to correction terms in critical behaviour. By comparing (2.99) with (2.98) we see that for a van der Waals fluid $\delta = 3$, $\beta = \frac{1}{2}$, and $h(x)$ is linear. After deriving any result for the general equation (2.98) we shall particularize to (2.99) for van der Waals (vdW).

Griffiths and Wheeler first point out that for this discussion it is advantageous to use the field variables π and τ in which the coexistence region is represented by a curve as in Figure 2.5. In this plane one direction is automatically selected – that of the tangent to the coexistence curve at T_c. Hence, we might expect critical behaviour to depend on the direction of approach to T_c, whether it is asymptotically parallel to the coexistence curve, or makes an angle with the coexistence curve (the numerical value of the angle is arbitrary since it depends on the units of measurement).

The quantities whose critical behaviour has been investigated, like C_P, K_T and C_v represent the second derivatives of the chemical potential μ (i.e. the Gibbs free energy) and Griffiths and Wheeler note that for the first two having a strong divergence

$$\frac{C_P}{T} = -\frac{\partial^2\mu}{\partial T^2}, \quad vK_T = -\frac{\partial^2\mu}{\partial P^2} \tag{2.100}$$

the directions of the derivatives make an angle with the coexistence curve; but C_v, which has a weak divergence, corresponds to the second derivative of μ in a direction parallel to the coexistence curve. This can best be seen from the thermodynamic identity giving the second derivative $d^2\mu/dT^2$ in an arbitrary direction dP/dT in the (P,T) plane,

$$-\frac{d^2\mu}{dT^2} = \frac{C_v}{T} + vK_T\left[\frac{dP}{dT} - \left(\frac{\partial P}{\partial T}\right)_v\right]^2. \tag{2.101}$$

The identity can be derived as follows. By standard partial differentiation we have

$$\begin{aligned} -\frac{d^2\mu}{dT^2} &= -\frac{\partial^2\mu}{\partial T^2} - 2\frac{\partial^2\mu}{\partial P\,\partial T}\frac{dP}{dT} - \frac{\partial^2\mu}{\partial P^2}\left(\frac{dP}{dT}\right)^2 \\ &= \frac{C_P}{T} - 2\left(\frac{\partial v}{\partial T}\right)_P\frac{dP}{dT} + vK_T\left(\frac{dP}{dT}\right)^2 \\ &= \frac{C_v}{T} + \frac{(\partial v/\partial T)_P^2}{vK_T} - 2\left(\frac{\partial v}{\partial T}\right)_P\frac{dP}{dT} + vK_T\left(\frac{dP}{dT}\right)^2 \end{aligned} \tag{2.102}$$

in which we have used the standard formula for $C_P - C_v$; other simple thermodynamic formulae (Pippard 1957) readily demonstrate the equality of (2.101) and (2.102).

From (2.101) it is clear that when dP/dT differs from the slope of the coexistence curve at T_c, $(\partial P/\partial T)_v$, $d^2\mu/dT^2$ has a strong divergence proportional to K_T; but when it is equal to this slope the second term vanishes and it has a weak singularity.

In regard to the second question, the actual form of the singularity at T_c, the direction of approach to T_c is important. Griffiths and Wheeler note that in a direction parallel to the coexistence curve K_T and C_P have a 'γ-divergence',

$$C_P \sim K_T \sim \tau^{-\gamma} \sim \pi^{-\gamma} \quad [\gamma = \beta(\delta - 1)] \tag{2.103}$$

whereas in a direction making an angle with the coexistence curve they have an 'ϵ-divergence',

$$C_P \sim K_T - \tau^{-\epsilon} \sim \pi^{-\epsilon} \quad (\epsilon = 1 - 1/\delta). \tag{2.104}$$

For quantities with weak divergences like C_v, along a path asymptotically parallel to the coexistence there is an 'α-divergence',

$$C_v \sim \tau^{-\alpha} \quad [\alpha = 2 - \beta(1 + \delta)]. \tag{2.105}$$

On a path making an angle with the coexistence curve there is a 'ξ-divergence',

$$C_v \sim \tau^{-\xi} \quad \xi = \alpha/\beta\delta. \tag{2.106}$$

Before demonstrating the above results for equation of state (2.98) attention must be drawn to the unusual character of the Legendre transformation from the $(\omega$–$\tau)$ to the $(\pi$–$\tau)$ plane in the neighbourhood of T_c. We have seen in §2.2

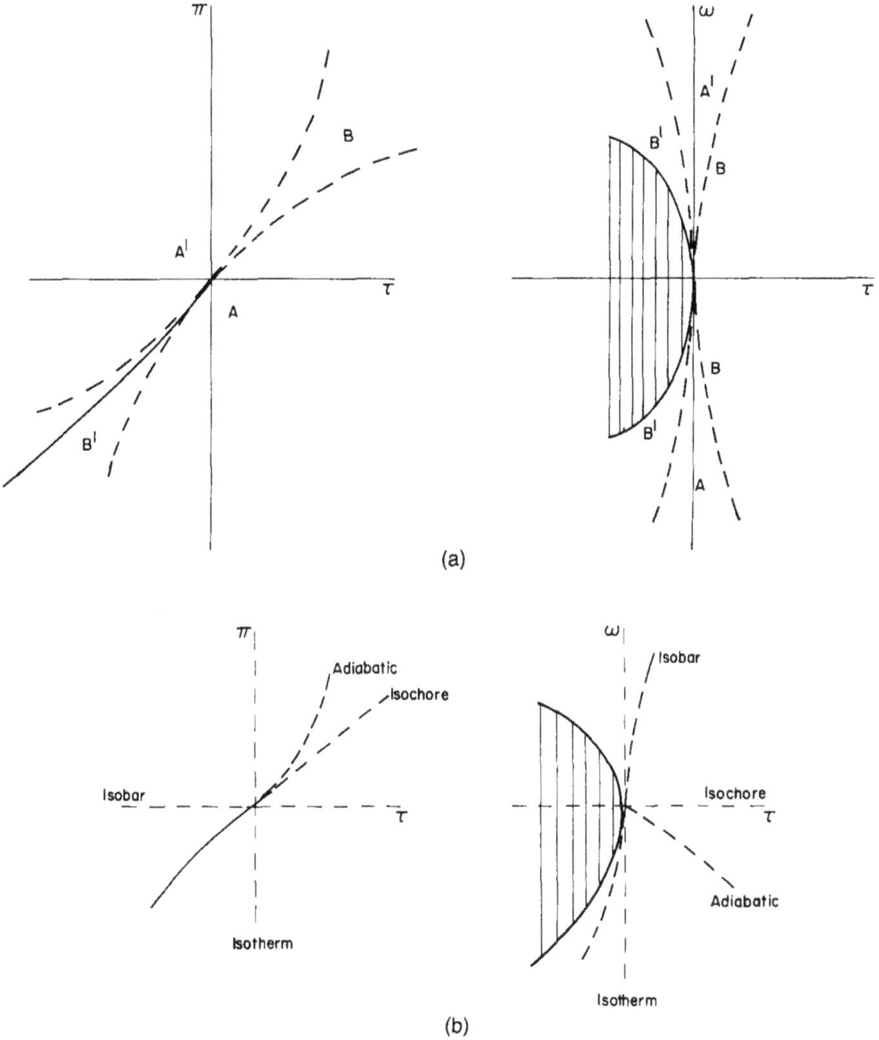

Figure 2.12 (a) Distortion of metric arising from the Legendre transformation near a critical point (following Griffiths and Wheeler 1970); (b) transformation of thermodynamic constraints from π–τ to ω–τ planes.

that this Legendre transformation is degenerate in the coexistence region of a first-order transition, and some features of the singular behaviour remain at T_c. These can be seen on reference to Figure 2.12(a). The coexistence curve is given by

$$a\tau - \pi = 0, \quad \tau'|\omega|^{-1/\beta} = b, \quad \omega = \pm B\tau'^{\beta} \tag{2.107}$$

where

$$h(-b) = 0 \quad \text{and} \quad B = b^{-\beta}$$
$$\omega = \pm 2\tau'^{1/2}(v\ \mathrm{d}W). \tag{2.108}$$

Any line making an angle with the coexistence curve in the (π–τ) plane has the equation

$$a\tau - \pi = c\tau \quad (c \neq 0)$$

and this corresponds to

$$c\tau = \omega|\omega|^{\delta-1}h(\tau|\omega|^{1/\beta})$$
$$\tau = \frac{\omega|\omega|^{\delta-1}}{c}\,h(0) \quad (\tau|\omega|^{-1/\beta} \sim \omega^{\delta-1/\beta} \to 0 \text{ as } \omega \to 0,$$
$$\text{since } \beta\delta > 1 \text{ for all physical models}) \tag{2.109}$$
$$\tau = \frac{3}{2c}\,\omega^3(v\ \mathrm{d}W).$$

Thus the extensive AA′ region in the (π–τ) plane transforms into a narrow wedge in the (ω–τ) plane.

Considering now the extensive BB′ region in the (ω–τ) plane given by

$$\tau|\omega|^{-1/\beta} = d \quad (-b < d < \infty) \tag{2.110}$$

this corresponds to

$$a\tau - \pi = D\omega|\omega|^{\delta-1} = \pm D|\tau|^{\beta\delta} \quad D = h(d)$$
$$4\tau - \pi = \pm D|\tau|^{3/2}(v\ \mathrm{d}W). \tag{2.111}$$

Hence it corresponds to a narrow wedge in the (π–τ) plane.

With this background we draw in Figure 2.12(b) the adiabatic, isochore, isotherm and isobar through T_c in both planes, and we see clearly that the first two are asymptotically parallel to the coexistence curve, whilst the last two make an angle with the coexistence curve.

Returning now to the calculation of K_T ($\tau > 0$), a typical quantity with a strong divergence, we find from (2.98) that

$$\frac{1}{K_T} \sim \left(\frac{\partial \pi}{\partial \omega}\right)_\tau = \delta\omega^{\delta-1}h(\tau\omega^{-1/\beta}) - \frac{\tau}{\beta}\,\omega^{\delta-1-1/\beta}h'(\tau\omega^{-1/\beta}). \tag{2.112}$$

For a path making an angle with the coexistence curve $\tau \sim \omega^\delta$ from (2.109), the first term in (2.112) dominates, and

$$K_T \sim \omega^{-(\delta-1)} \sim \tau^{-(\delta-1)/\delta}. \tag{2.113}$$

For a path parallel to the coexistence curve we have a relationship of the form (2.110), $\omega \sim \tau^\beta$, and both terms in (2.112) are of the same order giving

$$K_T \sim \tau^{-\beta(\delta-1)} = \tau^{-\gamma}. \tag{2.114}$$

To calculate C_v, a typical quantity with a weak divergence, we must first integrate (2.98) to determine the free energy, as we did for the van der Waals equation (2.37). We defer the calculation to Chapter 6, but we quote the result (6.56):

$$f(\tau,\omega) = A_0(\tau) + \omega^{\delta+1}k(\tau\omega^{-1/\beta}) \tag{2.115}$$

where $k(x)$ is related to $h(x)$. From this we find that

$$C_v = -T\left(\frac{\partial^2 f}{\partial \tau^2}\right)_\omega \sim \omega^{\delta+1-2/\beta}k''(\tau\omega^{-1/\beta}). \tag{2.116}$$

For a direction parallel to the coexistence curve $\omega \sim \tau^\beta$ and

$$C_v \simeq \tau^{-\alpha} \quad [\alpha = 2 - \beta(\delta + 1)]. \tag{2.117}$$

For a direction making an angle with the coexistence curve $\tau \sim \omega^\delta$ and

$$C_v \simeq \tau^{-\alpha/\beta\delta}. \tag{2.118}$$

Finally, Griffiths and Wheeler draw attention to the central role played by the matrix

$$\mathbf{M}(P,T) = \begin{vmatrix} -\partial^2\mu/\partial P^2 & -\partial^2\mu/\partial P\partial T \\ -\partial^2\mu/\partial P\partial T & -\partial^2\mu/\partial T^2 \end{vmatrix}. \tag{2.119}$$

Of the two eigenvalues of this matrix one diverges weakly and the other strongly, upon approaching the critical point, and the corresponding eigenvectors become respectively parallel and perpendicular to the coexistence curve. The determinant $D(P,T)$ of $\mathbf{M}(P,T)$ is given by

$$D(P,T) = vK_T C_v/T = vK_s C_P/T \tag{2.120}$$

as can be verified using elementary thermodynamic relations (Pippard 1957). Each of the products in (2.120) contains one strongly divergent and one weakly divergent quantity.

2.7 Capillarity and Surface Tension of a Liquid*

When a narrow tube open at both ends is placed vertically with its lower end immersed in a liquid, the level of the liquid in the tube rises. This *capillary effect*, which is a characteristic property of liquids, was first discovered by Leonardo de Vinci (1452–1519). The important idea of *surface tension* of a liquid, that the surface of a liquid is in a state of tension similar to that of a membrane stretched equally in all directions, was introduced by Segner in 1751, and he ascribed the tension to very short-range attractive forces. In 1805

* This subject has been given comprehensive treatment in a specialist text by Rowlinson and Widom (1982).

Thomas Young showed that surface tension was able to account for capillary phenomena.

The above facts are taken from a remarkable article by James Clerk Maxwell entitled *Capillary Action* and published in the 9th edition of the *Encyclopedia Brittanica* in 1876. The article could serve as a model of scientific writing, and Maxwell starts by reviewing concisely and accurately the major experimental and theoretical contributions to the development of the subject up to his own day. He tells us that among those who attempted to establish a theory of surface tension based on intermolecular attractive forces were Young, Laplace and Gauss. Laplace derived the formula for the pressure in a drop of radius l

$$P = K + H/l \tag{2.121}$$

where the first term represents the bulk contribution, the second term the surface tension, and K and H are integrals over the intermolecular forces.

If we wish to understand Laplace's result in terms of models of type (2.25) and (2.27) for intermolecular forces, we must be careful to specify that whilst the range of these forces is large compared with the mean spacing between molecules, it is small on the macroscopic scale. Thus λ must now be small but finite, J/λ representing the range of the forces. Laplace was concerned with the static problem, and the term K representing the mean energy corresponds to (2.30).

Let us divide the total attractive field seen by the ith molecule in a bulk liquid into layers distant z from $\mathbf{r}_i$ of thickness dz. Replacing the sum of these fields by an integral we find the value

$$2\pi \frac{N}{V} dz \int_0^\infty \phi[(t^2 + z^2)^{1/2}]t \, dt = \frac{N}{V} \psi(z) \, dz \tag{2.122}$$

where

$$\psi(z) = 2\pi \int_0^\infty \phi[(t^2 + z^2)^{1/2}]t \, dt.$$

The integration of (2.122) from $z = -\infty$ to ∞ yields (2.29), and we then sum over all i–j pairs to give (2.30). But we could alternatively sum over all pairs of *layers* at z and ζ to obtain the same result.

Let us follow the latter procedure for a *finite* system of large cross-sectional area A and height L. The interaction between the layer $(z, z + dz)$ and the layer $(\zeta, \zeta + d\zeta)$ will be, from (2.122),

$$\frac{N}{V} \psi(z - \zeta) \, dz \frac{N}{V} A \, d\zeta = \frac{N^2}{V^2} A\psi(z - \zeta) \, dz \, d\zeta. \tag{2.123}$$

This must be integrated over values of ζ from 0 to z, and z from 0 to L (Figure 2.13). Transforming to variables $u = z - \zeta$, $v = z + \zeta$, the integral over (2.123)

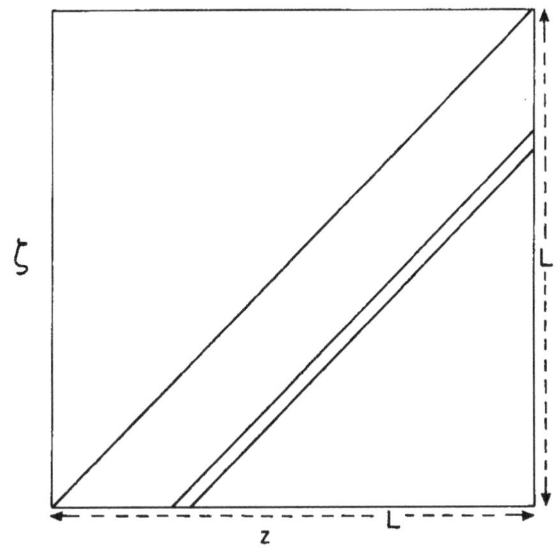

Figure 2.13 Transformation of integration variables from (z, ζ) to (u, v) $(u = z - \zeta, v = z + \zeta)$.

is transformed to

$$\frac{N^2}{V^2} A \int_0^L (L - u)\psi(u)\, du. \tag{2.123'}$$

The first term in the integral (2.123') corresponds to the bulk contribution (2.29) proportional to the volume AL; the factor $\frac{1}{2}$ arises since the integral of $\psi(u)$ goes from 0 to $L(\infty)$ instead of from $-\infty$ to ∞. The second term corresponds to a surface energy per unit area equal to

$$-\frac{1}{2}\frac{N^2}{V^2}\int_0^\infty u\psi(u)\, du = -\frac{\pi}{2}\frac{N^2}{V^2}\int_0^\infty r^3\phi(r)\, dr. \tag{2.124}$$

The factor $\frac{1}{2}$ arises here because there are two free surfaces. Replacing V by $\frac{4}{3}\pi l^3$ and A by $4\pi l^3$ we obtain Laplace's results (2.121). For the exponential law of force (2.27) the surface energy (2.124) is equal to

$$\frac{3}{8}\frac{N^2}{V^2}\frac{A}{\lambda}J \tag{2.125}$$

which is of order $1/\lambda L$ times the bulk energy (2.30).

A thermodynamic theory of capillarity and surface tension was developed by Gibbs (1875–8) in his classic paper 'On the equilibrium of heterogeneous substances'. This forms the basis for the discussion to be found in standard textbooks on thermodynamics (e.g. Zemansky 1968).

In 1893 van der Waals published a major paper in Dutch on this topic in the *Transactions of the Royal Academy of Sciences* in Amsterdam. The paper made a considerable impact, a German translation by Ostwald appeared in 1894, and an anonymous translation into French (possibly by van der Waals himself) in 1895. Van der Waals put forward two serious criticisms of the work of his predecessors: they had postulated a surface of discontinuity between the two phases whereas the true picture should be a continuously varying transition layer (a similar criticism of Laplace's work had been advanced by Poisson in 1831); and all attempts at a molecular theory had been static rather than thermodynamic.

Van der Waals' formulated a theory which attempted to remedy these defects. He needed first to tackle the difficult problem of generalizing the formulae of §2.3.1 to an inhomogeneous fluid; then using the criterion of minimization of free energy for equilibrium he was able to calculate the thickness and density profile of the transition layer. Finally, applying the calculations to the neighbourhood of the critical point, he found that the surface tension vanishes at T_c, and the thickness of the transition layer becomes infinite as T approaches T_c. This was in keeping with the experimental observations of Andrews (see §4.1).

Van der Waals work was badly neglected for several decades after 1914 (Rowlinson 1973). His results were re-derived by Cahn and Hilliard in 1958 in a more modern guise. Rowlinson (1979) has rendered a valuable service to the history of science by publishing an English translation of van der Waals' original paper.

2.7.1 *Van der Waals' Theory of Density Profile and Surface Tension*

Van der Waals assumed a transition layer of variable density ρ $(=N/V)$ between the liquid of density ρ_l at the bottom of a vessel and the gas of density ρ_g at the top. The density ρ is thus a function of height h, and it is quite easy to calculate the change in static energy arising as a result of the variation in density. The interaction energy of two layers separated by a distance h is now

$$A\rho(z)\rho(z+h)\psi(h)\,\mathrm{d}z\,\mathrm{d}h \qquad (2.126)$$

and this must be multiplied by $\frac{1}{2}$ to avoid double counting, and integrated over h and z to obtain the total energy of the system. We can expand $\rho(z+h)$ as a Taylor series and carry out the integration with respect to h; since $\psi(h)$ is an even function of h all odd terms will vanish. The zeroth-order term gives the energy corresponding to a homogeneous fluid of density ρ. The first correction term is

$$-\tfrac{1}{2}Ac_2\,\rho\,\frac{\mathrm{d}^2\rho}{\mathrm{d}z^2}\,\mathrm{d}z \qquad (2.127)$$

where

$$c_2 = -\int_{-\infty}^{\infty} \tfrac{1}{2}h^2\psi(h)\,\mathrm{d}h = -\frac{2\pi}{3}\int_0^{\infty} r^4\phi(r)\,\mathrm{d}r. \tag{2.128}$$

Van der Waals pointed out that this differs from the result (2.124) of the Laplace theory since it involves the second moment of the intermolecular force distribution instead of the first moment. Higher-order correction terms involve $\mathrm{d}^4\rho/\mathrm{d}z^4$ with a coefficient c_4, the fourth moment of the distribution, and so on. It is legitimate to take only the first correction term into account and ignore the remainder if ρ does not change significantly over the range of the intermolecular forces ($\mathrm{d}\rho/\mathrm{d}z \ll \lambda$). Van der Waals considered this a reasonable assumption to make. In addition, he assumed that the entropy of an inhomogeneous system depends on the density only and not on its gradient; hence (2.127) can be taken as the correction to the Helmholtz free energy arising from the inhomogeneity. Integration by parts with respect to z gives the alternative formula

$$\tfrac{1}{2}Ac_2\int\left(\frac{\mathrm{d}\rho}{\mathrm{d}z}\right)^2\mathrm{d}z \tag{2.129}$$

since $\mathrm{d}\rho/\mathrm{d}z$ tends to zero at the ends of the range of integration, in the bulk liquid and gas. This 'square gradient' term is retained in the modern theory, but has a different justification.

Consider now a fixed volume V of liquid and gas in equilibrium at a constant temperature T below T_c. There will be transition layer of profile $\rho(z)$ between the liquid and gas whose shape will be determined by minimizing the Helmholtz free energy. The external pressure which we will denote by $P_\infty(T)$ (corresponding to $z = \pm\infty$) is equal to

$$P_c[1 - \pi_0'(\tau')] \tag{2.130}$$

where $\pi_0'(\tau')$ is given by (2.75). The chemical potential of the bulk liquid and gas, denoted by $\mu_\infty(T)$, is given by (2.53′) with an appropriate value of r determined by ω_1 or ω_3 in (2.75).

Van der Waals assumed that the dependence on density of the free energy in the transition layer would be given by equation (2.37) with values of density ρ between ρ_g and ρ_l; even though these states are metastable and unstable in bulk matter, he argued that they would be stable in a transition layer where they are forced to interpolate between ρ_g and ρ_l. Hence he took for the total free energy per unit area

$$I = \int\left[\rho f + \tfrac{1}{2}c_2\left(\frac{\mathrm{d}\rho}{\mathrm{d}z}\right)^2\right]\mathrm{d}z = \int\left[a(\rho) + \tfrac{1}{2}c_2\left(\frac{\mathrm{d}\rho}{\mathrm{d}z}\right)^2\right]\mathrm{d}z \tag{2.131}$$

with $f(\rho)$ given by (2.37) and $a(\rho)$ by (2.39). This integral must be minimized using the standard calculus of variations treatment with the supplementary condition

$$\int \rho \, \mathrm{d}z = \text{constant.} \tag{2.132}$$

Introducing a Lagrange parameter λ, the Euler equations give (Margenau and Murphy 1956, Ch. 6)

$$\frac{\mathrm{d}a}{\mathrm{d}\rho} - c_2 \frac{\mathrm{d}^2\rho}{\mathrm{d}z^2} = \lambda \quad \text{or} \quad c_2 \frac{\mathrm{d}^2\rho}{\mathrm{d}z^2} = \mu(\rho,T) - \mu_\infty(T) \tag{2.133}$$

since $\mathrm{d}^2\rho/\mathrm{d}z^2$ becomes zero as $z \to \pm\infty$ in the bulk liquid and gas. Equation (2.133) determines the required profile $\rho(z)$.

To gain a better physical understanding of the content of equation (2.133), we differentiate at constant temperature and obtain

$$c_2 \, \mathrm{d}\left(\frac{\mathrm{d}^2\rho}{\mathrm{d}z^2}\right) = \mathrm{d}\mu = \frac{\mathrm{d}P}{\rho}$$

which can be integrated to give

$$P - P_\infty(T) = c_2 \int \rho \, \mathrm{d}\left(\frac{\mathrm{d}^2\rho}{\mathrm{d}z^2}\right) = c_2\left[\rho \frac{\mathrm{d}^2\rho}{\mathrm{d}z^2} - \frac{1}{2}\left(\frac{\mathrm{d}\rho}{\mathrm{d}z}\right)^2\right]. \tag{2.134}$$

The pressure throughout the closed vessel must be constant and equal to $P_\infty(T)$, and equation (2.134) shows how this pressure is derived from $P(T)$ for the homogeneous fluid at a given density ρ.

This calculation must now be related to the surface tension, and we define the surface free energy by the formula

$$\begin{aligned} \sigma &= \text{surface free energy} = \text{total free energy} - \text{bulk free energy} \\ &= \int \rho(f + P_\infty v - \mu_\infty) \, \mathrm{d}z = \int [\rho(\mu - \mu_\infty) + P_\infty - P] \, \mathrm{d}z. \end{aligned} \tag{2.135}$$

From (2.133) and (2.134) we deduce that

$$\begin{aligned} \sigma &= \tfrac{1}{2}c_2 \int_{-\infty}^{\infty} \left(\frac{\mathrm{d}\rho}{\mathrm{d}z}\right)^2 \mathrm{d}z = \tfrac{1}{2}c_2 \int_{\rho_g}^{\rho_l} \left(\frac{\mathrm{d}\rho}{\mathrm{d}z}\right) \mathrm{d}\rho \\ &= \int_{\rho_g}^{\rho_l} \{\tfrac{1}{2}c_2[\rho(\mu - \mu_\infty) + P_\infty - P]\}^{1/2} \, \mathrm{d}\rho. \end{aligned} \tag{2.136}$$

Equation (2.136) enables σ to be calculated from the equation of state (2.36) and the free energy (2.44).

To complete this theory van der Waals calculated the second-order term $\delta^2 I$ in (2.131), and showed that this is positive; hence he concluded that the solution which he derived in (2.133) is a genuine minimum and is stable.

2.7.2 *Critical Behaviour of Surface Tension and Density Profiles*

With the aid of the approximate formulae developed for the critical region the critical behaviour of the surface tension and of the density profile, and the average width of the transition layer, can readily be determined. From equation (2.75) we have

$$\frac{P_\infty}{P_c} - 1 = \pi_0 = -\pi_0' = -4\tau' + \frac{24}{5}\tau'^2 + \cdots \tag{2.137}$$

and from (2.71)

$$\frac{P}{P_c} - 1 = \pi = -\pi' = -4\tau' + 6\tau'\omega - 9\tau'\omega^2 - \frac{3}{2}\omega^3 + \frac{27}{2}\tau'\omega^3 + \frac{21}{4}\omega^4 - \frac{81}{5}\tau'\omega^4 - \frac{99}{8}\omega^5 + \cdots. \tag{2.138}$$

Substituting the relation

$$\omega = (1 + r)^{-1} = -r + r^2 - r^3 + r^4 \cdots \tag{2.139}$$

into (2.138) and retaining terms up to order r^4 or τ'^2 ($r \sim \tau'^{1/2}$ in the critical region) we obtain

$$\frac{P}{P_c} - 1 = -4\tau' - 6\tau' r + \frac{3}{2}r^3 - 3\tau' r^2 + \frac{3}{4}r^4. \tag{2.140}$$

From (2.53′) we have

$$\frac{\rho_c[\mu - \mu_0(T)]}{P_c} = -6\tau' r + \frac{3}{2}r^3 - \frac{3}{8}r^4 + \cdots. \tag{2.141}$$

Hence

$$\frac{\mu\rho}{P_c} = \frac{\mu_0(T)\rho}{P_c} - 6\tau' r + \frac{3}{2}r^3 - 6\tau' r^2 + \frac{9}{8}r^4 + \cdots. \tag{2.142}$$

To determine μ_∞ we take the value of ω_1 or ω_3 from (2.75) (both must give the same value for μ_∞), change to r_1 or r_3 by

$$r = (1 + \omega)^{-1} - 1 \tag{2.143}$$

and substitute in (2.141). We find after a little arithmetic

$$\mu_\infty \simeq 4\frac{4}{5}\tau'^2. \tag{2.144}$$

Hence, we deduce that

$$\rho(\mu - \mu_\infty) + P_\infty - P \simeq \frac{3}{8}(16\tau'^2 - 8\tau' r^2 + r^4) = \frac{3}{8}(4\tau' - r^2)^2. \tag{2.145}$$

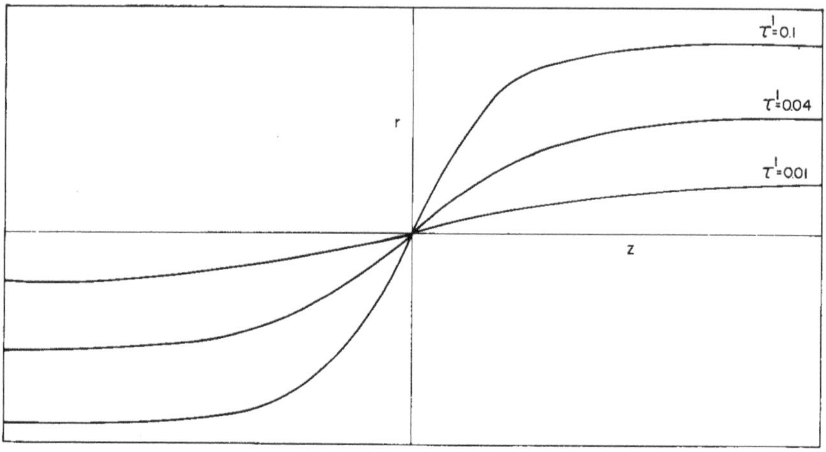

Figure 2.14 The van der Waals calculation of the density profile near T_c.

From (2.136) we can now calculate the surface free energy in the critical region,

$$\sigma = \int_{-2\tau'^{1/2}}^{2\tau'^{1/2}} \left(\frac{3}{16} c_2 P_c\right)^{1/2} (4\tau' - r^2)\rho_c \, \mathrm{d}r = \frac{8}{\sqrt{3}} \rho_c (c_2 P_c)^{1/2} \tau'^{3/2} \tag{2.146}$$

Thus the surface tension becomes zero at T_c with exponent 3/2. Modern experiments estimate this exponent as approximately 1.25; as in the bulk theory, van der Waals' theory provides a satisfactory qualitative explanation of critical behaviour, but the detailed quantitative aspects of the theory are incorrect.

The shape of the density profile can be determined from (2.133) and (2.134):

$$\begin{gathered} \tfrac{1}{2}c_2\left(\frac{\mathrm{d}\rho}{\mathrm{d}z}\right)^2 = \rho(\mu - \mu_\infty) - (P - P_\infty) = \frac{3}{8} P_c(4\tau' - r^2)^2 \\ \frac{\mathrm{d}r}{\mathrm{d}z} = m(4\tau' - r^2) \quad m = \frac{\sqrt{3}}{2\rho_c}\left(\frac{P_c}{c_2}\right)^{1/2}. \end{gathered} \tag{2.147}$$

This equation can be integrated to give

$$r = 2\tau'^{1/2} \tanh 2\tau'^{1/2} mz. \tag{2.148}$$

The density profile represented by (2.148) is illustrated in Figure 2.14.

Following Cahn and Hilliard (1958) we define the width of the boundary layer l as

$$l = \frac{\rho_l - \rho_g}{(\mathrm{d}\rho/\mathrm{d}z)_0} \simeq \frac{4\tau'^{1/2}}{4\tau' m} = m^{-1}\tau'^{-1/2}. \tag{2.149}$$

This becomes infinite at T_c with exponent 1/2.

2.8 Conclusion

With the aid of quite simple assumptions van der Waals was able to account for the most important qualitative features of the first-order transition between liquids and gases – the existence of a critical temperature, the continuity of state above the critical temperature, the characteristic discontinuities below the critical temperature, the behaviour of the surface tension, and the density profile of the boundary layer. This was a remarkable achievement. Further biographical and historical details relating to the origins of van der Waals' theory will be found in articles by de Boer (1974) and Klein (1974) in the proceedings of the conference in 1973 celebrating the centenary of the publication of van der Waals' thesis.

Notation for Chapter 2

$F(T,V)/V = \rho f(T,V) = a(T,\rho)$.
m mass, h Planck's constant, $\Lambda = h/(2\pi mkT)^{3/2}$.
P_c, T_c, V_c, ρ_c values of pressure, temperature, volume, density at the critical point.
a, b constants in the van der Waals equation.
$\pi = P/P_c - 1, \omega = V/V_c - 1, \tau = T/T_c - 1, r = \rho/\rho_c - 1$.
α coefficient of thermal expansion $= 1/v(\partial v/\partial T)$.
K_s adiabatic compressibility $= -1/v(\partial v/\partial P)_s$.
$\phi(r)$ intermolecular potential

$$\psi(h) = 2\pi \int_0^\infty \phi[(t^2 + h^2)^{1/2}]t \, dt \quad \text{(see (2.122))}$$

$$c_2 = -\int_{-\infty}^{\infty} \frac{1}{2} h^2 \psi(h) \, dh = -\frac{2\pi}{3} \int_0^\infty r^4 \phi(r) \, dr \quad \text{(see (2.128))}.$$

Exponents $\alpha, \delta, \epsilon, \zeta$: see (2.103) and (2.106).

References

CAHN, J. W. and HILLIARD, J. E. (1958) *J. Chem. Phys.* **28**, 258.
CALLEN, H. B. (1960) *Thermodynamics*, New York: John Wiley.
CLAUSIUS, R. (1850) *Ann. Phys. Chem.* **79**, 368, 500 (English translation by W. F. MAGIE (1899) in *The Second Law of Thermodynamics*, Harper and Brothers, New York, reprinted by Dover, New York, 1960, in *Reflections on the Motive Power of Fire*, Sadi Carnot, ed. E. B. MENDOZA).
DE BOER, J. (1974) *Physica* **73**, 1.
GIBBS, J. W. (1873) *Trans. Connecticut Acad.* **2**, 382; republished in *The Scientific Papers of J. Willard Gibbs*, Longmans Green, New York, 1906, reprinted by Dover, New York, 1961, Vol. 1, pp. 33–54.
GIBBS, J. W. (1875–8) *Trans. Connecticut. Acad.* **3**, 108, 343; republished in *The Scientific Papers of J. Willard Gibbs*, Longmans Green, New York, 1906, reprinted by Dover, New York, 1961, Vol. 1, pp. 219–331.

GRIFFITHS, R. B. (1967) *Phys. Rev.* **158**, 176.

GRIFFITHS, R. B. and WHEELER, J. C. (1970) *Phys. Rev. A* **2**, 1047.

HEMMER, P. C. and LEBOWITZ, J. L. (1976) DG **5b**, Ch. 2.

KAC, M., UHLENBECK, G. E. and HEMMER, P. C. (1963) *J. Math. Phys.* **4**, 216.

KLEIN, M. J. (1974) *Physica* **73**, 28.

MANDL, F. (1970) *Statistical Physics*, London: John Wiley.

MARGENAU, H. and MURPHY, G. M. (1956) *The Mathematics of Physics and Chemistry*, 2nd edition Ch. 6, Princeton, NJ: Van Nostrand.

MAXWELL, J. C. (1875) *Nature* **11**, 357, 374 (see *The Scientific Papers of J. C. Maxwell*, ed. W. D. NIVEN, Vol. 2, Cambridge, 1890, p. 418).

MAXWELL, J. C. (1876) *Capillary Action*, Encyclopedia Brittanica, 9th edition; reproduced in the *Scientific Papers of J. C. Maxwell*, ed. W. D. NIVEN, Vol. 2, Cambridge, 1890, p. 541 (reprinted Dover, New York, 1965).

MAXWELL, J. C. (1878) *Nature* **17**, 257.

PIPPARD, A. B. (1957) *The Elements of Classical Thermodynamics*, Cambridge.

POISSON, S. D. (1831) *Nouvelle Theorie de l'Action Capillaire*, Paris: Bachelier.

RAYLEIGH, LORD (1891) *Nature* **45**, 80 (*Scientific Papers*, 1902, **3**, 469).

ROWLINSON, J. S. (1973) *Nature* **244**, 414.

ROWLINSON, J. S. (1979) *J. Stat. Phys.* **20**, 197.

ROWLINSON, J. S. and WIDOM, B. (1982) *Molecular Theory of Capillarity*, Oxford.

RUSHBROOKS, G. S. (1949) *Statistical Mechanics*, Oxford.

SEGNER, F. (1751) *Comment Soc. Reg. Gottingen* **1**, 301.

THOMSON, JAMES (1871) *Proc. R. Soc.* **20**, 1 (reprinted in *Collected Papers in Physics and Engineering*, Cambridge, 1912, p. 228).

UHLENBECK, G. E. (1968) *Fundamental Problems in Statistical Mechanics*, Vol. 2, ed. E. G. D. COHEN, Amsterdam: North-Holland, p. 13.

VAN DER WAALS, J. D. (1893) *Verhandel. Kon. Acad. Weten, Amsterdam* (*Sect. 1*) **1**, NO. 8, 56.

VAN KAMPEN, N. G. (1964) *Phys. Rev. A* **135**, 362.

YOUNG, T. (1805) *Philos. Trans. R. Soc.* **95**, 65.

ZEMANSKY, M. W. (1968) *Heat and Thermodynamics*, 5th edition, New York: McGraw-Hill.

3

Magnets: Classical Theory

3.1 Thermodynamics of Magnetism

Following the distinction in §2.1 between extensive and intensive variables, we note that for magnetic systems the extensive variable is the magnetization, M, and the intensive variable is the magnetic field, H. In order to find the appropriate analogue of equation (2.3), an analysis must be undertaken of the external work done in a magnetization process. The problem is tackled from first principles by Pippard (1957, pp. 24–6), who demonstrates conclusively that the correct analogue is

$$dQ = T\,dS = dU - \mathbf{H}\,d\mathbf{M} \tag{3.1}$$

when the magnetization $\mathbf{M}$ and field $\mathbf{H}$ are not in the same direction, and

$$dQ = T\,dS = dU - H\,dM \tag{3.2}$$

in the more usual case when the magnetization and field are in the same direction. The fundamental relation corresponding to (2.7) is then

$$dU = T\,dS + H\,dM. \tag{3.3}$$

The first Legendre transformation changes the extensive variable S into the intensive variable T; it has become common practice to denote the corresponding free energy by $A(T,M)$:

$$A(T,M) = U - TS \tag{3.4}$$

$$dA = -S\,dT + H\,dM. \tag{3.5}$$

The second Legendre transformation changes the extensive variable M into the intensive variable H. For reasons related to the canonical ensemble in statistical mechanics, which we will discuss shortly, the corresponding free energy is denoted by $F(T,H)$:

$$F(T,H) = A - HM \tag{3.6}$$

$$\mathrm{d}F = -S\,\mathrm{d}T - M\,\mathrm{d}H. \tag{3.7}$$

Thus, $F(T,H)$ *for magnets is the analogue of* $G(T,P)$ *for fluids*; this is quite confusing, but the notation has become so widespread that it is difficult to advocate any change.

In §1.2 we discussed the introduction of the canonical ensemble for any macroscopic body, and the partition function was defined in (1.7). The energies of the macroscopic assemblies were expresssed in terms of the volume V, an extensive variable which is the natural independent variable for fluids. For magnetic assemblies the natural independent variable is intensive, the magnetic field H, and the different possible energy levels of the assembly are functions of H. The magnetization is defined by

$$M = -\frac{1}{Z_N}\sum_n \exp(-\beta E_n)\frac{\partial E_n}{\partial H} = \frac{1}{\beta}\frac{\partial}{\partial H}(\ln Z_N). \tag{3.8}$$

Thus the free energy $F(T,H,N)$ defined as in (1.8) by

$$F = -kT \ln Z_N \tag{3.9}$$

satisfies the first two differential relations of (1.9) as well as (3.8) and hence

$$\mathrm{d}F = -S\,\mathrm{d}T - M\,\mathrm{d}H. \tag{3.10}$$

This free energy is therefore identified with $A - HM$ of (3.6).

Introducing densities as in §2.1, and denoting M/N by m and A/N by a, we find that

$$\mathrm{d}a(T,m) = -s\,\mathrm{d}T + H\,\mathrm{d}m. \tag{3.11}$$

The analogue of m is ρ rather than v, since m increases with increasing H, and ρ increases with increasing P, whilst v decreases with increasing P. On comparing (3.11) with the function $a(T, \rho)$ introduced in (2.16), and satisfying

$$\mathrm{d}a(T,\rho) = -s\,\mathrm{d}T + \mu\,\mathrm{d}\rho \tag{3.12}$$

we find that the chemical potential μ is the analogue of the magnetic field H.

3.2 Statistical Mechanics of Paramagnetism

We illustrate the above discussion by considering in detail the statistical mechanics of paramagnetism. Consider, first, an assembly of N non-interacting *classical* dipoles of magnetic moment μ_0 in a magnetic field H. The energy of a dipole making an angle θ with the field is $-\mu_0 H \cos\theta$. Hence the classical partition function per dipole is

$$Z = \frac{1}{4\pi}\int_0^{2\pi} d\phi \int_0^{\pi} \exp(\beta\mu_0 H \cos\theta)\sin\theta \, d\theta$$
$$= \sinh(\beta\mu_0 H)/\beta\mu_0 H \quad (\beta = 1/kT). \tag{3.13}$$

By Boltzmann's relation (1.1) the average magnetic moment in the direction of H is given by

$$\langle\mu_0 \cos\theta\rangle = \frac{1}{4\pi}\int_0^{2\pi} d\phi \int_0^{\pi} \exp(\beta\mu_0 \mathrm{H} \cos\theta)\mu_0 \cos\theta \sin\theta \, d\theta$$
$$= \frac{1}{\beta}\frac{\partial}{\partial H}(\ln Z). \tag{3.14}$$

Our definition (3.8) thus identifies the magnetization m with the average magnetic moment, and this is very reasonable. From (3.13) and (3.14) we deduce that

$$m = \mu_0 L(\beta\mu_0 H) \tag{3.15}$$

where L is the Langevin function defined by

$$L(x) = \coth x - 1/x. \tag{3.16}$$

The free energy $f(T,H)$ is defined by (3.9),

$$f = -kT[\ln \sinh(\beta\mu_0 H) - \ln(\beta\mu_0 H)]. \tag{3.17}$$

If, however, we wish to calculate the free energy $a(T,m)$, we must calculate $f + Hm$ and express it as a function of m. This involves inverting the relation (3.15) to obtain

$$\beta\mu_0 H = L^{-1}\left(\frac{m}{\mu_0}\right) \tag{3.18}$$

the details of which we will discuss shortly. We now consider an assembly of N non-interacting *quantum* spins each of which has magnetic moment $g\beta_B s = \mu_0$, and whose component in the field direction is $\mu_0 s_z/s$; g is the gyromagnetic ratio, β_B the Bohr magneton and the maximum magnetic moment has been normalized so that we recover the above classical result when $s \to \infty$. The spin quantum number s_z can take values $s, s-1, s-2, \ldots, -s$, and s can have an integral or half-integral value from $s = \frac{1}{2}$ upwards. The partition function per spin is now given by

$$Z = \exp(\beta\mu_0 H)\left[1 + \exp\left(-\frac{\beta\mu_0 H}{s}\right) + \exp\left(-\frac{2\beta\mu_0 H}{s}\right) + \cdots\right.$$
$$\left. + \exp\left(-\frac{2s\beta\mu_0 H}{s}\right)\right] = \frac{\sinh\{\beta\mu_0 H[(2s+1)/2s]\}}{\sinh(\beta\mu_0 H/2s)}. \tag{3.19}$$

As before, the magnetization is identical with the average magnetic moment and is given by

$$m = kT\frac{\partial}{\partial H}(\ln Z) = \mu_0 B_s(\beta\mu_0 H) \tag{3.20}$$

where $B_s(x)$, called the Brillouin function, is defined by

$$B_s(x) = \left(\frac{2s+1}{2s}\right)\coth\left(\frac{2s+1}{2s}\right)x - \frac{1}{2s}\coth\left(\frac{x}{2s}\right). \tag{3.21}$$

Note that

$$B_{1/2}(x) = \tanh x \tag{3.22}$$

$$B_\infty(x) = L(x) \tag{3.23}$$

so that we recover the classical theory by allowing s to tend to infinity.

In the case of $s = \frac{1}{2}$, we can obtain an explicit form of the free energy $a(T,m)$. We can invert (3.20) to give

$$\beta\mu_0 H = \tanh^{-1}(m/\mu_0) = \tfrac{1}{2}\ln[(1 + m/\mu_0)/(1 - m/\mu_0)]. \tag{3.24}$$

Hence

$$\begin{aligned} a(T,m) = f(H,T) + Hm &= -kT\ln(\cosh\beta\mu_0 H) + \frac{m}{2\beta\mu_0}\ln\left(\frac{1 + m/\mu_0}{1 - m/\mu_0}\right) \\ &= -kT\ln[2(1 - m^2/\mu_0^2)^{-1/2}] + \frac{m}{2\beta\mu_0}\ln\left(\frac{1 + m/\mu_0}{1 - m/\mu_0}\right). \end{aligned} \tag{3.25}$$

In the case of general s, even though we cannot find a closed-form expression of type (3.25), we can derive series expansions from which any desired properties can be calculated. In Figure 3.1 $B_s(x)$ is plotted as a function of x for a variety of values of s. The general pattern of behaviour is the same; for small x, $B_s(x)$ is linear, and for sufficiently large x the function approaches the asymptotic value 1. The linear region gives rise to the initial susceptibility in zero field, and the asymptotic region to saturation in high fields.

The Taylor expansion of $B_s(x)$ for small x can be written in the form

$$y = B_s(x) = a_1(s)x + a_3(s)x^3 + a_5(s)x^5 + \cdots \tag{3.26}$$

where

$$a_1(s) = \frac{s+1}{3s}, \quad a_3(s) = -\frac{(s+1)[(2s+1)^2+1]}{180s^3}$$

$$a_5(s) = \frac{(s+1)[(2s+1)^4 + (2s+1)^2 + 1]}{7560s^4}.$$

This can be inverted to give

$$x = B_s^{-1}(y) = b_1(s)y + b_3(s)y^3 + b_5(s)y^5 + \cdots \tag{3.27}$$

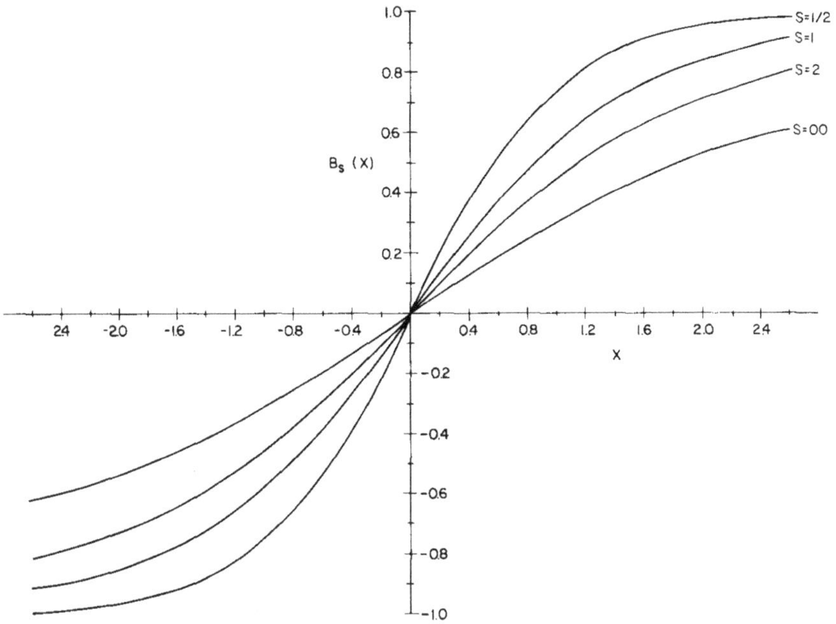

Figure 3.1 Brillouin function $B_s(x)$ vs. x (equation (3.21)).

where

$$b_1(s) = \frac{3s}{s+1}, \quad b_3(s) = \frac{9s(s^2 + s + \frac{1}{2})}{5(s+1)^3}$$

$$b_5(s) = \frac{27s[11(2s+1)^4 + 32(2s+1)^2 + 11]}{2800(s+1)^5}.$$

To determine the form of $a(T,m)$ for general s we must obtain an expression for ln Z, and it is convenient to use (3.20) in the form

$$\frac{\mathrm{d}}{\mathrm{d}x}(\ln Z) = B_s(x) \quad (x = \beta\mu_0 H). \tag{3.28}$$

Hence

$$\ln Z = \int B_s(x)\,\mathrm{d}x = \int y[B_s^{-1}(y)]'\,\mathrm{d}y = \ln(2s+1) + \tfrac{1}{2}b_1(s)y^2 + \tfrac{3}{4}b_3(s)y^4 + \tfrac{5}{6}b_5(s)y^6 + \cdots \quad (y = m/\mu_0) \tag{3.29}$$

where the constant term has been determined from the random state $m = 0$. We thus find generally that

$$\begin{aligned} a(T,m) &= -(\ln(2s+1) + \tfrac{1}{2}b_1(s)y^2 + \tfrac{3}{4}b_3(s)y^4 + \tfrac{5}{6}b_5(s)y^6 + \cdots) \\ &\quad + y[b_1(s)y + b_3(s)y^3 + b_5(s)y^5 + \cdots] \\ &= -\ln(2s+1) + \tfrac{1}{2}b_1(s)y^2 + \tfrac{1}{4}b_3(s)y^4 + \tfrac{1}{6}b_5(s)y^6 + \cdots. \end{aligned} \tag{3.30}$$

3.3 Experimental Characteristics of Ferromagnets

The magnetic properties of loadstone were known to the Ancient Greeks perhaps 2800 years ago. The first compass in the Western world was produced in the twelfth century, and because of potential commercial importance, experimental investigations of the properties of magnets were undertaken very early in the history of science. (An interesting and detailed survey of the early history of magnetism is given in Mattis (1965).) In 1600 William Gilbert published his classic treatise, *De Magnete*, which contained the observation that magnets lose their magnetism in a furnace. It took nearly 300 years more to clarify precisely how this magnetism, or magnetization as we call it nowadays, disappears. As we have mentioned in §1.4.2, among those who contributed to progress in the nineteenth century were Faraday (1845), Barrett (1874), Bauer (1880) and particularly John Hopkinson (1889), a professor of engineering at King's College, London. Hopkinson was a gifted mathematician and experimenter who was elected to the Royal Society at the age of 29 and received a Royal Medal in 1890 at the age of 40. Unfortunately he died tragically with three of his children in a climbing accident in 1898. Hopkinson was the first to use the term 'critical temperature' in connection with a ferromagnet; in his 1889 paper he refers to 'the temperature at which the magnetism disappears which we may appropriately call the critical temperature'.

We have referred in §1.4.2 to Curie's classic paper of 1895 in which he investigated the thermodynamic behaviour of ferromagnets. Of this paper Stoner (1934) wrote:

> All the observations had been in a sense isolated. A comprehensive experimental survey of a wide range of substances under widely varied conditions was needed before much further progress could be made in elucidating the significance of magnetic phenomena. It is such a survey which was provided by Curie's work, which is worthy to rank among the great classical experimental researches .... The whole investigation may be said to form the experimental foundation for modern theoretical work on magnetism.

In §1.4.2 we mentioned the analogy between magnets and fluids discovered by Curie, which led Pierre Weiss to advance his molecular field hypothesis. In the next section we shall discuss the mathematics of the Weiss theory in detail. The conventional form of m–H isothermals illustrated in Figure 3.2 results from a synthesis of the experimental work of Curie and the theoretical investigations of Weiss.

The term *Curie point* for the temperature at which ferromagnetism disappears was introduced incidentally in a footnote to a paper by Weiss and Kamerlingh Onnes in 1910. We have discussed elsewhere (Domb 1971) the injustice to Hopkinson latent in this terminology, and the time which elapsed before the term was accepted. But it is now used universally, and cannot really be changed.

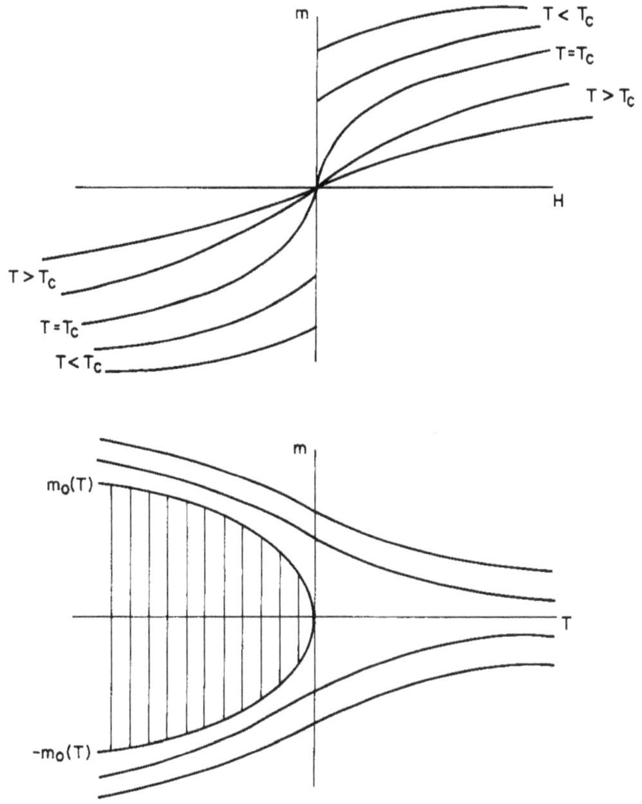

Figure 3.2 Thermodynamic behaviour of an ideal ferromagnet.

The major features of the isothermals of Figure 3.2 are the existence of T_c, the characteristic discontinuity from $-m_0(T)$ to $+m_0(T)$ in the magnetization for $T < T_c$, the behaviour of $m_0(T)$ as a function of T for $T < T_c$, the form of the critical m–H isothermal at $T = T_c$ and the behaviour of the initial magnetic susceptibility $\chi_0(T) = (\partial m/\partial H)_{H=0}$ for $T > T_c$.

3.4 The Weiss Molecular Field Hypothesis

In 1907 Pierre Weiss suggested that in a ferromagnet the interactions of the elementary molecular magnets combine to provide an internal field which must be added to the external field to obtain the true field acting on an elementary magnetic dipole. This field must depend on the degree of orientation of the elementary dipoles, and Weiss assumed empirically that it is proportional to the magnetization. Thus the true field seen by the magnetic dipole is

$H + \theta m$, where θ is a coupling constant. For a ferromagnet, instead of the Langevin relation (3.15), we should have

$$m = \mu_0 L[\beta\mu_0(H + \theta m)] \tag{3.31a}$$

and from this equation of state we should be able to derive the characteristic experimental features described in the previous section.

If the elementary molecular magnet is a quantum mechanical spin s having $(2s + 1)$ possible orientations, the same empirical assumption would lead to the equation

$$m = \mu_0 B_s[\beta\mu_0(H + \theta m)] \tag{3.31b}$$

where $B_s(x)$ is the Brillouin function defined in (3.21). Equation (3.31a) is a special case of (3.31b) with $s = \infty$.

We now examine (3.31b) to see if it gives rise to a spontaneous magnetization at sufficiently low temperatures. A solution is required having $m_0 \neq 0$ when $H = 0$. It is convenient to introduce the variable $\xi = \beta\mu_0\theta m$, and equation (3.31b) with $H = 0$ then takes the form

$$B_s(\xi) = \xi/\beta\mu_0^2\theta. \tag{3.32}$$

The left-hand side is a fixed curve for a given s starting linearly with slope $(s + 1)/3s$ and saturating to the value 1, as shown in Figure 3.1. It is clear from the form of the curves in Figure 3.1 that if the slope of the right-hand side is greater than $(s + 1)/3s$ the only solution is $\xi = 0$, but if the slope is less, i.e.

$$\frac{kT}{\mu_0^2\theta} < \frac{(s+1)}{3s}, \tag{3.33}$$

then there are three solutions, $\xi = 0$ and $\xi = \pm\xi_0$. The latter correspond to a spontaneous magnetization, and we will shortly show that they give rise to a lower free energy than the $\xi = 0$ solution and therefore correspond to the stable state. Hence we may write for the Curie temperature T_c

$$kT_c = \frac{\mu_0^2\theta(s+1)}{3s}. \tag{3.34}$$

The form of the $m_0(T)$ curve for $s = \frac{1}{2}$ and $s = \infty$ is plotted in Figure 3.3.

3.4.1 Mean Field Approximation

The above theory is empirical and does not use a microscopic model involving elementary interactions. In 1928 Heisenberg suggested that the origin of the large interactions giving rise to ferromagnetism lay in the quantum mechanical exchange forces which can be represented by a vector coupling of spins,

$$J\mathbf{s}_i \cdot \mathbf{s}_j = J(s_{xi}s_{xj} + s_{yi}s_{yj} + s_{zi}s_{zj}). \tag{3.35}$$

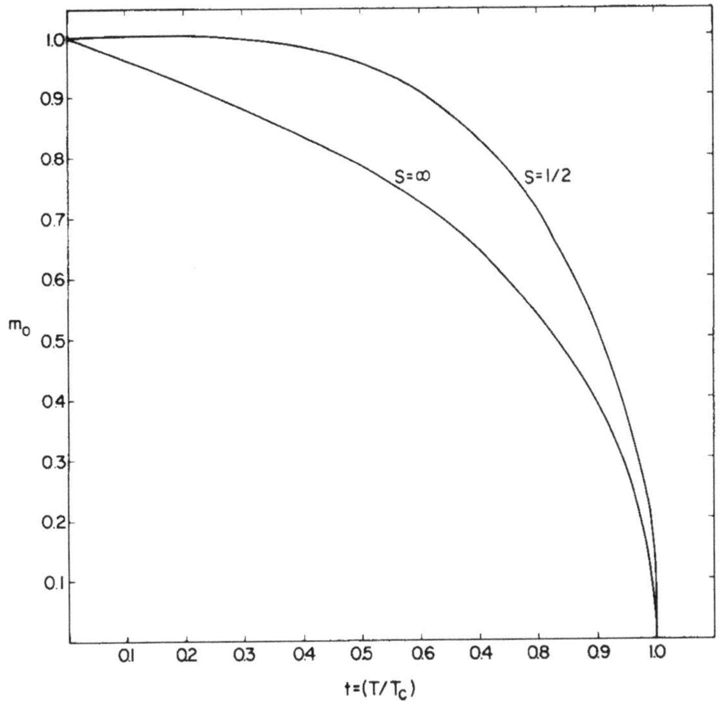

Figure 3.3 Spontaneous magnetization vs. reduced temperature in the mean field approximation.

Here s_x, s_y, s_z are quantum mechanical non-commuting operators representing the components of spin.

Because of the non-commutation of the spin operators the quantum mechanical Heisenberg model presents special difficulties which are not relevant to the approximation with which we are currently concerned. If we allow s to become infinite we derive the *classical Heisenberg model* with an interaction

$$J_i \boldsymbol{\sigma}_0 \cdot \boldsymbol{\sigma}_i \tag{3.36}$$

between spins at lattice sites 0 and i, where $\boldsymbol{\sigma}_0$, $\boldsymbol{\sigma}_i$ are unit vectors which take on all possible orientations. The internal field acting on molecule 0 as a result of all its interactions is then

$$H_{\text{int}} = \sum_{i=1}^{r} \frac{J_i}{\mu_0} \cos \theta_i \tag{3.37}$$

where θ_i is the angle between the spins. This is a fluctuating field as all the θ_i vary, and if we replace the field by its mean value we obtain a *mean field approximation* precisely analogous to the van der Waals approximation for fluids. We should expect this to be valid when a large number of neighbours contribute significantly to the interaction, i.e. for long-range forces. In fact the results have been shown to be rigorously correct by Kac and Helfand (1963) for a force of the form $\mathrm{Lt}_{\lambda\to 0}\, J\lambda \exp(-\lambda r)$ in one dimension, which is the same force as was used for the van der Waals theory, (1.45).

Replacing (3.37) by its mean value, we have

$$H_{\text{int}} = \sum_{i=1}^{r} \frac{J_i}{\mu_0} \langle \cos \theta_i \rangle. \tag{3.38}$$

But the definition of magnetization in §3.1 led to the conclusion (3.14) that $\langle \cos \theta_i \rangle$ is equal to m/μ_0. Hence we derive the Weiss hypothesis

$$H_{\text{int}} = \sum_{i=1}^{r} \frac{J_i}{\mu_0^2} m = \theta m \tag{3.39}$$

with

$$\theta = \frac{1}{\mu_0^2} \sum_{i=1}^{r} J_i$$

and for classical dipoles we will be led to equation (3.31a).

An alternative classical type of interaction is provided by the anisotropic *Ising interaction* of spin s (see §4.8),

$$\frac{J_i s_{z0} s_{zi}}{s^2} \tag{3.40}$$

where the s_{zi} can take values $s, s-1, s-2, \ldots, -s$. (The normalization has been chosen so that the maximum interaction is independent of s.) For this interaction the equivalent mean field is

$$H_{\text{int}} = \frac{1}{\mu_0} \sum_{i=1}^{r} \frac{J_i \langle s_{zi} \rangle}{s} = \frac{m}{\mu_0^2} \sum_{i=1}^{r} J_i \tag{3.41}$$

so that relation (3.39) remains valid. However, since the elementary molecular magnet has spin s, we will be led to equation (3.31b). From (3.34) we derive a formula for T_c in terms of the interactions J_i:

$$kT_c = \frac{(s+1)}{3s} \sum_{i=1}^{r} J_i. \tag{3.42}$$

3.4.2 *Equation of State of a Ferromagnet*

Writing $T/T_c = t = 1 + \tau$, and $m/\mu_0 = m^*$, we have from (3.31b) and (3.34)

$$m^* = B_s\left(\beta \mu_0 H + \frac{3s}{s+1} t^{-1} m^*\right). \tag{3.43}$$

Two important particular cases are

$$m^* = \tanh(\beta\mu_0 H + t^{-1}m^*) \quad (s = \tfrac{1}{2})$$
$$m^* = L(\beta\mu_0 H + 3t^{-1}m^*) \quad (s = \infty). \tag{3.44}$$

The above forms are implicit and it is much more convenient to invert and obtain $H(m,T)$,

$$\beta\mu_0 H = B_s^{-1}(m^*) - \frac{3s}{s+1}\, t^{-1}m^* \tag{3.45}$$

where the inverse function B_s^{-1} is defined in (3.27). For the particular case $s = \frac{1}{2}$ we can obtain the explicit formula

$$\beta\mu_0 H = \tanh^{-1}(m^*) - t^{-1}m^* = \tfrac{1}{2}\ln\left(\frac{1+m^*}{1-m^*}\right) - t^{-1}m^*. \tag{3.46}$$

But for general s we must use the power series definition (3.27),

$$\beta\mu_0 H = b_1(s)m^*(1 - t^{-1}) + b_3(s)m^{*3} + b_5(s)m^{*5} \cdots \tag{3.47}$$

and for $s = \infty$, the classical case,

$$\beta\mu_0 H = 3m^*(1 - t^{-1}) + \frac{9}{5}m^{*3} + \frac{297}{175}m^{*5} + \cdots. \tag{3.48}$$

3.4.3 *Critical Behaviour near T_c*

From equation (3.47) we can readily calculate critical behaviour near $t = 1$ for general s.

3.4.3.1 *Initial susceptibility ($T \rightarrow T_{c+}$)*

$$\frac{\chi_0}{\mu_0} = \mathop{\mathrm{Lt}}_{H\to 0} \frac{m^*}{H} = \frac{\beta\mu_0}{b_1(s)}\frac{1}{1 - t^{-1}}$$
$$\chi_0 = \frac{s+1}{3s}\frac{\mu_0^2}{k(T - T_c)} \quad \text{(Curie–Weiss law)} \tag{3.49}$$

(cf. fluids $K_T \sim \tau^{-1}$).

3.4.3.2 *Critical isotherm ($T = T_c$)*

Since $t = 1$ the first term vanishes and we find

$$\beta\mu_0 H = b_3(s)m^{*3}. \tag{3.50}$$

From (3.27)

$$b_3(\tfrac{1}{2}) = 1/3, \quad b_3(\infty) = 9/5$$

(cf. fluids $\pi \sim \omega^3 \sim r^3$).

3.4.3.3 *Spontaneous magnetization* ($T \to T_{c-}$)

The first term is now negative, and we must find a non-zero m^* satisfying $H = 0$. We easily find

$$m_0^{*2} \simeq \frac{b_1(s)}{b_3(s)} \tau' \tag{3.51}$$

$$\begin{aligned} m_0^* &\simeq \pm\sqrt{3}\,\tau'^{1/2} \quad (s = \tfrac{1}{2}) \\ m_0^* &\simeq \pm\sqrt{5/3}\,\tau'^{1/2} \quad (s = \infty). \end{aligned} \tag{3.52}$$

(cf. fluids coexistence boundary $\omega \sim r \sim \pm\tau'^{1/2}$).

3.4.3.4 *Initial susceptibility* ($T \to T_{c-}$ *at* $m = m_0^*$)

We have

$$\beta\mu_0 \frac{\mathrm{d}H}{\mathrm{d}m^*} = -b_1(s)\tau' + 3b_3(s)m^{*2} = -b_1(s)\tau' + 3b_1(s)\tau' = 2b_1(s)\tau'$$

Thus,

$$\chi_0 \simeq \frac{s+1}{6s} \frac{\mu_0^2}{k(T - T_c)} \tag{3.53}$$

which is exactly half of the high-temperature susceptibility, as for fluids.

3.4.4 Free Energy

The equation of state is not a fundamental thermodynamic relation, and we must derive the free energy from it if we wish to calculate the specific heat, and to confirm that the state with non-zero spontaneous magnetization is indeed the stable thermodynamic state. We shall integrate the thermodynamic relation $H = (\partial a/\partial m)_T$ to obtain $a(T,m)$.

We start with equation (3.46) for $s = \frac{1}{2}$, since in this case the integration can be carried out in closed form, to give

$$\beta a(T,m) = m^* \tanh^{-1}(m^*) + \tfrac{1}{2}\ln(1 - m^{*2}) - \tfrac{1}{2}t^{-1}m^{*2} + a_0(T). \tag{3.54}$$

Comparing with (3.25) and allowing T_c to become zero, we find that

$$a_0(T) = -\ln 2. \tag{3.55}$$

Hence we can write

$$a(T,m) = kT[m^* \tanh^{-1}(m^*) - \ln 2(1 - m^{*2})^{-1/2}] - \tfrac{1}{2}kT_c m^{*2}. \tag{3.56}$$

In the case of general s we must use the series expansion (3.48) and compare with (3.31a). We find

$$\beta a(T,m) = -\ln(2s + 1) + [\tfrac{1}{2}b_1 m^{*2}(1 - t^{-1}) + \tfrac{1}{4}b_3 m^{*4} + \tfrac{1}{6}b_5 m^{*6} + \cdots]. \tag{3.57}$$

If we examine the behaviour of (3.57) as a function of m, we find that it is an even function which becomes large as m becomes large. Since we have already established that there are exactly three values for which $(\partial a/\partial m)_T = 0$, the two values m_0 must be minima, and the value $m = 0$ a maximum.

3.4.5 *Specific Heat*

In order not to confuse the entropy with the spin number we shall denote the former by s_e in the remaining sections of this chapter. From (3.57) we derive a formula for the entropy,

$$s_e = k[\ln(2s + 1) - \tfrac{1}{2}b_1 m^{*2} - \tfrac{1}{4}b_3 m^{*4} - \tfrac{1}{6}b_5 m^{*6} \cdots] \tag{3.58}$$

from which we immediately deduce that.

$$C_m = T\left(\frac{\partial s_e}{\partial T}\right)_m = 0 \tag{3.59}$$

anywhere in the one-phase region. This corresponds precisely to the behaviour of C_v for fluids ((2.40) and (2.57)).

However, when we come to calculate C_H, we no longer find the strong divergence that characterized C_P for fluids. We have

$$\begin{aligned} C_H = T\left(\frac{\partial s_e}{\partial T}\right)_H &= T\left(\frac{\partial s_e}{\partial T}\right)_m + T\left(\frac{\partial s_e}{\partial m}\right)_T\left(\frac{\partial m}{\partial T}\right)_H \\ &= T(\partial H/\partial T)_m^2/(\partial H/\partial m)_T \end{aligned} \tag{3.60}$$

and since $(\partial H/\partial T)_m$ is zero when $m = 0$, C_H is also zero in this case. When m is different from zero in the one-phase region, C_H differs from zero and is equal to

$$\frac{k(b_1 m^* + b_3 m^{*3} + \cdots)^2}{[b_1(1 - t^{-1}) + 3b_3 m^{*2} + 5b_5 m^{*4} + \cdots]}. \tag{3.61}$$

But the expression (3.61) remains finite however we approach T_c. For a path in the $(H\text{–}T)$ plane of the form $H = a\tau$, the corresponding path in the $(m\text{–}\tau)$ plane is $m^* \sim \tau^{1/2}$ and the first terms in the numerator and denominator are dominant. For a path of the form $m^* \sim \tau^{\theta}$ ($\theta < \frac{1}{2}$) the second term in the denominator dominates and again the result is finite.

The general theory of strong and weak divergences developed in §2.6 is still appropriate; C_H is weakly divergent (in contrast to C_p for fluids) because it is a derivative in a direction parallel to the coexistence curve; but $\chi_0 = (\partial m/\partial T)_H$ is strongly divergent because it is a derivative in a direction perpendicular to the coexistence curve. The matrix $\mathbf{M}(H,T)$ corresponding to (2.119) again has two eigenvalues, one strongly and the other weakly divergent, and its determinant, corresponding to (2.120), is equal to $\chi_0 C_m/T = \chi_{s_e} C_H/T$. But since C_m and C_H are now both weakly divergent, χ_0 and χ_{s_e} must both be strongly divergent.

In the two-phase region we proceed as for fluids, writing

$$\begin{aligned} m^* &= \theta' m + (1 - \theta')m \\ s_e(T,m) &= \theta' s_e(T,m_+) + (1 - \theta')s_e(T,m_-) \end{aligned} \tag{3.62}$$

where m_+ is equal to $+m_0(T)$ and m_- to $-m_0(T)$. Since $s_e(T,m)$ depends only on m^2 (3.58), it is the same function of T for m_+ and m_-, and is therefore the same function of T for the whole of the coexistence region. We then have

$$C_m = T\,\frac{\mathrm{d}s_e}{\mathrm{d}T} \tag{3.63}$$

which can be conveniently evaluated at $m = m_0(T)$. In order to get a detailed picture of the behaviour of C_m near T_c we must evaluate an additional term in the expansion of (3.51)

$$m_0^{*2} = \frac{b_1}{b_3}\tau' + \left(\frac{b_1}{b_3} - \frac{b_5 b_1^2}{b_3^3}\right)\tau'^2 \tag{3.64}$$

and must then evaluate $s_e(T,m)$ as far as the term in τ'^2. We find that near T_c, C_m is of the form

$$C_m \sim k\Delta_s[1 - A(s)\tau'] \tag{3.65}$$

where

$$\Delta_s = \frac{b_1^2}{2b_3} = \frac{5s(s+1)}{2(s^2 + s + \frac{1}{2})}$$

$$A(s) = \frac{2b_1 b_5}{b_3^2} - 2 = \frac{11(2s+1)^4 + 32(2s+1)^2 + 1}{56(s^2 + s + \frac{1}{2})^2} - 2.$$

This has the particular values

$$\frac{3}{2}k\left(1 - \frac{8}{5}\tau'\right) \quad (s = \tfrac{1}{2}) \qquad \frac{5}{2}k\left(1 - \frac{8}{7}\tau'\right) \quad (s = \infty). \tag{3.66}$$

Table 3.1

Magnets	Fluids
Spontaneous magnetization $m_0 \simeq \pm\tau'^{1/2}$	Coexistence boundary $r_0 \sim \pm\tau'^{1/2}$
Critical isotherm $H \sim m^3$	Critical isotherm $\pi \sim r^3$
Initial susceptibility ($H = 0$, $T \to T_{c+}$) $\chi_0 \sim \tau^{-1}$	Isothermal compressibility on critical isochore ($T \to T_{c+}$)$K_T \sim \tau^{-1}$
Initial susceptibility ($m = \pm m_0$, $T \to T_{c-}$) $\frac{1}{2}\chi_0(T \to T_{c+})$	Isothermal compressibility on coexistence boundary $\frac{1}{2}K_T(T \to T_{c+})$
Specific heat $C_m \sim \Delta_s k$ ($\tau < 0$)	Specific heat $C_v - \frac{3}{2}k \sim \frac{9}{2}k$ ($\tau < 0$)
$= 0$ ($\tau > 0$)	$= 0$ ($\tau > 0$)

Since $H = 0$ in the whole of the coexistence region, and $s_e(T,m)$ remains the same function of T, we can conclude that C_H in this region is not infinite, but has exactly the same value as C_m.

3.5 Magnet–Fluid Analogy

Even though the classical equations of state of fluids and magnets are very different we have noted that the pattern of critical behaviour is the same. This is summarized in Table 3.1.

For magnets we have found that critical exponents do not depend on s (universality), but critical amplitudes and higher correction terms do. The symmetry in magnets is exact, whereas in fluids it is only approximate. Magnets do not manifest any latent heat. This is because the (H–T) coexistence curve has zero slope, and

$$-\frac{\mathrm{d}H}{\mathrm{d}T} = \frac{\Delta s_e}{\Delta m} = 0. \tag{3.67}$$

Since $\Delta m \neq 0$, $\Delta s_e = 0$. Also, from thermodynamics

$$\Delta u - T\Delta s_e - H\Delta m = 0 \tag{3.68}$$

so that $\Delta u = 0$; the difference between the behaviour of C_H and C_P arises because for magnets the coexistence curve is parallel to the axis.

3.6 Equation of State in the Critical Region

In applying statistical mechanics to specific magnetic models it is usually convenient to use H as an independent variable. The equation of state (3.47), which we have derived, expresses H as a function of m, whereas from statistical mechanics we would derive m as a function of H. If we invert (3.47) we find that

$$m^* = c_1(\beta\mu_0 H) + c_3(\beta\mu_0 H)^3 + c_5(\beta\mu_0 H)^5 \cdots \tag{3.69}$$

where

$$\begin{aligned} c_1 &= b_1^{-1}(1-t^{-1})^{-1} \sim \tau^{-1} \quad c_3 = b_3 b_1^{-1}(1-t^{-1})^{-4} \sim \tau^{-4} \\ c_5 &= -b_5 b_1^{-1}(1-t^{-1})^{-6} + 3b_3^2 b_1^{-7}(1-t^{-1})^{-7} \sim \tau^{-7}. \end{aligned} \tag{3.70}$$

Thus,

$$\frac{dm}{dH} \sim \tau^{-1}, \quad \frac{d^3m}{dH^3} \sim \tau^{-4}, \quad \frac{d^5m}{dH^5} \sim \tau^{-7} \cdots, \tag{3.71}$$

i.e. there is a constant 'gap' exponent $\Delta = 3/2$ such that

$$\frac{d^{2r+1}m}{dH^{2r+1}} \sim \tau^{-1-2r\Delta}. \tag{3.72}$$

Retaining only the dominantly divergent terms in (3.69), we can write

$$m^* = H\tau^{-1}F(H\tau^{-3/2})$$

or more conveniently

$$m^*\tau^{-1/2} = \Phi(H\tau^{-3/2}). \tag{3.73}$$

Inverting (3.73) gives

$$H\tau^{-3/2} = \Psi(m^*\tau^{-1/2}) \tag{3.74}$$

and when we compare this with our original equation (3.47) we find that only two terms of the expansion have been retained, and

$$\Psi(u) = (\beta\mu)^{-1}(b_1 u + b_3 u^3) \quad (u \approx m^*\tau^{-1/2}). \tag{3.75}$$

In fact (3.75) has retained only terms needed to describe the leading order behaviour in the critical region, and the remaining terms in (3.47) give rise to corrections to this behaviour. The form (3.75) is not convenient if we need to pass to negative τ because of the square root. We therefore change variables and write

$$H = \tau^{3/2}\Psi(m^*\tau^{-1/2}) = m^{*3}h(m^{*-2}\tau) \tag{3.76}$$

where it is easy to verify that

$$h(x) = (\beta\mu_0)^{-1}(b_3 + b_1 x) \quad (x = \tau m^{*-2}). \tag{3.77}$$

The full equation (3.47) can then be written

$$\begin{aligned}\beta\mu_0 H &= m^*b_1(\tau - \tau^2 + \tau^3 - \tau^4 \ldots) + b_3 m^{*3} + b_5 m^{*5} + b_7 m^{*7} \cdots \\ &= m^{*3}(b_3 + b_1\tau m^{*-2}) + m^{*5}(b_5 - b_1\tau^2 m^{*-4}) \\ &\quad + m^{*7}(b_7 - b_1\tau^3 m^{*-6}) + \cdots\end{aligned} \tag{3.78}$$

the second and third terms representing the corrections mentioned above. Likewise if we return to the (μ–r) equation for fluids, (2.53′), we can write it as

$$\frac{\mu - \mu_0(T)}{P_c v_c} = \frac{3}{2} r^3(1 + 4\tau r^{-2}) - \frac{3}{8} r^4 + \frac{3}{2} r^5(\tau r^{-2}) \tag{3.79}$$

where the leading term is identical in form with the leading term of (3.78), but the correction terms are different.

The critical behaviour for $\tau < 0$ can readily be determined from (3.77). The spontaneous magnetization corresponds to a zero of $h(x)$

$$\begin{aligned}x_0 &= -b_3/b_1 = \tau m_0^{*-2} \\ m_0^* &= \left(\frac{b_1}{b_3}\tau'\right)^{1/2}\end{aligned} \tag{3.80}$$

as before. We can find higher derivatives at $m = m_0^*$ as follows:

$$\left(\frac{\mathrm{d}H}{\mathrm{d}m^*}\right)_{m_0^*} = 3m_0^{*2}h(-x_0) - 2\tau h'(-x_0) = 2\tau' b_1 \tag{3.81}$$

as before,

$$\begin{aligned}\left(\frac{\mathrm{d}^2H}{\mathrm{d}m^{*2}}\right)_{m_0^*} &= -6m_0^{*-1}\tau h'(-x_0) = 6m_0^{*-1}\tau' b_1 \sim \tau'^{1/2} \\ \left(\frac{\mathrm{d}^2m^*}{\mathrm{d}H^2}\right)_{m_0^*} &= -\frac{(\mathrm{d}^2H/\mathrm{d}m^{*2})}{(\mathrm{d}H/\mathrm{d}m^*)^3} \sim \tau'^{-5/2}\end{aligned} \tag{3.82}$$

and similarly for higher derivatives. Hence there is also a gap exponent $\Delta' = 3/2$ on the low-temperature side, and $\Delta' = \Delta$.

3.7 Surface Effects and Domain Walls

The discussion of §2.7 should have its analogue for a magnetic system. For example, formula (2.124) for the static surface energy near a free surface remains valid for a magnetic system. But if we wish to consider the analogue of the transition layer between a liquid and a gas, which is the domain wall between regions magnetized in different directions, different features enter which have no parallel for fluids. A van der Waals type of theory has only limited application and is largely of theoretical interest.

A fluid is homogeneous, and the sample size is determined by the material available. When a liquid and gas coexist each occupies its own homogeneous region with a single transition layer between them. However, a ferromagnet splits up into a large number of domains which are magnetized in different directions. The size of the domains and their direction of magnetization depend on a number of factors: on the nature of the main microscopic magnetic interaction mentioned briefly in §3.4.1; on dipolar forces which are usually much weaker than the main magnetic interaction but which have a long range; on the anisotropy energy which is determined by the orientation of the elementary magnets to the crystalline axes of the solid (for a discussion see Ashcroft and Mermin (1976, pp. 718–22)).

Because of the complexity of the problem little work was done on the critical behaviour of domain walls during the classical period.

Notation for Chapter 3

μ_0 magnetic moment of spin.
$m(H,T)$ magnetization per spin, $m_0(T)$ spontaneous magnetization per spin.
$m/\mu_0 = m^*$, $m_0/\mu_0 = m_0^*$.
$\mathbf{s}_i$ vector spin with components (s_{xi}, s_{yi}, s_{zi}).
$\boldsymbol{\sigma}_i$ unit vector $(\sigma_{xi}, \sigma_{yi}, \sigma_{zi})$ representing spin taking all orientations.
J_{ij} interaction energy between spins i, j.
$B_s(x)$ Brillouin function for spin s

$$= \left(\frac{2s+1}{2s}\right)\coth\left(\frac{2s+1}{2s}\right)x \; \frac{1}{2s}\coth\frac{x}{2s} \quad \text{(see (3.21))}.$$

$L(x)$ Langevin function $= B_\infty(x) = \coth x - 1/x$.
s_e entropy per spin (to avoid confusion with spin s)
χ_{s_e} adiabatic susceptibility $= (\partial M/\partial H)_{s_e}$.

References

ASHCROFT, N. W. and MERMIN, N. D. (1976) *Solid State Physics*, New York: Holt, Rinehart and Winston.

BARRETT, W. F. (1874) *Philos. Mag.* **47**, 51.

BAUER, C. (1880) *Wiedemann Ann. Phys. Chem.* **11**, 394.

CURIE, P. (1895) *Ann. Chim. Phys.* **5**, 289.

DOMB, C. (1971) *Statistical Mechanics at the Turn of the Decade*, ed. E. G. D. COHEN, New York: Marcel Dekker.

FARADAY, M. (1845) *Experimental Researches in Electricity*, §2343–7 (Vol. 3, pp. 54–6).

GILBERT, W. (1600) *De Magnete* (republished by the Gilbert Club, London, 1900; revised edition by Basic Books, New York, 1958).

HEISENBERG, W. (1928) *Z. Phys.* **49**, 619.

HOPKINSON, J. (1889) *Philos. Trans. R. Soc. A* **180**, 443.
KAC, M. and HELFAND, E. (1963) *J. Math. Phys.* **4**, 1078.
MATTIS, D. C. (1965) *The Theory of Magnetism*, New York: Harper and Row.
PIPPARD, A. B. (1957) *The Elements of Classical Thermodynamics*, Cambridge.
STONER, E. C. (1934) *Magnetism and Matter*, London: Methuen.
WEISS, P. (1907) *J. Phys.* **6**, 661.
WEISS, P. and KAMERLINGH ONNES, H. (1910) *J. Phys.* **9**, 555.

4

Light Scattering and Correlations: Classical Theory

4.1 Critical Opalescence

In the opening remarks of his 1869 Bakerian lecture Andrews quoted from an earlier communication in 1863 (which was published in the 3rd edition of Miller's *Chemical Physics*, p. 328). In it he said:

> On partially liquefying carbonic acid by pressure alone, and gradually raising at the same time the temperature to 88° Fahr., the surface of demarcation between the liquid and gas became fainter, lost its curvature, and at last disappeared. The space was then occupied by a homogeneous fluid, which exhibited, when the pressure was suddenly diminished or the temperature slightly lowered, a peculiar appearance of moving or flickering striae throughout its entire mass.

This phenomenon, subsequently termed critical opalescence, attracted the attention of a number of experimentalists in the late nineteenth and early twentieth centuries. Avenarius (1874) described the experimental results in great detail noting striking changes of colour and the onset of turbidity in carbon disulfide, ether, carbon dichloride and acetone. Altschul (1893), Wesendonck (1894), Travers and Usher (1906) and Young (1906) investigated the phenomenon more precisely, although none of them refers to the observations of Andrews or Avenarius. It is indeed striking to observe a colourless transparent fluid suddenly becoming opaque and changing colour in a narrow band of temperatures around T_c. As the temperature is lowered, the fluid splits into colourless liquid and gas with a meniscus separating them. In addition to the fluids mentioned by Avenarius, the phenomenon was observed for sulfur dioxide, pentane, isopentane, hexane and octane, so that it could reasonably be regarded as universal. Doubts were expressed as to whether the simple van der Waals theory was capable of accounting for the observations.

Smoluchowski (1908, 1912) and Einstein (1910) were the first to identify the source of the opalescent bands. Fluctuations in the density of the fluid give rise to fluctuations in the refractive index, and hence to light scattering. Einstein had shown how to calculate these fluctuations, and the conclusion was drawn that they would increase dramatically in the neighbourhood of T_c.

4.2 Thermodynamic Fluctuations

The formula

$$S = k \ln \Omega \tag{4.1}$$

relating the thermodynamic entropy of a given state to its statistical probability is usually associated with the name of Boltzmann, and is even engraved on his tombstone; in fact it was first enunciated by Planck, although Boltzmann had developed all of the background ideas. Einstein was unaware of Boltzmann's work and obtained the result independently, but Einstein's approach to the formula differed significantly from that of Boltzmann, and was really complementary to it (for details see Pais (1982, pp. 60–75)).

Boltzmann's aim was to establish a theory of matter based on probability and statistics, and formula (4.1) served as the bridge by which he could get back to thermodynamics. Einstein in his studies on Brownian movement had been concerned with fluctuations and their experimental observation. Rewriting (4.1) in the form

$$\Omega = \exp(S/k) \tag{4.2}$$

and making use of thermodynamic information expressing the entropy as a function of state, it is possible to calculate the probability of deviations from equilibrium values. He applied this to the calculation of fluctuations in temperature, density, pressure, etc., in a finite system.

Consider a small region of a large reservoir, which we can consider to be attached to it in equilibrium, as illustrated diagrammatically in Figure 4.1. The suffix 0 will be used to refer to the reservoir, and we can assume that the temperature T_0 and pressure P_0 of the reservoir are constant; for convenience we keep the number of molecules N in the small region constant, allowing the volume to fluctuate. Consider any fluctuation of the small region of temperature change ΔT, volume change ΔV, entropy change ΔS, and internal energy change ΔU. The probability of this fluctuation, w, is given from (4.2) by

$$w \propto \exp(\Delta S + \Delta S_0)/k \tag{4.3}$$

where the total system, reservoir + small region, is isolated at constant energy and volume. Thus,

$$\Delta U + \Delta U_0 = 0, \quad \Delta V + \Delta V_0 = 0. \tag{4.4}$$

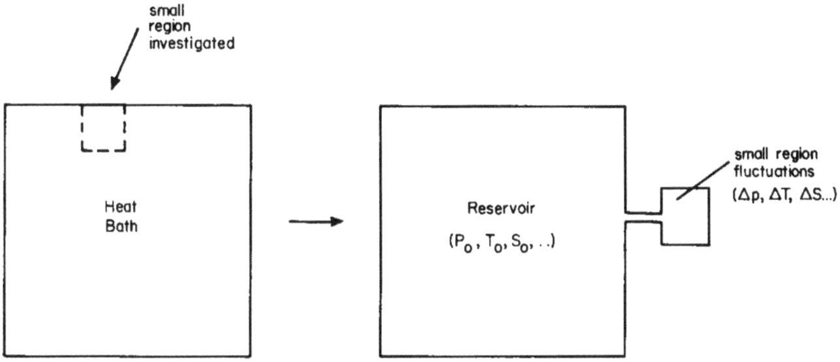

Figure 4.1 Thermodynamic fluctuations (following Einstein).

But

$$\Delta S_0 = \left(\frac{\Delta Q}{T_0}\right)_{\text{reservoir}} = \frac{\Delta U_0 + P_0 \Delta V_0}{T_0} = -\frac{\Delta U + P_0 \Delta V}{T_0}.$$

Hence,

$$w \propto \exp[-\beta(\Delta U - T_0 \Delta S + P_0 \Delta V)]. \tag{4.5}$$

To determine the fluctuations of any pair of thermodynamic variables the exponent in (4.5) must be expanded as far as second-order terms. If the variables of interest are ΔT and ΔV, it is convenient to use the Helmholtz free energy ΔF, and we readily find that

$$w \propto \exp[-\beta(\Delta F + S_0 \Delta T + P_0 \Delta V + \Delta S \Delta T)]. \tag{4.6}$$

Note that this formula is exact. Expanding to second order,

$$\Delta F = -S_0 \Delta T - P_0 \Delta V - \frac{1}{2}\left(\frac{C_V}{T}(\Delta T)^2 + 2\left(\frac{\partial S}{\partial V}\right)_T \Delta T \Delta V + \left(\frac{\partial P}{\partial V}\right)_T (\Delta V)^2\right)$$

$$\Delta S = \frac{C_V}{T}\Delta T + \left(\frac{\partial S}{\partial V}\right)_T \Delta V$$

we deduce that

$$w \propto \exp\left[-\frac{C_V}{2kT^2}(\Delta T)^2 + \frac{1}{2kT}\left(\frac{\partial P}{\partial V}\right)_T (\Delta V)^2\right]. \tag{4.7}$$

Hence the fluctuations follow a Gaussian distribution, and

$$\begin{aligned} \langle \Delta T^2 \rangle &= kT^2/C_V \\ \langle \Delta V^2 \rangle &= -kT(\partial V/\partial P)_T = kTVK_T. \end{aligned} \tag{4.8}$$

The fluctuations in volume are thus proportional to the compressibility and are large near T_c.

For convenience N has been kept constant and V allowed to fluctuate. Since the density ρ is equal to N/V, we find that the fluctuations in density are given by

$$\langle \Delta\rho^2 \rangle = \frac{N^2}{V^4} \langle \Delta V^2 \rangle = \frac{\rho^2 kT}{V} K_T. \tag{4.9}$$

It would have been equally correct to have calculated $\langle \Delta\rho^2 \rangle$ by keeping V constant and varying N, with

$$\langle \Delta\rho^2 \rangle = \langle \Delta N^2 \rangle / V^2. \tag{4.9'}$$

By comparing (4.9) and (4.9′) we deduce that

$$\langle \Delta N^2 \rangle = \rho^2 kTVK_T = \rho kTV \left(\frac{\partial \rho}{\partial P} \right)_T \quad \left[K_T = \frac{1}{\rho} \left(\frac{\partial \rho}{\partial P} \right)_T \right]. \tag{4.10}$$

This could have also been deduced directly from (4.6) by taking ΔN as one of the variables, and using the appropriate thermodynamic function.

4.3 Theory of Einstein and Smoluchowski*

Smoluchowski and Einstein assumed that the change in density is accompanied by a change in refractive index according to the Clausius–Mossotti relation (Ashcroft and Mermin 1976, p. 542)

$$\frac{\epsilon - 1}{\epsilon + 2} = A\rho \tag{4.11}$$

so that

$$\langle \Delta\epsilon^2 \rangle = \langle \Delta\rho^2 \rangle \left(\frac{\mathrm{d}\epsilon}{\mathrm{d}\rho} \right)^2 = \frac{kTK_T}{9V} (\epsilon - 1)^2 (\epsilon + 2)^2 \tag{4.12}$$

using (4.9). Assuming that the scattering from an individual element followed Rayleigh's law, and that scattering from different elements added randomly, they arrived at the formula (see Callen 1960, p. 180)

$$\frac{I_\theta}{I_0} = \frac{\pi^2 V^2 \langle \Delta\epsilon^2 \rangle}{2\lambda_0^4} \frac{1 + \cos^2\theta}{r^2} \tag{4.13}$$

Here θ is the angle between the incident and scattered beams, I_θ is the intensity of the scattered wave and I_0 of the incident wave for light of wavelength λ_0; the first factor on the right-hand side represents the Rayleigh scattering from the inhomogeneities, and the second factor is the polar diagram arising from the random addition of the intensities of the two polarizations.

* The respective contributions of Einstein and Smoluchowski are outlined in Pais (1982, pp. 100–3).

The scattering intensity is thus proportional to K_T/λ_0^4, and it was assumed that critical opalescence was associated with the large increase in K_T near T_c. At T_c the scattering remains finite experimentally even though K_T becomes infinite; it was assumed that since the coefficient of $(\Delta V)^2$ in (4.7) becomes zero at T_c, higher-order terms of the exponent would be needed to deal with the situation at this temperature, and $\langle \Delta N^2 \rangle$ in (4.10) would be proportional to V^b with $b > 1$.

In their classic paper of 1914, Ornstein and Zernike showed that such a conclusion was logically untenable if (as had been assumed) correlations between different elements of fluid were ignored. In fact they suggested that such correlations could not be ignored, and played a major role in determining critical behaviour; as mentioned in §1.4.3, they introduced the concept of the molecular pair distribution function which has figured centrally in all subsequent work on liquids.

4.4 Correlation in Fluids

Let $\nu(\mathbf{dr})$ be a random variable representing the number of particles in a volume $\mathbf{dr}$ centred at $\mathbf{r}$. Since $\mathbf{dr}$ is small the probability of occupation by more than one particle is of order $\mathbf{dr}^2$ (the distribution of $\nu(\mathbf{dr})$ will be Poisson). Hence we can write

$$\langle \nu(\mathbf{dr}) \rangle = n_1(\mathbf{r})\ \mathbf{dr} \tag{4.14}$$

$$\langle \nu^2(\mathbf{dr}) \rangle = n_1(\mathbf{r})\ \mathbf{dr} \tag{4.15}$$

where $n_1(\mathbf{r})$ defines the density ρ which we take to be constant.

For the correlation between particles at $\mathbf{r}_1$ and $\mathbf{r}_2$ we define the average

$$\langle \nu(\mathbf{dr}_1)\nu(\mathbf{dr}_2) \rangle = n_2(\mathbf{r}_1,\mathbf{r}_2)\ \mathbf{dr}_1\ \mathbf{dr}_2 \tag{4.16}$$

and $n_2(\mathbf{r}_1,\mathbf{r}_2)$ is called the *pair distribution function.* It is usual to normalize by defining

$$g(\mathbf{r}_1,\mathbf{r}_2) = n_2(\mathbf{r}_1,\mathbf{r}_2)/n_1(\mathbf{r}_1)n_1(\mathbf{r}_2) \tag{4.17}$$

which would equal 1 if there were no correlation (since we should then have $\langle \nu(\mathbf{dr}_1)\nu(\mathbf{dr}_2) \rangle = \langle \nu(\mathbf{dr}_1) \rangle \langle \nu(\mathbf{dr}_2) \rangle$). For a homogeneous fluid

$$g(\mathbf{r}_1,\mathbf{r}_2) = g(|\mathbf{r}_1 - \mathbf{r}_2|) = g(\mathbf{r}). \tag{4.18}$$

To represent deviations from randomness we write

$$h(\mathbf{r}) = g(\mathbf{r}) - 1 \tag{4.19}$$

so that $h(\mathbf{r}) = 0$ for random distributions. $h(\mathbf{r})$ is called the *pair correlation function.*

Now consider a fixed volume V containing a fluctuating number N of particles. Dividing into microscopic cells of volume $d\mathbf{r}_i$,

$$N = \sum_i \nu(d\mathbf{r}_i)$$

from which we deduce that

$$\langle N \rangle = \sum_i \langle \nu(d\mathbf{r}_i) \rangle = \sum_i n_1(\mathbf{r}_i)\, d\mathbf{r}_i = \rho V. \tag{4.20}$$

Likewise,

$$\begin{aligned}\langle N^2 \rangle &= \left\langle \sum_i \nu^2(d\mathbf{r}_i) \right\rangle + \sum_{i \neq j} \langle \nu(d\mathbf{r}_i)\nu(d\mathbf{r}_j)\rangle = \rho V + \iint n_2(\mathbf{r}_2,\mathbf{r}_2)\, d\mathbf{r}_1\, d\mathbf{r}_2 \\ &= \rho V + \rho^2 V \int g(\mathbf{r})\, d\mathbf{r}.\end{aligned} \tag{4.21}$$

Hence

$$\begin{aligned}\langle \Delta N^2 \rangle &= \langle N^2 \rangle - \langle N \rangle^2 = \rho V - \rho^2 V^2 + \rho^2 V \int g(\mathbf{r})\, d\mathbf{r} \\ &= \rho V + \rho^2 V \int [g(\mathbf{r}) - 1]\, d\mathbf{r}.\end{aligned} \tag{4.22}$$

Thus from (4.10) we deduce the relation

$$kT\rho K_T = 1 + \rho \int [g(\mathbf{r}) - 1]\, d\mathbf{r} = 1 + \rho \int h(\mathbf{r})\, d\mathbf{r}. \tag{4.23}$$

This is often referred to as the *fluctuation theorem*.

4.5 Terminology for Light Scattering (Figure 4.2)

Assume an incident wave $\mathbf{E}_0 \exp(-i\omega_0 t + \mathbf{k}_0 \mathbf{R})$ and a scattered wave from each molecule $(A\mathbf{E}_0/R) \exp(-i\omega_0 t + \mathbf{k}_s \mathbf{R})$. The difference in phase between incident and scattered waves for a molecule P at position $\mathbf{r}$ is $(\mathbf{k}_0 - \mathbf{k}_s) \cdot \mathbf{r}$; $(\mathbf{k}_0 - \mathbf{k}_s)$, which has magnitude $2|\mathbf{k}_0|\sin(\theta/2)$ and is inversely proportional to λ_0, will be denoted by $\mathbf{q}$.

Summing the scattering over all N molecules of the body we have

$$\mathbf{E}_s = \frac{A\mathbf{E}_0}{R} \sum_{j=1}^{N} \exp(i\mathbf{q} \cdot \mathbf{r}_j). \tag{4.24}$$

Denote the intensity of the scattered light by $I(\mathbf{q})$. Then

$$I(\mathbf{q}) = \langle \mathbf{E}_s \mathbf{E}_s^* \rangle = \frac{|A|^2 |\mathbf{E}_0^2|}{R^2} \left\langle \sum_{j=1}^{N} \sum_{l=1}^{N} \exp[i\mathbf{q}(\mathbf{r}_j - \mathbf{r}_l)] \right\rangle$$

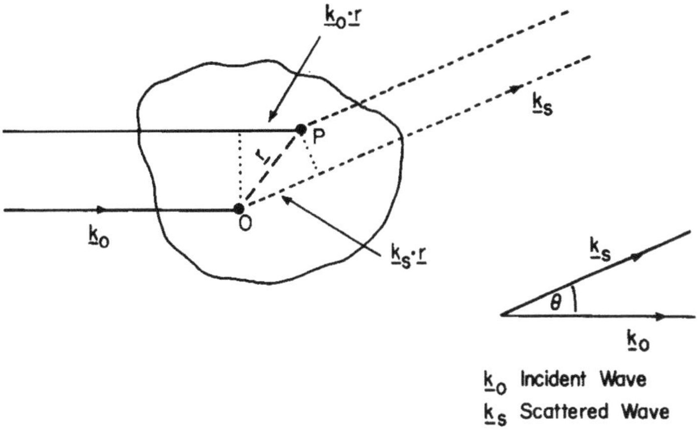

Figure 4.2 Light scattering terminology.

and

$$S(\mathbf{q}) = \frac{I(\mathbf{q})}{I_0(\mathbf{q})} = \frac{1}{N}\left\langle \sum_{j=1}^{N} \sum_{l=1}^{N} \exp[i\mathbf{q}(\mathbf{r}_j - \mathbf{r}_l)] \right\rangle \tag{4.25}$$

where $I_0(\mathbf{q}) = N|A|^2|\mathbf{E}_0^2|/R^2$ refers to random scatterers. $S(\mathbf{q})$ is sometimes called the structure factor, and is equal to 1 for random scatterers:

$$\begin{aligned} S(\mathbf{q}) &= 1 + \frac{1}{N}\left\langle \sum_{j\neq l} \exp[i\mathbf{q}(\mathbf{r}_j - \mathbf{r}_l)] \right\rangle \\ &= 1 + \frac{1}{N}\iint d\mathbf{r}_1\, d\mathbf{r}_2 \exp[i\mathbf{q}(\mathbf{r}_2 - \mathbf{r}_1)]\mathbf{n}_2(\mathbf{r}_2, \mathbf{r}_1) \\ &= 1 + \rho\int \exp(i\mathbf{q}\cdot\mathbf{r})g(\mathbf{r})\, d\mathbf{r} \end{aligned} \tag{4.26}$$

using (4.16)–(4.18).

Now $g(\mathbf{r})$ is nearly 1 for large $\mathbf{r}$ and $\int \exp(i\mathbf{q}\cdot\mathbf{r})\, d\mathbf{r} = \delta(\mathbf{q})$. If we define

$$\chi(\mathbf{q}) = 1 + \rho\int \exp(i\mathbf{q}\cdot\mathbf{r})h(\mathbf{r})\, d\mathbf{r} \tag{4.27}$$

we have

$$S(\mathbf{q}) = \chi(\mathbf{q}) + \rho\delta(\mathbf{q}) \tag{4.28}$$

and, from (4.23),

$$\operatorname*{Lt}_{q\to 0} \chi(q) = kT\rho K_T \quad \text{(small-angle scattering).} \tag{4.29}$$

Because of (4.27) it is natural to define Fourier transforms in d dimensions,

$$\hat{h}(\mathbf{q}) = \int h(\mathbf{r}) \exp(\mathrm{i}\mathbf{q} \cdot \mathbf{r})\, \mathrm{d}\mathbf{r}$$

$$h(\mathbf{r}) = (2\pi)^{-d} \int \hat{h}(\mathbf{q}) \exp(-\mathrm{i}\mathbf{q} \cdot \mathbf{r})\, \mathrm{d}\mathbf{q} \qquad (4.30)$$

the integrals being taken over the whole of d-dimensional $\mathbf{r}$ and $\mathbf{q}$ space.

In three dimensions for an isotropic system,

$$\hat{h}(q) = 2\pi \int_0^\infty r^2\, \mathrm{d}r h(r) \int_0^\infty \exp[\mathrm{i}(qr \cos\theta)] \sin\theta\, \mathrm{d}\theta$$

$$= 4\pi \int_0^\infty \frac{\sin qr}{qr} h(r) r^2\, \mathrm{d}r. \qquad (4.31)$$

We then have

$$\chi(\mathbf{q}) = 1 + \rho\hat{h}(\mathbf{q}). \qquad (4.32)$$

4.6 Theory of Ornstein and Zernike (O–Z): Direct and Indirect Correlation*

O–Z divided the correlation between two molecules 1 and 2 into two parts, a *direct* part arising from the 1–2 molecular interaction, and an *indirect* part arising from the intervention of other molecules (Figure 4.3). They envisaged that the direct correlation function $c(\mathbf{r})$ would be of short range (e.g. $-\beta\phi(r)$,

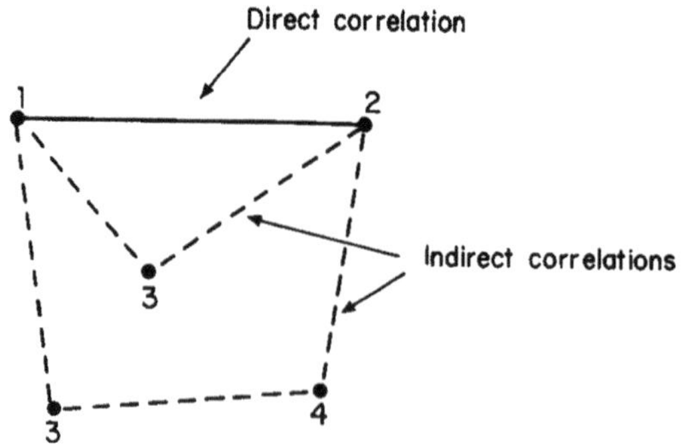

Figure 4.3 Basic idea of Ornstein–Zernike theory.

* This section is based largely on Fisher (1964).

where $\phi(r)$ is the intermolecular potential) even if the total correlation function $h(\mathbf{r})$ were long ranged. In fact, they wrote down the equation

$$\begin{aligned} h(\mathbf{r}_{12}) &= c(\mathbf{r}_{12}) + \int c(\mathbf{r}_{13})c(\mathbf{r}_{32})\rho \, d\mathbf{r}_3 \\ &\quad + \iint c(\mathbf{r}_{13})c(\mathbf{r}_{34})c(\mathbf{r}_{42})\rho^2 \, d\mathbf{r}_3 \, d\mathbf{r}_4 + \cdots \\ &= c(\mathbf{r}_{12}) + \rho \int c(\mathbf{r}_{13})h(\mathbf{r}_{32}) \, d\mathbf{r}_3 \end{aligned} \tag{4.33}$$

which is usually known as the O–Z integral equation. Without any theory we can regard (4.33) as defining a new function $c(\mathbf{r})$ related to $h(\mathbf{r})$. We can write (4.33) rather more conveniently in the form

$$h(\mathbf{r}) = c(\mathbf{r}) + \rho \int c(\mathbf{R} - \mathbf{r})h(\mathbf{R}) \, d\mathbf{R}. \tag{4.34}$$

Taking Fourier transforms in (4.34) we have

$$\hat{h}(\mathbf{q}) = \hat{c}(\mathbf{q}) + \rho\hat{h}(\mathbf{q})\hat{c}(\mathbf{q}), \tag{4.35}$$

or

$$\begin{aligned} \hat{h}(\mathbf{q}) &= \hat{c}(\mathbf{q})/[1 - \rho\hat{c}(\mathbf{q})] \\ \hat{c}(\mathbf{q}) &= \hat{h}(\mathbf{q})/[1 + \rho\hat{h}(\mathbf{q})]. \end{aligned} \tag{4.36}$$

Also

$$\chi(\mathbf{q}) = 1 + \rho\hat{h}(\mathbf{q}) = [1 - \rho\hat{c}(\mathbf{q})]^{-1} \tag{4.37}$$

and, from (4.23),

$$\chi(0) = 1 + \rho\hat{h}(0) = kT_{\rho}K_T. \tag{4.38}$$

All the considerations so far have been exact. The assumption of O–Z was that $c(\mathbf{r})$ is sufficiently short ranged for its moments to exist, and hence that $\hat{c}(\mathbf{q})$ should have a Taylor expansion in q. This will be the case, for example, if $c(\mathbf{r})$ has a finite cut-off, or an exponential tail. For isotropic systems in three dimensions

$$\hat{c}(\mathbf{q}) = 4\pi \int_0^{\infty} \frac{\sin qr}{qr} c(r)r^2 \, dr \tag{4.39}$$

and following the O–Z assumption, one can write

$$\hat{c}(q) = \hat{c}(0) - L^2q^2 + O(q^4) \tag{4.40}$$

where

$$\hat{c}(0) = 4\pi \int_0^\infty r^2 c(r)\, dr, \quad L^2 = \frac{2\pi}{3}\int_0^\infty c(r) r^4\, dr \tag{4.41}$$

and L has the dimensions of a length. Noting that $1/\chi(0)$ is equal to $1 - \rho\hat{c}(0)$ from (4.37), we find that $\hat{c}(0)$ is a slowly varying function of T finite at T_c, and we expect similar behaviour for L^2. We then have

$$\hat{h}(q) \sim \hat{c}(0)/\{[1 - \rho\hat{c}(0)] + \rho L^2 q^2\} = \frac{\hat{c}(0)\rho^{-1}L^{-2}}{q^2 + \kappa^2} \tag{4.42}$$

where

$$\kappa^2 = [1 - \rho\hat{c}(0)]/\rho L^2. \tag{4.43}$$

To identify κ, O–Z used the van der Waals result that $K_T \sim \tau^{-1}$ near T_c. Hence from (4.38) and (4.42),

$$\kappa^2 \sim \tau, \quad \kappa \sim \tau^{1/2}. \tag{4.44}$$

From (4.42) and (4.44) we note that at T_c, $\chi(q)$ is no longer infinite but

$$\chi(q) \sim 1/q^2. \tag{4.45}$$

The infinity occurs only in the limit of very small angles and is no longer observable. However, the total scattering in all directions is of the order

$$\int_0^\pi \frac{2\pi \sin\theta\, d\theta}{4\pi^2 \sin^2(\theta/2)} = \frac{1}{\pi}\int_0^\pi \cot\frac{\theta}{2}\, d\theta \tag{4.46}$$

which diverges. Thus the O–Z theory is not completely consistent, and we shall find that the deficiency is corrected in modern theories.

From (4.25) and (4.42) the dependence on wavelength of the scattered light is

$$\frac{1}{\lambda_0^4}\frac{1}{(q^2 + \kappa^2)} = \frac{1}{\lambda_0^2[\lambda_0^2\kappa^2 + 4\pi^2\sin^2(\theta/2)]}. \tag{4.47}$$

At T_c the scattering is no longer Rayleigh but $\sim 1/\lambda_0^2$. Near T_c it varies between $1/\lambda_0^2$ and $1/\lambda_0^4$, moving towards Rayleigh scattering as the angle θ becomes smaller.

Equation (4.42) is a 'Lorentzian' scattering curve for the variation of scattering amplitude with q. If we plot $1/\chi(q)$ as a function of q^2 near T_c using (4.37) and (4.40), we should obtain a series of straight lines which go through the origin when $T = T_c$ (Figure 4.4), and whose intercept on the y axis gives $\kappa^2(\tau)$.

To calculate $h(r)$ in three dimensions we take Fourier transforms in (4.42) and using (4.30) and (4.31) we find that

$$h(r) \sim \int_0^\infty \frac{\sin qr}{qr}\frac{q^2\, dq}{(q^2 + \kappa^2)} = \frac{1}{r} f(\kappa r) \tag{4.48}$$

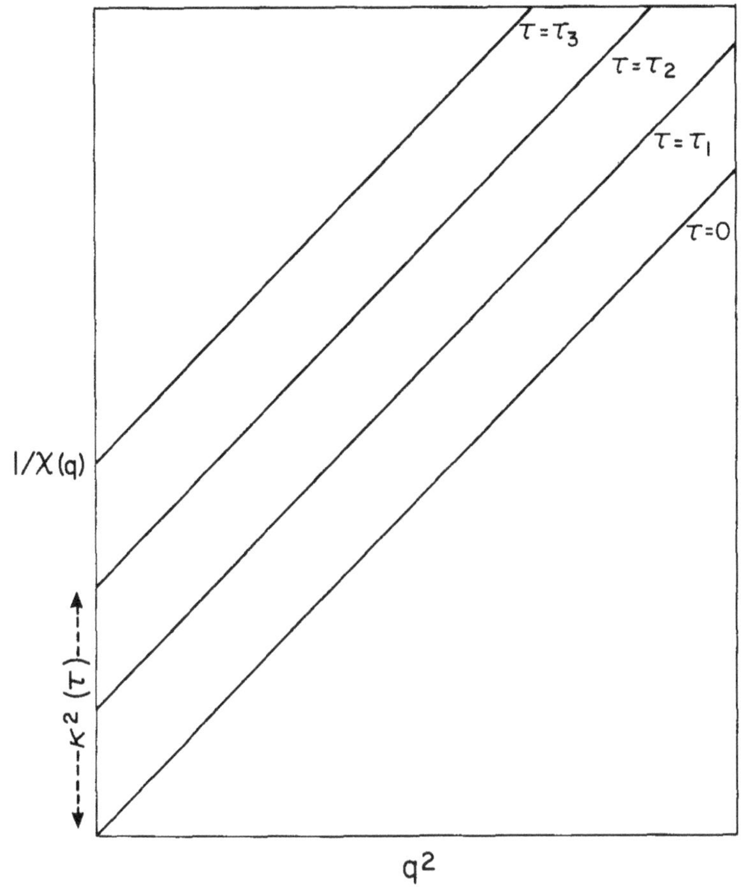

Figure 4.4 Plot of inverse scattering amplitude vs. q^2 (following Ornstein–Zernike theory).

where

$$f(y) = \int_0^\infty \frac{u \sin u \, du}{u^2 + y^2} = \exp(-y).$$

Hence

$$h(r) \sim \frac{1}{r} \exp(-\kappa r). \tag{4.49}$$

Where τ is significantly different from zero the correlation drops off exponentially and is short ranged. But at $\tau \to 0$ it becomes long ranged $\sim 1/r$. We can define a 'coherence length' l by

$$l^2 \sim \int r^4 h(r)\, dr \Big/ \int r^2 h(r)\, dr \sim \kappa^{-2} \sim \tau^{-1} \tag{4.50}$$

and this goes to infinity at $\tau \to 0$. (The integrals in (4.50) can be evaluated exactly from $\hat{h}(q)$ by differentiating w.r.t. q.)

4.6.1 Generalization to d Dimensions

The discussion of the previous section generalizes without difficulty to d dimensions. The major formulae remain unchanged, only auxiliary formulae like (4.39) requiring appropriate modifications. We note briefly for the record that the formula for transforming to d-dimensional hyperspherical polar coordinates is

$$\begin{aligned} x_1 &= r \sin\theta_1 \cos\theta_2 \ldots \cos\theta_{d-1} \\ x_2 &= r \sin\theta_1 \cos\theta_2 \ldots \sin\theta_{d-1} \\ x_3 &= r \sin\theta_1 \cos\theta_2 \ldots \sin\theta_{d-2} \\ x_4 &= r \sin\theta_1 \cos\theta_2 \ldots \sin\theta_{d-3} \\ &\vdots \\ x_{d-1} &= r \sin\theta_1 \sin\theta_2 \\ x_d &= r \cos\theta_1. \end{aligned} \tag{4.51}$$

The volume element in d-dimensional polar coordinates is

$$r^{d-1} \sin^{d-2}\theta_1 \sin^{d-3}\theta_2 \ldots \sin\theta_{d-2}\, dr\, d\theta_1\, d\theta_2 \ldots d\theta_{d-1}. \tag{4.52}$$

The limits of integration are 0 to π for θ_1, and 0 to 2π for $\theta_2, \ldots, \theta_{d-1}$. The volume element for a system with hyperspherical symmetry is obtained by integrating over θ_1 to θ_{d-1}, and is $2\pi^{d/2} r^{d-1}/\Gamma(d/2)$. We find instead of (4.46) that for all $d \geq 4$ the integral which gives the total scattering converges; O–Z theory is therefore completely consistent for $d \geq 4$.

To calculate the function $h(r)$ in d dimensions from $h(q)$ in (4.42) it is convenient to use Cartesian coordinates as in (4.30). We need to evaluate

$$\int_{-\infty}^{\infty} \cdots \int_{-\infty}^{\infty} dq_1\, dq_2 \ldots dq_d \frac{\exp(-i\mathbf{q}\cdot\mathbf{r})}{q^2 + \kappa^2} \tag{4.53}$$

and we can apply the elegant method used by Elliott Montroll for a similar integral in the theory of lattice Green's functions (see e.g. Montroll and Weiss 1965). Using the identity

$$\int_0^{\infty} \exp(-tx)\, dx = \frac{1}{t} \tag{4.54}$$

we can write (4.53) as

$$\int_0^\infty dt \int_{-\infty}^{\infty} \cdots \int_{-\infty}^{\infty} dq_1 \ldots dq_d \exp[-t(q^2 + \kappa^2) - i\mathbf{q} \cdot \mathbf{r}]. \tag{4.55}$$

Now

$$\begin{aligned} \int_{-\infty}^{\infty} \exp[-(tq_1^2 + iq_1x_1)]\, dq_1 &= \int_{-\infty}^{\infty} \exp\left[-t\left(q_1 + \frac{ix_1}{2t}\right)^2\right] \exp\left(-\frac{x_1^2}{4t}\right) dq_1 \\ &= \left(\frac{\pi}{t}\right)^{1/2} \exp\left(-\frac{x_1^2}{4t}\right). \end{aligned} \tag{4.56}$$

Hence (4.55) reduces to

$$\int_0^\infty dt \left(\frac{\pi}{t}\right)^{d/2} \exp[-(\kappa^2 t + r^2/4t)]. \tag{4.57}$$

Using the standard form (Gradzshtyn and Ryzhik 1965, p. 340, formula (9))

$$\int_0^\infty x^{\nu-1} \exp\left(-\frac{\beta}{x} - \gamma x\right) dx = 2\left(\frac{\beta}{\gamma}\right)^{\nu/2} K_\nu[2(\beta\gamma)^{1/2}] \tag{4.58}$$

where K_ν is the Bessel function ($K_\nu = K_{-\nu}$), we find that in d dimensions,

$$h(\mathbf{r}) \sim \frac{1}{r^{d-2}} (\kappa r)^{d/2-1} K_{d/2-1}(\kappa r) = \frac{1}{r^{d-2}} f(\kappa r) \tag{4.59}$$

where

$$f(u) = u^{d/2-1} K_{d/2-1}(u). \tag{4.60}$$

We therefore find that at $T = T_c$ ($\kappa = 0$) the long-range correlation decays as $1/r^{d-2}$, and $f(\kappa r)$ has the asymptotic expansion for large κr,

$$f(\kappa r) = (\kappa r)^{d/2-1-1/2} \exp(-\kappa r) = (\kappa r)^{(d-3)/2} \exp(-\kappa r). \tag{4.61}$$

Away from T_c at a fixed temperature T, the decay of $h(\mathbf{r})$ has the asymptotic form, for large r,

$$r^{(-1/2)(d-1)} \exp(-\kappa r). \tag{4.62}$$

Let us finally look at the form of $f(u)$ for initial values of d:

$d = 2$: $f(u) = K_0(u)$ diverges logarithmically at $u = 0$. Sensible results can be derived only for $T \neq T_c$.
$d = 3$: $f(u) = u^{1/2} K_{1/2}(u) \sim \exp(-u)$ as before.
$d = 4$: $f(u) = uK_1(u) \sim A + B u \ln u$ at $u = 0$.
$d = 5$: $f(u) = u^{3/2} K_{3/2}(u) \sim (1 + u) \exp(-u)$.

4.7 Order in Solids

So far the question of symmetry has not entered into the discussion. There is no difference in symmetry between the liquid and vapour phases, and the properties of a ferromagnet, in which there is a clear difference of symmetry between the magnetized and unmagnetized phases, were derived by analogy with a fluid. But in considering the phenomena of order–disorder transitions in binary alloys, which were briefly mentioned in §1.4.4, the important role which symmetry plays was manifest at the outset. In the low-temperature phase there is a regular arrangement of atoms which gives rise to superlattice lines in the X-ray diffraction picture; these lines are not present in the high-temperature phase.

There are two types of atom, A and B, which in the perfectly ordered state occupy a and b sites respectively. We shall confine attention to an AB type of alloy (like β-brass) on a lattice which can be decomposed into two equivalent sublattices (e.g. simple cubic, body-centred cubic). Early theoretical treatments attempted to deal with the more general case of $A_r B_{1-r}$, and did not appreciate how much more complicated the problem is when A and B sites are non-equivalent. A state of partial order is defined by a parameter t:

$$t = \frac{(\text{A on } a)_t - (\text{A on } a)_{\text{random}}}{(\text{A on } a)_{\text{perfect}} - (\text{A on } a)_{\text{random}}} = \frac{(\text{A on } a)_t - \frac{1}{2}}{1 - \frac{1}{2}} \tag{4.63}$$

where $(\text{A on } a)_t$ represents the fraction of A atoms on a sites.

Thus

$$(\text{A on } a)_t = \tfrac{1}{2}(1 + t), \quad (\text{A on } b)_t = \tfrac{1}{2}(1 - t)$$

and by symmetry (4.64)

$$(\text{B on } b)_t = \tfrac{1}{2}(1 + t), \quad (\text{B on } a)_t = \tfrac{1}{2}(1 - t).$$

For a perfectly ordered alloy $t = 1$, and for a random mixture $t = 0$.

Bragg and Williams (1934) had in mind the kinetic picture of interchanges taking place continuously and dynamic equilibrium being maintained. They focused attention on one particular atom A in equilibrium spending part of its time on an a site and part on a b site, the ratio at temperature T being determined by Boltzmann's relation (1.1). Hence they were led to an equation for t as a function of the energy of an interchange and T. The method is short and neat, but it is difficult to see the nature of the approximation involved. We shall therefore adopt the alternative approach of other authors who attempted to construct the free energy $U - TS$ as a function of t, and minimized it to obtain $t(T)$.

The entropy S as a function of t is easy to derive by simple combinational arguments. The number of ways of organizing the A atoms is

$$\frac{(N/2)!}{[(N/2)\frac{1}{2}(1+t)]!\,[(N/2)\frac{1}{2}(1-t)]!} \tag{4.65}$$

with an identical factor for the B atoms. Hence

$$\begin{aligned}\frac{S(t)}{k} &= 2\ln\frac{N}{2}! - 2\ln\left(\frac{N}{4}(1+t)\right)! - 2\ln\left(\frac{N}{4}(1-t)\right)! \\ &= N\ln 2 - \left(\frac{N}{2}(1+t)\ln(1+t) + (1-t)\ln(1-t)\right).\end{aligned} \tag{4.66}$$

To calculate the energy of a given configuration it was assumed that interatomic forces decrease rapidly with distance so that a model taking account of only nearest-neighbour interactions is reasonable. Let us denote the interactions of an AA, AB and BB pair of neighbours by $-\epsilon_{AA}$, $-\epsilon_{AB}$ and $-\epsilon_{BB}$ respectively. Consider a configuration in which there are N_{AA}, N_{AB} and N_{BB} such pairs. It is easy to establish the relations

$$\begin{aligned} 2N_{AA} + N_{AB} &= qN_A \\ N_{AB} + 2N_{BB} &= qN_B \end{aligned} \tag{4.67}$$

by enumerating all q bonds which originate in each atom of the lattice (q being the coordination number of the lattice). There is therefore only one free parameter if N_A, N_B are fixed, and the energy of the configuration is

$$\begin{aligned} w &= -N_{AA}\epsilon_{AA} - N_{AB}\epsilon_{AB} - N_{BB}\epsilon_{BB} \\ &= -\tfrac{1}{2}q(N_A\epsilon_{AA} + N_B\epsilon_{BB}) - N_{AB}\epsilon \quad [\epsilon = \epsilon_{AB} - \tfrac{1}{2}(\epsilon_{AA} + \epsilon_{BB}) > 0]. \end{aligned} \tag{4.68}$$

The values of N_{AB} will follow a probability distribution, and correspondingly the different possible energies w. If we replace w by its average value, we will be using a mean field type of approximation discussed in §3.4.1. The mean value of N_{AB} is readily calculated from relations (4.64), starting with the A atom and considering separately the cases when it is on an a site or a b site. We find

$$\langle N_{AB}\rangle = \frac{N}{2}\cdot\tfrac{1}{2}(1+t)\cdot\tfrac{1}{2}(1+t) + \frac{N}{2}\cdot\tfrac{1}{2}(1-t)\cdot\tfrac{1}{2}(1-t) = \frac{N}{4}(1+t^2). \tag{4.69}$$

We therefore write for the internal energy

$$U(t) = -\tfrac{1}{2}q(N_A\epsilon_{AA} + N_B\epsilon_{BB}) - \frac{N}{4}(1+t^2). \tag{4.70}$$

The equilibrium value is obtained by minimizing the free energy $U - TS$ from (4.66) and (4.70) as a function of t, and is given by

$$\ln\frac{1+t}{1-t} = t\epsilon/kT.$$

This is identical with the equation for the spontaneous magnetization in the classical magnetic theory for $s = \frac{1}{2}$ (equation (3.31b) with $H = 0$). Bragg and Williams observed:

> The general conclusion that the order sets in abruptly below a critical temperature T_c has a case analogy in ferromagnetism, and there are, in fact, many points of similarity between the present treatment and the classic equations of Langevin and Weiss. We may compare the degree of order in the alloy with the intrinsic field of a ferromagnetic. The average orientation of the atomic magnets corresponds to the degree of order in the alloy, and the difference in potential energy for the parallel and antiparallel positions to the $V(w)$ we have considered above. There is a similar dependence of the one on the other, and on the temperature, in both cases. Hence the alloy has a critical temperature above which no order or superlattice exists, just as the ferromagnetic has a *Curie point*.

4.8 Ising Model

The analogy between alloys and magnets to which Bragg and Williams drew attention can be seen more clearly in relation to the Ising model, which has been mentioned in §3.4.1. In 1925 W. Lenz, who had been looking for a simple model which might serve to explain ferromagnetism, put forward the following suggestion to his graduate student E. Ising. Assume that each atom possesses a spin and hence a magnetic moment, μ_0, which can orient either parallel or antiparallel to an external magnetic field H. There is an interaction between nearest-neighbour spins in the lattice, parallel spins having an interaction $-J$, and antiparallel spins $+J$ ($J > 0$). Lenz hoped that the model might manifest a non-zero spontaneous magnetization, i.e. that as $H \to 0$ the ratio

$$(N_2 - N_1)/(N_2 + N_1) \tag{4.71}$$

might tend to a non-zero value. Here N_1, N_2 are the numbers of spins antiparallel and parallel to the field H.

Ising was able to solve the problem in one dimension only where the statistical problem is quite simple, and the solution did not possess a spontaneous magnetization at any temperature above zero. He concluded wrongly that the model would not give rise to a spontaneous magnetization in higher dimensions. In 1936 Peierls was the first to demonstrate convincingly that for a two-dimensional Ising model, at sufficiently low but non-zero temperatures, the ratio (4.71) does tend to a non-zero value as $H \to 0$.

Ising published no other papers in this field and by now thousands of papers have appeared discussing the properties of the Ising model. To give proper credit to the true initiator of the model some have adopted the terminology *Lenz–Ising model*, but this practice has unfortunately not become very widespread. (For an interesting historical review of the model see Brush (1967).)

The Hamiltonian for the Ising model can be conveniently written in the form

$$\mathscr{H} = -J\sum_{\langle ij\rangle}\sigma_i\sigma_j - \mu_0 H \sum_i \sigma_i \tag{4.72}$$

where the sum i is taken over all N spins of the lattice, and $\langle ij\rangle$ over all of the nearest-neighbour pairs of the lattice, counting each pair once only. Each σ_i takes on the values ∓ 1 according as the spin i is pointing in the direction 1 or 2. Ising considered only a model of spin $\frac{1}{2}$; the natural generalization to spin s is

$$\mathscr{H} = -\frac{J}{s^2}\sum_{\langle ij\rangle} s_{zi}s_{zj} - \frac{\mu_0 H}{s}\sum_i s_{zi} \tag{4.73}$$

where s_{zi} are standard quantum spins taking on the values s, $(s-1)$, ..., $-(s-1)$, $-s$. The normalization is chosen to keep the maximum interaction between nearest-neighbour spins, and the maximum interaction with the external field, independent of s. The term Ising model without qualification usually refers to the model of spin $\frac{1}{2}$; it is sometimes called the simple Ising model for greater clarity and will be denoted by $I(\frac{1}{2})$; the more general model (4.73) will be denoted by $I(s)$.

Let us now consider the partition function for the $I(\frac{1}{2})$ model. Any configuration of spins characterized by the numbers N_1, N_2 of spins pointing in directions 1 and 2, and N_{11}, N_{12}, N_{22} different types of nearest-neighbour pair, will have energy

$$\mu_0 H(N_1 - N_2) + J(N_{12} - N_{11} - N_{22}). \tag{4.74}$$

But by analogy with (4.67) we can write

$$\begin{aligned} 2N_{11} + N_{22} &= qN_1 \\ 4N_{12} + 2N_{22} &= qN_2. \end{aligned} \tag{4.75}$$

For convenience, to avoid surface effects, we adopt the usual solid state physics device of cyclic boundary conditions, so that all lattice sites are equivalent. Also we know that

$$N_1 + N_2 = N. \tag{4.76}$$

Hence (4.74) can be written

$$-(\mu_0 H + \tfrac{1}{2}qJ)N + 2\mu_0 HN_1 + 2JN_{12}. \tag{4.77}$$

Of the five numbers introduced only two are free, and these are conveniently taken to be N_1, N_{12}.

The major difficulty in the calculation of the partition function is the combinatorial problem of determining how many states there are on a lattice with a given (N_1,N_{12}). If we denote this by $g(N;N_1,N_{12})$, the partition function can be written as

$$Z_N^1 = \langle \exp(-\beta\mathscr{H}) \rangle = y^{-N/2} z^{-qN/4} \sum_{N_1, N_{12}} g(N; N_1, N_{12}) y^{N_1} z^{N_{12}} \tag{4.78}$$

$$[y = \exp(-2\beta\mu_0 H) \quad z = \exp(-2\beta J) \quad \beta = 1/kT].$$

The direct analogue of an Ising ferromagnet in the realm of alloys is a *solid solution* with N_A atoms of type A, and N_B atoms of type B. Equations (4.67) and (4.68) remain valid, but $\varepsilon < 0$ (cf. equation (4.68)) so that the lowest energy state corresponds to the minimum of N_{AB}, i.e. to two separated phases of A and B atoms. The partition function (p.f.) for the solid solution is therefore

$$Z_N^s = z_A^{(1/2)qN_A} z_B^{(1/2)qN_B} \sum_{N_{AB}} g(N; N_A, N_{AB}) z^{N_{AB}} \tag{4.79}$$

$$[z_A = \exp(\beta\epsilon_{AA}) \quad z_B = \exp(\beta\epsilon_{BB}) \quad z = \exp(\beta\epsilon)].$$

We have differentiated between N_A, N_B for the alloy model and N_1, N_2 for the magnetic model to remind us that the former are fixed and the latter have to be summed over to derive the p.f.

On comparing (4.79) and (4.78) we see that except for ground state terms the p.f. for the alloy is obtained by picking out the coefficient of y^{N_A} in the p.f. for the magnet. Writing

$$\Lambda_N(y, z) = \sum_{N_1, N_{12}} g(N; N_1, N_{12}) y^{N_1} z^{N_{12}} \tag{4.80}$$

we can obtain the required coefficient by the standard method of statistical mechanics and saddle-point technique (Fowler 1936). The coefficient is given by

$$\frac{1}{2\pi i} \oint_c \frac{\Lambda_N(y, z)}{y^{N_A+1}} \, dy \tag{4.81}$$

and the saddle point by

$$\frac{\delta}{\delta y} (\ln \Lambda_N - N_A \ln y) = 0. \tag{4.82}$$

We readily find that for the magnet the p.f. is given by

$$\ln Z_N^1 = -N/2 \ln y - qN/4 \ln z + \ln \Lambda_N(y, z) \tag{4.83}$$

whereas for the alloy

$$\ln Z_N^s = N/4 \ln(z_A z_B) + \ln \Lambda_N(y, z) - N_A \ln y \tag{4.84}$$

at the value of y given by (4.82).

An alternative and more fundamental way of looking at the relationship between alloy and magnet problems is to note that (4.80) defines a grand p.f. for the alloy, y playing the role of the activity. The result (4.84) can then be readily derived from standard p.f. theory (Domb 1960).

The magnetic p.f. is unchanged by a reversal of magnetic field. Hence from (4.78)

$$y^{-N/2}\Lambda_N(y, z) = y^{N/2}\Lambda_N(y^{-1}, z). \tag{4.85}$$

Taking logarithms, differentiating, and putting $H = 0$ ($y = 1$), we find that

$$\left(\frac{\delta}{\delta y} \ln \Lambda_N(y, z)\right)_{y=1} = \frac{N}{2}. \tag{4.86}$$

Hence when $N_A = \frac{1}{2}$ the p.f. for the alloy is effectively identical with that for the magnet.

The order–disorder problem discussed in the previous section corresponds to the Ising model of an antiferromagnet ($J < 0$), and for simplicity we will restrict attention to 'even' lattices like the SQ, SC and BCC which can be decomposed into two equivalent sublattices.

The relationship between the magnet and alloy problems is as before, and in zero magnetic field the two are effectively identical. Finally it is easy to see that in zero field there is a precise correspondence of configurations between ferro- and antiferro-magnet. The analogy noted by Bragg and Williams does not depend on their particular approximation but is exact.

4.8.1 *One-dimensional Assemblies: Transfer Matrix*

For a one-dimensional assembly of systems each having t possible energy levels a general treatment is possible which reduces the evaluation of the p.f. to the determination of the largest eigenvalue of a matrix. Let $U(x_i, x_{i+1})$ denote the energy between any two systems x_i and x_{i+1}, each x_i taking t possible values. For simplicity we shall deal with a ring of systems in which the Nth is connected cyclically to the first. The p.f. for N systems is then given by

$$Z_N = \sum_{x_i} \exp\{-\beta[U(x_1,x_2) + U(x_2,x_3) + \cdots + U(x_N,x_1)]\}$$

$$= \sum_{i,j,\ldots,k=1}^{t} V_{ij} V_{jk} \cdots V_{kl} V_{li} \tag{4.87}$$

where $V_{ij} = \exp(-\beta U_{ij})$ corresponds to the interaction between neighbouring systems in states i and j. The elements V_{ij} thus form a $t \times t$ matrix $\mathbf{V}$ and the continued product (4.87) will readily be identified as the trace of $\mathbf{V}^N$. Thus if $\lambda_1, \lambda_2, \ldots, \lambda_t$ are the eigenvalues of $\mathbf{V}$

$$Z_N = \lambda_1^N + \lambda_2^N + \cdots + \lambda_t^N. \tag{4.88}$$

For sufficiently large N only the largest eigenvalue, λ_1, is significant, and we may identify Z_N with λ_1^N. It should be noted that if the interaction U_{ij} is symmetric in i and j the matrix $\mathbf{V}$ is symmetric. Also the ith diagonal element of $\mathbf{V}^N$ is proportional to the probability that a particular system is in state i.

The above method of treating one-dimensional assemblies was suggested independently by Kramers and Wannier (1941), Lassettre and Howe (1941) and Montroll (1941). It constituted the first preliminary step towards modern theories of critical behaviour, focusing attention on the exact calculation of the p.f. The matrix **V** is usually called the *transfer matrix*.

Let us apply the method to the one-dimensional simple Ising model. To preserve the symmetry we can split the interaction with the magnetic field between i and $(i + 1)$, and write

$$U(x_i, x_{i+1}) = -J\sigma_i\sigma_{i+1} + -\tfrac{1}{2}\mu_0 H(\sigma_i + \sigma_{i+1}). \tag{4.89}$$

Hence we find for **V** the 2×2 matrix

$$\mathbf{V} = \begin{vmatrix} y^{-1/2}z^{-1/2} & z^{1/2} \\ z^{1/2} & y^{1/2}z^{-1/2} \end{vmatrix} \tag{4.90}$$

so that

$$Z_N = y^{-N/2}z^{-N/2}(\Lambda_1^N + \Lambda_2^N) \tag{4.91}$$

with

$$\left.\begin{matrix}\Lambda_1\\ \Lambda_2\end{matrix}\right\} = \tfrac{1}{2}(1 + y) \pm [(1 - y)^2 + 4yz^2]^{1/2}. \tag{4.92}$$

This is the solution found by Ising (1925).

For large N we can ignore Λ_2, and since Λ_1 is a smooth analytic function of (y, z) there are no singularities. There is not spontaneous magnetization for $T > 0$; the only manifestation of singular behaviour occurs at $T = 0$ where the susceptibility χ_0 takes on the value

$$\chi_0 \to \beta\mu_0^2 \exp(\beta J) \tag{4.93}$$

which becomes infinite. The physical properties of the one-dimensional Ising chain are derived in Domb (1960).

The method outlined above can be applied to a finite strip of width l, and the number of different states t to be taken into account for the transfer matrix is then 2^l. Similarly, for a three-dimensional block of dimensions $l \times m$ the transfer matrix is of order 2^{lm}. As long as the transfer matrix is finite, and the largest eigenvalue is non-degenerate, it is easy to show that it has no singularities (Domb 1960, §3.2.3), and a theorem of Perron (1907) and Frobenius (1908) ensures that this eigenvalue is indeed non-degenerate since all its elements are positive. To achieve singular behaviour the matrices must be infinite, and this fits in with the demonstration of Peierls (1936).

In 1941 Kramers and Wannier examined the matrices corresponding to finite strips when $H = 0$, and using a unitary transformation, discovered a relation between the properties of the p.f. at high and low temperatures. It can be written in the form

$$\Lambda_N(1, z) = \frac{2}{(1+v)^2}\Lambda_N(1, v) \quad \left(v = \frac{(1-z)}{(1+z)}\right) \tag{4.94}$$

and provides a one-to-one correspondence between values of Λ_N at z and those at v. As $z \to 0$, $v \to 1$ and vice versa; knowing the p.f. from $z = 0$ to $z = z_c = (\sqrt{2} - 1) = (1 - z_c)/(1 + z_c)$ the values from $z = z_c$ to $z = 1$ could automatically be deduced. Kramers and Wannier thus concluded that if any singularity occurred as $l \to \infty$ for $z < z_c$ it would be repeated for $z > z_c$, and if there was only one singularity it must occur at $z = z_c$. Hence they conjectured that the Curie point of the quadratic lattice occurs at $z = z_c = (\sqrt{2} - 1)$. By evaluating the specific heat at this point for finite strips of increasing width, Kramers and Wannier also conjectured that the specific heat increases logarithmically with the width of the strip. Both of these conjectures were borne out by Onsager's exact solution (1944) to be discussed in the next chapter.

4.8.2 Correlation in Solids

The ideas of Ornstein and Zernike for fluids discussed in §4.4 can usefully be applied to solids, and can be extended to take account of long-range order. We start with the alloy problem for which the fluid language can be used with only minor changes. We denote the densities of A and B atoms by α_A, α_B ($\alpha_A \leq \alpha_B$):

$$\alpha_A = N_A/N, \quad \alpha_B = N_B/N. \tag{4.95}$$

Let $\alpha_{AA}(\mathbf{r})$, $\alpha_{BB}(\mathbf{r})$, $\alpha_{AB}(\mathbf{r})$ denote the mean fractions of occupation of a pair of sites a distance $\mathbf{r}$ apart by A–A, B–B, A–B atoms respectively. Then it is clear that the following relations hold:

$$\alpha_{AA}(\mathbf{r}) + \alpha_{AB}(\mathbf{r}) = \alpha_A \qquad \alpha_{AB}(\mathbf{r}) + \alpha_{BB}(\mathbf{r}) = \alpha_B. \tag{4.96}$$

Hence α_{AA}, α_{AB}, α_{BB} depend on only one parameter which can be used to define the order at a distance $\mathbf{r}$. If there were no correlation in the occupation of the sites, we should have

$$\alpha_{AA}(\mathbf{r}) = \alpha_A^2, \quad \alpha_{AB}(\mathbf{r}) = \alpha_A\alpha_B, \quad \alpha_{BB}(\mathbf{r}) = \alpha_B^2. \tag{4.97}$$

In general we shall write

$$\alpha_{AA}(\mathbf{r}) = \alpha_A^2[1 + \theta(\mathbf{r})] \tag{4.98}$$

where for convenience α_A is taken to correspond to the smaller component, $\alpha_A \leq \frac{1}{2}$. From (4.96) we find

$$\alpha_{AB}(\mathbf{r}) = \alpha_A[\alpha_B - \alpha_A\theta(\mathbf{r})] \qquad \alpha_{BB}(\mathbf{r}) = \alpha_B^2 + \alpha_A^2\theta(\mathbf{r}). \tag{4.99}$$

$\theta(\mathbf{r})$ is the analogue of the pair correlation function $h(\mathbf{r})$ in §4.4. Clearly the formula (4.99) takes on a much simpler form when $\alpha_A = \alpha_B = \frac{1}{2}$.

For a solid solution $\theta(\mathbf{r})$ behaves monotonically, whereas for an alloy it alternates in sign as one moves from a to b sites; we will then focus attention on its absolute value. If $\theta(\mathbf{r})$ tends to a non-zero value as $r \to \infty$ we say that there is long-range order in the assembly. To identify with the treatment in §4.7 we take

$$\theta(\mathbf{r}) = t(\mathbf{r})^2 \tag{4.100}$$

and refer to $t(\mathbf{r})$ as the degree of order at distance $\mathbf{r}$, and to its limiting value t as $r \to \infty$ as the long-range order.

We thus see that the long-range order is only one parameter amongst many needed to provide a proper description of the state of an alloy. After the work of Bragg and Williams attempts were indeed made to take account of short-range order, the best known being that of Bethe (1935), who introduced a second parameter for this purpose. However, the basic pattern of critical behaviour remained unchanged. More sophisticated *closed-form approximations* were devised with the same result; they produced refinements in the details, e.g. the location of T_c and the magnitude of the specific heat jump, but no change in the classical pattern summarized in Table 3.1 (see Burley (1972) for a review). Subsequently Onsager's solution indicated that it would be necessary to take an infinite number of parameters into account to achieve a breakaway from classical behaviour.

4.8.3 Correlations in the Ising Model

For the simple Ising model it is convenient to use spin averages in defining correlations. We first note that

$$\langle\sigma_0\rangle = \frac{\sum \sigma_0 \exp(-\beta\mathscr{H})}{Z_N} \tag{4.101}$$

can be identified with the reduced magnetization, $m^* = m/\mu_0$, since

$$M = -\frac{\partial F}{\partial H} = \frac{1}{\beta}\frac{\partial}{H}(\ln Z_N) = \mu_0 \sum_{i=1}^{N} \langle\sigma_i\rangle. \tag{4.102}$$

If two spins distance $\mathbf{r}$ apart were uncorrelated we would have

$$\langle\sigma_0\sigma_\mathbf{r}\rangle = \langle\sigma_0\rangle\langle\sigma_\mathbf{r}\rangle \tag{4.103}$$

where

$$\langle\sigma_0\sigma_\mathbf{r}\rangle = \frac{\sum \sigma_0\sigma_\mathbf{r} \exp(-\beta\mathscr{H})}{Z_N}. \tag{4.104}$$

We therefore define the *spin–spin correlation function* as

$$\Gamma(\mathbf{r}) = \langle \sigma_0 \sigma_{\mathbf{r}} \rangle - \langle \sigma_0 \rangle \langle \sigma_{\mathbf{r}} \rangle. \tag{4.105}$$

In zero magnetic field the second term on the right of (4.105) vanishes. For an antiferromagnet $\Gamma(\mathbf{r})$ alternates in sign as $\mathbf{r}$ moves from odd to even sites.

The magnetic susceptibility is obtained from the second derivative of the p.f. with respect to the magnetic field H,

$$\chi = \frac{\partial M}{\partial H} = \frac{1}{\beta} \frac{\partial^2}{\partial H^2} (\ln Z_N) = \frac{1}{\beta} \left[\frac{\partial^2 Z_N/\partial H^2}{Z_N} - \left(\frac{\partial Z_N/\partial H}{Z_N} \right)^2 \right], \tag{4.106}$$

i.e.

$$\frac{\chi}{N\beta\mu_0^2} = 1 + \sum_{r \neq 0} \Gamma(\mathbf{r}). \tag{4.107}$$

Relation (4.107) is the magnetic analogue of the fluctuation theorem (4.23) for fluids. $N\beta\mu_0^2$ is the susceptibility of an assembly of non-interacting magnets, and corresponds to $N\beta\rho^{-1}$, the compressibility of an ideal gas.

For the Ising model of spin s it is convenient to replace $\langle \sigma_0 \rangle$ by $\langle s_{z0} \rangle / s$ and $\langle \sigma_0 \sigma_{\mathbf{r}} \rangle$ by $\langle s_{z0} s_{zr} \rangle / s^2$ in (4.101) and (4.105). Relation (4.107) remains valid but the $N\beta\mu_0^2$ must be multiplied by a factor $(s+1)/3s$ which is now the susceptibility of an assembly of ideal magnets.

4.8.4 X-ray and Neutron Scattering in Solids

We now wish to adapt the discussion of §4.5 to scattering from solids having a regular crystalline structure. For simplicity we consider only Bravais lattices with a unit cell given by vectors $(\mathbf{a}_1, \mathbf{a}_2, \mathbf{a}_3)$ so that the general lattice point $\mathbf{l}$ is given by

$$\mathbf{l} = l_1 \mathbf{a}_1 + l_2 \mathbf{a}_2 + l_3 \mathbf{a}_3. \tag{4.108}$$

Define the reciprocal lattice by

$$\boldsymbol{\tau}_1 = \frac{2\pi}{v_0} (\mathbf{a}_2 \times \mathbf{a}_3), \quad \boldsymbol{\tau}_2 = \frac{2\pi}{v_0} (\mathbf{a}_3 \times \mathbf{a}_1), \quad \boldsymbol{\tau}_3 = \frac{2\pi}{v_0} (\mathbf{a}_1 \times \mathbf{a}_2) \tag{4.109}$$

a general lattice point being given by

$$\boldsymbol{\tau} = t_1 \boldsymbol{\tau}_1 + t_2 \boldsymbol{\tau}_2 + t_3 \boldsymbol{\tau}_3; \tag{4.110}$$

v_0 is the volume of the unit cell of the original lattice, i.e. $[\mathbf{a}_1, \mathbf{a}_2, \mathbf{a}_3]$. From these definitions we have

$$\mathbf{l} \cdot \boldsymbol{\tau} = 2\pi(l_1 t_1 + l_2 t_2 + l_3 t_3). \tag{4.111}$$

For fluids we were concerned with problems in which all the molecules could be treated as identical. The corresponding problem for a regular solid will lead to the standard theory of X-ray diffraction. In (4.24) all the terms will be consistent in sign when $\mathbf{q}$ is a point of the reciprocal lattice, whereas the phases will randomize if $\mathbf{q}$ is any other point. Hence in (4.25) $S(\mathbf{q})$ will be proportional to

$$\sum_{\boldsymbol{\tau}} \delta(\mathbf{q} - \boldsymbol{\tau}). \tag{4.112}$$

Very large intensities will be found in the specific directions given by points of the reciprocal lattice, and very small intensities in other directions.

However, in alloys and magnets we are concerned with problems in which individual molecules scatter the radiation differently. Let us take a particular model of a binary alloy in which the A atoms scatter with intensity v_A and B atoms v_B. Instead of (4.25) we now have

$$S(\mathbf{q}) = \frac{1}{N}\left\langle \sum_{j=1}^{N} \sum_{k=1}^{N} v_j v_k \exp[i\mathbf{q}(\mathbf{l}_j - \mathbf{l}_k)] \right\rangle. \tag{4.113}$$

Confining attention to a particular distance $l_j - l_k = \mathbf{r}$, we substitute for the proportion of A–A, A–B, B–B pairs from (4.98) and (4.99), and find for the total $v_j v_k$ coefficient

$$v_A^2 \alpha_{AA}(\mathbf{r}) + 2v_A v_B \alpha_{AB}(\mathbf{r}) + v_B^2 \alpha_{BB}(\mathbf{r}) = (v_A \alpha_A + v_B \alpha_B)^2 + \alpha_A^2 \theta(\mathbf{r})(v_A - v_B)^2. \tag{4.114}$$

We now sum (4.113) over all distances, and the $1/N$ cancels because of the double summation. The first term in (4.114) gives a standard X-ray term of the form (4.112) usually termed *coherent scattering*. The second term gives a contribution to $S(\mathbf{q})$ equal to

$$\alpha_A^2(v_A - v_B)^2 \left| 1 + \sum_{\mathbf{r} \neq 0} \theta(\mathbf{r}) \exp(i\mathbf{q} \cdot \mathbf{r}) \right|. \tag{4.115}$$

When there is no long-range order present and $\theta(\mathbf{r})$ decays to zero for large r, the sum in (4.115) parallels the integral in (4.27) with $\theta(\mathbf{r})$ analogous to $h(\mathbf{r})$. This is generally termed *diffuse scattering*. However, when there is long-range order present and $\theta(\mathbf{r})$ tends to a non-zero value as $\mathbf{r} \to \infty$, there will clearly be an additional coherent contribution of the form (4.112) proportional to $\theta(\infty)$. If $\theta(\infty)$ alternates in sign on moving from odd to even sites the size of the reciprocal lattice must be halved so that odd multiples of π are obtained in (4.111). The intensity of the *superlattice* spots is then proportional to $\theta(\infty)$.

The above discussion applies also to the scattering of neutrons from binary alloys, although additional diffuse scattering may arise from isotopic effects. However, an important new feature of neutron diffraction is the interaction of neutrons with the magnetic moments of the molecules, and the dependence of the scattering intensity on their orientation. As a result neutrons can be used as an experimental tool for the investigation of magnetic order. A detailed account of how this operates in practice can be found in specialist books on neutron diffraction (see e.g. Marshall and Lovesey 1971). We shall content ourselves with a simple illustrative example for the Ising model, assuming that spins oriented in directions 1 and 2 have scattering intensities v_1 and v_2 respectively.

By analogy with the alloy we define α_1, α_2 as the mean fraction of spins pointing in directions 1, 2 ($\alpha_1 \leq \alpha_2$) and $\alpha_{11}(\mathbf{r})$, $\alpha_{12}(\mathbf{r})$, $\alpha_{22}(\mathbf{r})$ as the mean fractions of corresponding pairs at lattice sites separated by a distance $\mathbf{r}$. Then

$$\begin{aligned} \alpha_{11} + \alpha_{12} &= \alpha_1 \\ \alpha_{12} + \alpha_{22} &= \alpha_2 . \end{aligned} \tag{4.116}$$

By analogy with (4.98) we define

$$\alpha_{11}(\mathbf{r}) = \alpha_1^2[1 + \theta(\mathbf{r})] \tag{4.117}$$

and find that

$$\begin{aligned} \alpha_{12}(\mathbf{r}) &= \alpha_1[\alpha_2 - \alpha_1\theta(\mathbf{r})] \\ \alpha_{22}(\mathbf{r}) &= \alpha_2^2 + \alpha_1^2\theta(\mathbf{r}). \end{aligned} \tag{4.118}$$

We also have

$$\langle\sigma_0\sigma_{\mathbf{r}}\rangle = \alpha_{11} + \alpha_{22} - 2\alpha_{12} \tag{4.119}$$

and

$$\langle\sigma_0\rangle = \alpha_2 - \alpha_1. \tag{4.120}$$

Hence after a little algebra it is easy to show from (4.106) that

$$\Gamma(\mathbf{r}) = 4\alpha_1^2\theta(\mathbf{r}). \tag{4.121}$$

The intensity coefficient $v_j v_k$ in (4.113) is now

$$\begin{aligned} (v_1\alpha_1 + v_2\alpha_2)^2 + (v_1 - v_2)^2\alpha_1^2\theta(\mathbf{r}) = [\tfrac{1}{2}(v_1 + v_2) + m^*(v_1 - v_2)]^2 \\ + \tfrac{1}{4}(v_1 - v_2)^2\Gamma(\mathbf{r}). \end{aligned} \tag{4.122}$$

Here m^* is the reduced magnetization

$$m^* = M/N\mu_0 = \alpha_2 - \alpha_1. \tag{4.123}$$

We thus see that there is a coherent scattering contribution to $S(\mathbf{q})$ from which the magnetization can be determined, and a diffuse scattering contribution from which $\Gamma(\mathbf{r})$ can be determined. The contribution to $S(\mathbf{q})$ giving rise to diffuse scattering can be written in the form

$$\tfrac{1}{4}(v_1 - v_2)^2\left(1 + \sum_{\mathbf{r}\neq 0} \Gamma(\mathbf{r}) \exp(i\mathbf{q}\cdot\mathbf{r})\right) = \tfrac{1}{4}(v_1 - v_2)^2[1 + \hat{\Gamma}(\mathbf{q})] \tag{4.124}$$

where

$$\hat{\Gamma}(\mathbf{q}) = \sum_{\mathbf{r}\neq 0} \Gamma(\mathbf{r}) \exp(i\mathbf{q}\cdot\mathbf{r}) \tag{4.125}$$

is the analogue of the Fourier transforms in (4.31). It is sometimes useful in relation to (4.107) and (4.108) to interpret the sum in larger brackets in (4.124) as a wave-number-dependent susceptibility in the spatially varying field

$$H(\mathbf{r}) = H \exp(i\mathbf{q}\cdot\mathbf{r}). \tag{4.126}$$

4.9 Landau Theory

In the opening paragraph of his classic 1937 paper Landau emphasized that the continuity of state between liquid and gas at sufficiently high temperatures is possible only because the liquid and gas phases have the same symmetry. A transition between two phases of different symmetry cannot be continuous; elements of symmetry are either present or absent and no intermediate case is possible.

It is surprising that this simple and convincing argument was by-passed for at least 15 years, with discussion continuing on the possibility that the solid–fluid melting transitions might end in a critical point (see e.g. Domb 1951, Munster 1952 and discussion). A summary of Landau's major ideas had appeared in English in a letter to *Nature* in 1936, so there was really no valid reason for ignoring them.

Landau proceeded to a detailed analysis of the role of symmetry in λ-point and Curie-point transitions, which he developed further in his book with Lifshitz (1958). The ordered low-temperature phase always has lower symmetry than the disordered phase and its symmetry group contains elements which are not invariant under transformations of the higher-symmetry group of the disordered phase. The extra density of the ordered phase can be expressed in terms of appropriate irreducible representations, and with their help an *order parameter*, $\boldsymbol{\eta}$, can be defined which will be zero in the disordered phase, and non-zero in the ordered phase. In simple cases $\boldsymbol{\eta}$ will be a scalar, but in more complex cases it can have components η_i.

Landau assumed that the free energy Φ can be expanded as a power series in $\boldsymbol{\eta}$ about the critical point; since any stable state must correspond to a minimum in Φ, the linear term must be absent, i.e.

$$\Phi = \Phi_0 + \eta^2 A(P,T) + \eta^3 \sum_{\alpha} C_\alpha(P,T) f_\alpha^3(\gamma_i) + \eta^4 \sum_{\alpha} B_\alpha(P,T) f_\alpha^4(\gamma_i) + \cdots \tag{4.127}$$

Here $\gamma_i = \eta_i/\eta$, and group theoretical considerations were used to show that the square term does not involve γ_i. Landau argued that the Curie temperature T_c corresponds to $A(P,T_c) = 0$, A being positive for $T > T_c$ and negative for $T < T_c$; this allows for a solution of minimum free energy with $\eta \neq 0$ for $T < T_c$. Hence near T_c we can write

$$A(P,T_c) = e(P)(T - T_c). \tag{4.128}$$

For simplicity we will continue the argument with η a scalar; the generalization to $\boldsymbol{\eta}$ with several components is straightforward. We then have

$$\Phi = \Phi_0 + \eta^2 A(P,T) + \eta^3 C(P,T) + \eta^4 B(P,T) + \cdots. \tag{4.129}$$

Even when $A(P,T_c) = 0$, Φ is not a minimum at $\eta = 0$ unless the cubic coefficient $C(P,T)$ is also zero, and $B(P,T) > 0$. Landau differentiated between the case when $C(P,T)$ is identically zero for reasons of symmetry (e.g. a ferromagnet), and when it is zero because it satisfies the two equations

$$A(P,T) = C(P,T) = 0. \tag{4.130}$$

The former he identified with Curie points, whilst he showed that the latter were isolated points corresponding to the meeting of three phases.

Let us take the ferromagnet as an example with η representing the magnetization, and Φ the magnetic free energy,* so that

$$\Phi = \Phi_0 + AM^2 + BM^4 + \cdots. \tag{4.131}$$

Using the condition for equilibrium $\partial\Phi/\partial M = 0$, we find that $M = 0$ in the symmetric high-temperature phase, and

$$M^2 = -A/2B = \frac{e}{2B}(T_c - T) \tag{4.132}$$

in the unsymmetric low-temperature phase.

The magnetic equation of state is given by

$$H = \partial\Phi/\partial M = 2AM + 4BM^3 + \cdots. \tag{4.133}$$

Hence the initial magnetic susceptibility is given by

$$\frac{1}{\chi} = \left(\frac{\partial H}{\partial M}\right)_0 = 2A. \tag{4.134}$$

The entropy is given by

$$S = -\partial\Phi/\partial T = S_0 - M^2\partial A/\partial T \ldots. \tag{4.135}$$

This is equal to S_0 for the symmetric phase, and to

$$S_0 = \frac{e^2}{2B}(T_c - T) \tag{4.136}$$

* We denoted this by $A(M,T)$ in Chapter 3, but we wish to avoid confusion with the coefficient A.

for the unsymmetric phase. S is thus continuous at T_c but the specific heat

$$C_M = T(\partial S/\partial T)_M \tag{4.137}$$

undergoes a discontinuity, and drops by $e^2/2B$ on passing through T_c.

We see therefore that the Landau theory has reproduced all of the results of the classical treatment given in Chapter 3, but has suggested that the pattern of behaviour is the same for all assemblies governed by an order parameter with this type of symmetry.

4.9.1 Critical Fluctuations

An alternative approach to that of Ornstein and Zernike for the theory of critical fluctuations attempts to derive the free energy of a heterogeneous assembly in the region where deviations from homogeneity and gradients of density are not too large. The method was pioneered by Rocard (1933) who took two important steps. He used the virial theorem of Clausius (see §1.4.1) to obtain an expansion for the pressure in a heterogeneous assembly in terms of the density and its derivatives, and analyzed the deviations from homogeneity into Fourier components. Rocard was led to a formula which differed somewhat from that of Ornstein and Zernike, but the differences are of minor significance and need not concern us here.

In 1949 Klein and Tisza formulated a general theory of heterogeneous thermodynamic fluctuations by dividing a canonical ensemble into equivalent cells and allowing thermodynamic quantities to vary from cell to cell. The results of O–Z and Rocard were derived as particular cases of their treatment. Again the differences from O–Z achieved by their method are not very significant in the light of subsequent developments (a simple exposition of their ideas is given by Fierz (1960)).

We shall base our discussion on the exposition given by Landau (see Landau and Lifshitz 1958, §116) since a subsequent modification by Ginzburg and Landau (1950) has played a key role in the modern theory. Consider an assembly whose temperature and volume are constant, so that ΔT and ΔV in §4.1 are zero. The probability of a given fluctuation is given by

$$w \propto \exp(-\beta \Delta F). \tag{4.138}$$

The non-zero value of ΔF arises from variations in density; for a magnetic system it would arise from variations in magnetization, and in other systems from variations in the appropriate order parameter. We can divide the total free energy ΔF into contributions from different spatial regions,

$$\Delta F = \int \Delta a(\mathbf{r})\, \mathrm{d}V \tag{4.139}$$

where $a(T,\rho)$ is the free energy per unit volume defined in (2.16). $\Delta a(\mathbf{r})$ depends on $\delta\rho(\mathbf{r})$ and combinations of the derivatives of ρ which are spherically symmetrical. In an expansion of $\Delta a(\mathbf{r})$ the first derivatives of ρ can enter only as $(\text{grad } \rho)^2$, and the second derivatives as $\nabla^2\rho$. Terms in $\nabla^2\rho$ can be integrated to give a surface contribution which can be ignored, and terms in $\rho\nabla^2\rho$ to give a bulk contribution of the form $(\text{grad } \rho)^2$. We are thus led in the first order to a 'square gradient' term of the type discussed in §2.7.1 in relation to the theory of the interface. Also the linear term in $\delta\rho$ gives zero on integration over the volume, since the total number of particles is constant. We can thus write

$$\Delta a(\mathbf{r}) = \tfrac{1}{2}c\delta\rho^2 + \tfrac{1}{2}f(\nabla\rho)^2 + \cdots \tag{4.140}$$

where

$$c = (\partial^2 a/\partial\rho^2)_T = (\partial\mu/\partial\rho)_T = \frac{1}{\rho}(\partial P/\partial\rho)_{T>0} = 1/\rho^2 K_T \tag{4.141}$$

so that c is zero at $T = T_c$. f must be greater than zero for $\Delta a(\mathbf{r})$ to correspond to a minimum at T_c.

If $\delta\rho$ is now broken down into Fourier components

$$\delta\hat{\rho}_{\mathbf{q}} = V^{-1}\int \exp(\mathrm{i}\mathbf{q}\cdot\mathbf{r})\delta\rho(\mathbf{r})\,\mathrm{d}\mathbf{r} \tag{4.142}$$

on substituting in (4.140) and (4.139) we find for the total fluctuation

$$\Delta F = \tfrac{1}{2}v\sum_{\mathbf{q}}(c + fq^2)|\delta\hat{\rho}_{\mathbf{q}}|^2 \tag{4.143}$$

which is a sum of contributions from independent normal modes. The absence of interaction between the modes is due to the truncation of the expansions (4.140) at the square term.

Each mode $\mathbf{q}$ has an independent Boltzmann distribution with average energy kT, so that

$$\langle|\delta\hat{\rho}_{\mathbf{q}}|^2\rangle = kT/V(c + fq^2). \tag{4.144}$$

From (4.142) we find that

$$\begin{aligned}\langle|\delta\hat{\rho}_{\mathbf{q}}^2|\rangle = \langle\delta\rho_{\mathbf{q}}\delta\rho_{-\mathbf{q}}\rangle &= V^{-2}\iint \exp[\mathrm{i}\mathbf{q}\cdot(\mathbf{r}_1 - \mathbf{r}_2)]\langle\delta\rho(\mathbf{r}_1)\delta\rho(\mathbf{r}_2)\rangle\,\mathrm{d}\mathbf{r}_1\,\mathrm{d}\mathbf{r}_2 \\ &= V^{-1}\int \exp(\mathrm{i}\mathbf{q}\cdot\mathbf{r})\langle\delta\rho(0)\,\delta\rho(\mathbf{r})\rangle\,\mathrm{d}\mathbf{r}.\end{aligned} \tag{4.145}$$

From (4.17), (4.18) and (4.19) we can write

$$\langle\delta\rho(0)\delta\rho(\mathbf{r})\rangle = \rho^2[g(r) - 1] + \rho\delta(\mathbf{r}) \tag{4.146}$$

where the last term is included to take account of the possibility of $r = 0$ (cf. equation (4.21)). It follows from (4.144) and (4.145) that

$$1 + \rho\hat{h}(\mathbf{q}) = kT/\rho(c + fq^2) \tag{4.147}$$

which is the O–Z result (4.42).

The approximation of stopping the gradient terms in the expansion in (4.140) is reasonable for long-wavelength fluctuations which are responsible for critical behaviour. However, since c is zero at T_c it is reasonable to assume that a higher-order term $d(\delta\rho)^4$ is needed to get a valid picture in the critical region. This was the suggestion of Ginzburg and Landau (1950) for the order parameter of the superconducting transition. Its adaption complicates the above treatment very considerably, and has far-reaching consequences.

Notation for Chapter 4

w probability of a fluctuation.
I_0 intensity of incident wave, I_θ intensity of scattered wave at angle θ.
λ_0 wavelength of incident wave, $\mathbf{r}$ local distance vector, R large distance from origin.
$\mathbf{k}_0$ wave vector of incident wave, $\mathbf{k}_s$ wave vector of scattered wave ($k = 2\pi/\lambda$), $\mathbf{q} = \mathbf{k}_0 - \mathbf{k}_s$.
$g(\mathbf{r}_1, \mathbf{r}_2)$ pair distribution function (normalized).
$g(r)$ pair distribution function (normalized) where there is spherical symmetry.
$h(r) = g(r) - 1$ pair correlation function, $c(r)$ direct correlation function defined by (4.33).
$\hat{h}(q) =$ Fourier transform of $h(r)$ (^ denotes Fourier transform).
$S(\mathbf{q})$ structure factor $= \chi(\mathbf{q}) + \rho\delta(\mathbf{q})$ (see (4.28)).
L length defined by (4.41), $\kappa = \kappa(\tau)$ defined by (4.43).
Order in solids defined by parameter t in (4.63).
q coordination number of lattice (= number of nearest neighbours (n.n.)).
$\epsilon_{AA}, \epsilon_{AB}, \epsilon_{BB}$ interactions between AA, AB and BB atoms.
$\epsilon = \epsilon_{AB} - \frac{1}{2}(\epsilon_{AA} + \epsilon_{BB})$.

Ising model

Simple Ising model = Ising model of spin $\frac{1}{2}$ with n.n. interactions.
$I(s) =$ Ising model of spin s with n.n. interactions.
$y = \exp(-2\beta\mu_0 H)$, $z = \exp(-2\beta J)$, $v = (1 - z)/(1 + z)$.
N_A, N_B numbers of atoms of types A, B.
N_{AA}, N_{AB}, N_{BB} numbers of different types of n.n. pairs.
$g(N; N_A, N_{AB})$ combinatorial factor.
$z_A = \exp(\beta\epsilon_{AA})$, $z_B = \exp(\beta\epsilon_{BB})$, $z = \exp(\beta\epsilon)$.
$\Lambda_N(y, z)$ defined in (4.80).
Z_N^I Ising model p.f., Z_N^s alloy or solid solution p.f.
$\mathbf{V}$ transfer matrix (defined in (4.90)).
$\theta(\mathbf{r})$ defined in (4.98) is analogue for solids of $h(\mathbf{r})$ for fluids.
$t(\mathbf{r})$ degree of order at distance $\mathbf{r}$ (4.100).
$\langle\sigma_0\sigma_r\rangle$ pair correlation of spins in Ising model.
$\Gamma(\mathbf{r}) = \langle\sigma_0\sigma_r\rangle - \langle\sigma_0\rangle\langle\sigma_r\rangle$.
Unit cell of lattice $(\mathbf{a}_1, \mathbf{a}_2, \mathbf{a}_3)$, volume of unit cell $v_0 = [\mathbf{a}_1, \mathbf{a}_2, \mathbf{a}_3]$, reciprocal lattice $(\boldsymbol{\tau}_1, \boldsymbol{\tau}_2, \boldsymbol{\tau}_3)$ defined in (4.109).
v_A scattering intensity of A atoms, v_B scattering intensity of B atoms.

Landau Theory

Order parameter $\boldsymbol{\eta} = (\eta_1, \eta_2, \eta_3)$.
Gibbs free energy $\Phi(P,T)$.
$A(P,T)$, $B(P,T)$, $C(P,T)$ coefficients in expansion of Φ in powers of η (see (4.129)).
$A(P,T_c) = e$.
Magnetic free energy $\Phi(M,T)$ expansion in (4.131).
c, f coefficients in expansion of free energy of a heterogeneous assembly (see (4.140)).

References

ALTSCHUL, M. (1893) *Z. Phys. Chem.* **11**, 578.

ANDREWS, T. (1869) *Philos. Trans. R. Soc.* **159**, 575.

ASHCROFT, N. W. and MERMIN, N. D. (1976) *Solid State Physics*, New York: Holt, Rinehart and Winston.

AVENARIUS, M. (1874) *Ann. Phys. Chem.* **151**, 306.

BETHE, H. A. (1935) *Proc. R. Soc. A* **216**, 45.

BRAGG, W. L. and WILLIAMS, E. J. (1934) *Proc. R. Soc. A* **145**, 699.

BRUSH, S. G. (1967) *Rev. Mod. Phys.* **39**, 883.

BURLEY, D. M. (1972) DG **2**, Ch. 9.

CALLEN, H. B. (1960) *Thermodynamics*, New York: John Wiley.

DOMB, C. (1951) *Philos. Mag.* **42**, 1316.

DOMB, C. (1960) *Adv. Phys.* **9**, 149.

EINSTEIN, A. (1910) *Ann. Phys., Leipzig* **33**, 127.

FIERZ, M. (1960) *Theoretical Physics in the 20th Century*, ed. M. FIERZ and V. F. WEISSKOPF, New York: Interscience, p. 175.

FISHER, M. E. (1964) *J. Math. Phys.* **5**, 944.

FOWLER, R. H. (1936) *Statistical Mechanics*, Cambridge.

FROBENIUS, S. B. (1908) *Preuss. Akad. Wiss.* 471.

GINZBURG, V. L. and LANDAU, L. D. (1950) *Sov. Phys. JETP* **20**, 1064.

GRADZSHTYN, I. S. and RYZHIK, I. M. (1965) *Tables of Integrals*, New York and London: Academic Press.

ISING, E. (1925) *Z. Phys.* **31**, 253.

KLEIN, M. J. and TISZA, L. (1949) *Phys. Rev.* **76**, 1861.

KRAMERS, H. A. and WANNIER, G. H. (1941) *Phys. Rev.* **60**, 252, 263.

LANDAU, L. D. (1936) *Nature* **138**, 840.

LANDAU, L. D. (1937) *Phys. Z. Sow.* **11**, 26, 545 (English translation (1965) in *Collected Papers of L. D. Landau*, ed. D. TER HAAR, Oxford: Pergamon Press).

LANDAU, L. D. and LIFSHITZ, G. M. (1958) *Statistical Physics*, Oxford: Pergamon Press.

LASSETRE, E. N. and HOWE, J. P. (1941) *J. Chem. Phys.* **9**, 747.

MARSHALL, W. and LOVESEY, S. W. (1971) *Theory of Thermal Neutron Scattering*, Oxford.

MONTROLL, E. W. (1941) *J. Chem. Phys.* **9**, 706.

MONTROLL, E. W. and WEISS, G. H. (1965) *J. Math. Phys.* **6**, 167.

MUNSTER, A. (1952) *CR Le Reunion Chimie Physique*, Paris p. 21.

ONSAGER, L. (1944) *Phys. Rev.* **65**, 117.

Ornstein, L. S. and Zernike, F. (1914) *Proc. Akad. Sci. Amsterdam* **17**, 783.
Ornstein, L. S. and Zernike, F. (1916) *Proc. Akad. Sci. Amsterdam* **18**, 1520 (reproduced in *The Equilibrium Theory of Classical Fluids*, ed. A. L. Frisch and J. L. Lebowitz, Reading, MA: Benjamin, 1964).
Pais, A. (1982) *Subtle is the Lord, The Science and Life of Albert Einstein*, Oxford.
Peierls, R. E. (1936) *Proc. Camb. Philos. Soc.* **32**, 477.
Perron, O. (1907) *Math. Ann.* **64**, 248.
Rocard, Y. (1933) *J. Phys. Radium* **4**, 165.
Smoluchowski, M. S. (1908) *Ann. Phys, Leipzig* **25**, 205.
Smoluchowski, M. S. (1912) *Philos. Mag.* **23**, 165.
Travers, M. W. and Usher, F. L. (1906) *Proc. R. Soc. A* **78**, 247.
Wesendonck, K. von (1894) *Z. Phys. Chem.* **15**, 262.
Young, S. (1906) *Proc. R. Soc. A* **78**, 262.

5

The Onsager Revolution

5.1 Onsager's Contributions to Ising Model Theory

We have mentioned in §1.5 the shattering blow which Onsager's exact calculations for a two-dimensional Ising model delivered to the classical theories, and the challenge it posed in relation to critical behaviour. But Onsager's work had much wider ramifications since detailed aspects of his calculations stimulated important new areas of research and progress. His topological interpretation of the duality transformation discovered by Kramers and Wannier greatly extended its scope, and soon led, in conjunction with his formulation of the star–triangle transformation, to exact solutions for other two-dimensional lattices (see e.g. Domb 1960 pp. 211–20). The star–triangle transformation itself initiated a search which revealed other useful transformations of a similar kind (see e.g. Syozi 1972). The simplification of the calculations by the use of spinors led to the discovery of other simplified methods of calculation each of which had its own specific advantages (Temperley 1972, Schultz, Mattis and Lieb 1964). More remotely the existence of exact solutions to the Ising model in two dimensions suggested that other realistic models might give rise to exact solutions. Over 20 years elapsed before a breakthrough occurred (Lieb 1967). Subsequently this area of research has generated results of great interest and importance (Baxter 1982).

The calculation of the specific heat of a finite crystal stimulated a deeper analysis of finite size and boundary effects in two-dimensional models (see e.g. Watson 1972). The deviation of the boundary tension from the classical van der Waals theory of surface tension led to the re-examination and reformulation of this theory (see e.g. Widom 1972, Rowlinson and Widom 1982).

Because of their seminal historical importance, and because of the fundamental nature of the concepts introduced, we shall describe Onsager's contributions in some detail.

5.1.1 Historical Notes

The solution to the Ising model for the SQ lattice in zero field was first publicly presented in a discussion remark following a paper by Wannier at a meeting of the New York Academy of Science on February 28, 1942 (Shedlovsky and Montroll 1963). Onsager relished cryptic announcements of this kind, and the background story was forthcoming only 28 years later at the fifth Battelle Colloquium on Materials Science, which was devoted to the subject of critical phenomena (Mills, Ascher and Jaffee 1971). In the course of some autobiographical remarks Onsager told of a young man named Elliott Montroll who, early in 1940, had been strongly recommended by Joe and Maria Mayer. After considerable efforts Onsager had succeeded in securing a position for Montroll to occupy.

> I was very glad I had made that effort. And incidentally, so is everybody else in the Chemistry Department at Yale. I think everybody found Elliott very interesting to talk to .... In the Fall of 1940 Montroll brought the startling news that Wannier (working with Kramers) had located the critical point by way of a transformation .... I did not do anything until I had a chance to talk to Wannier some time between Christmas and New Year, and he gave me a very vague idea how it was done. In the course of the spring I started to reconstruct this dual transformation from imperfect information, and actually very soon it turned out to be something much more general .... It was a sort of investigation where you got a good lead, and certainly you had to pursue that; and before you reached the end of that lead up opened another, and this was if anything even more fascinating. Then one kept on going in several stages; one beautiful lead opening up after the other, and every one much too good to abandon. Before I got to the end of it, it looked like a very good guess what the answer might be; it was something that showed up so regular that it was easy to extrapolate it all the way to an infinite crystal. It took a few months, though, to verify the guess, but it was doable; and that was how the Ising model was solved. And that, for those who are not aware of all the details, was a first model of anything, any kind of system, that has what you call a continuous transition, that is one which occurs at a definite temperature but in such a manner that nothing changes by a finite amount. Every step is small. This was the first model of that sort that had ever been computed exactly, and the outcome demonstrated, once and for all, that the results of several more primitive theories had features that could not possibly be preserved in an exact computation. And for the rest, at the very least, it's taught us what *not* to do; what so stay away from . . .

At the time of the New York announcement the formula for the two-dimensional Ising model was still a conjecture, but Wannier laid a 4–1 bet that it was right. The complete solution was despatched to *Physical Review* in October 1943 and was published in February 1944.

The calculation was confined to the free energy of the model, and the leading eigenvalues and eigenvectors were calculated by the method of Lie algebras. A simplification which led to substantial further progress was introduced by Bruria Kaufman, who showed that the algebra required for the determination of the eigenvalues and eigenvectors was that of the spin representations of the orthogonal group, the general theory of which had been formulated previously by Brauer and Weyl (1935). The background was again described by Onsager at the Battelle Conference.

> Well, after I got out a paper on the Ising Model, there was a young lady, Bruria Kaufman, and I had no idea what she might be like; first of all she insisted on working with me. Well, I thought we might arrange a way to do that. Secondly, she insisted on working on the Ising model. Now, that was the kind of task that I would never want to impose on a student. So I did my best to talk her out of it. But no, she insisted. All right then, I tried to dream up some problems that would be significant but reasonably safe . . .
>
> By the summer of 1945 I had a proof for the conjectured full spectrum of the transfer matrix . . . . The approach was somewhat laborious . . . . Before long, however, Bruria Kaufman had developed a much better strategy . . . . I suggested that she explore the structure of W (the transfer matrix) as well as the effect of joining crystal ends on a torus with a twist etc. She made good progress; but as she got her bearings she was more intrigued by the ubiquitous trigonometric relations and decided to look for a possible connection with spinor theory. Why not? In fact it seemed like very good sense, and so it was . . . . By the summer of 1946 she had a beautifully compact computation of the partition function, bypassing all tedious detail.
>
> By itself that was only a more elegant derivation of an old result; but the approach looked powerful enough to produce a few new ones. Very well, how about correlations? That indeed required everything that had been developed before. Now we had to specify explicitly the planes and the angles of the commuting rotations; this proved a bit tedious but the road was known . . . . The successful performance of this task was another proof of her excellent intuition . . .

The resulting correlations were expressed in the form of special types of recurrent determinant known as Toeplitz determinants which could be evaluated. They assumed a particularly simple form for correlations along a diagonal. But the big question remaining was the evaluation of the correlation for a pair of sites whose separation tends to infinity, i.e. the long-range order.

In another cryptic discussion remark on 19 May 1949 after a paper by Rushbrooke at the Florence Conference on Statistical Mechanics, Onsager announced the result of his calculation for the long-range order. He never published any details of his solution. The same formula was derived for the spontaneous magnetization (which is equivalent to the long-range order) by Yang in 1952. At the Battelle Conference the curtain was lifted on the technique which Onsager had used, and on why he had not published his calculations.

> Then I tried to look at the infinite one; in a little while it dawned on me that this was just, in difference equation, the same as the Milne equation in an integral equation; and so I tried the same technique (Wiener-Hopf) which solves the Milne equation. It solves the difference equation too. Now, the reason that was not published the way it was computed was that, before I got around to it, I tried to find the general formula for the evaluation of determinants of the same type, and lo and behold I found it. It was a general formula for the evaluation of Toeplitz matrices. The only thing I did not know was how to fill out the holes in the mathematics and show the epsilons and deltas and all of that, and the limiting processes; I did not know just how it should be done, and what mathematicians really knew about limiting processes in that ball park.
>
> Later, six years ago, at a meeting of the Mathematics Society, I talked to Hirschman, who had the best theorem to date. It turned out that if I had just gone ahead the way I had started and then used a theorem of Wiener's, that I had not known about . . . I would have had exactly the same conditions as Hirschman's . . . the mathematicians got there first.

5.1.2 Duality Transformation

We have mentioned the low–high temperature relations discovered by Kramers and Wannier who applied a unitary transformation to the transfer matrix. Onsager gave a topological interpretation to this transformation which provided a clearer understanding of its nature, and enabled its application to be extended considerably. This was first published in a review paper by Wannier in 1945.

For any two-dimensional net we define a dual net by placing spins into every elementary polygon and drawing lines of interaction between them so that each old line is crossed by a new one. The duality relation is clearly a reciprocal one; the number of polygons in one net equals the number of vertices in the other, while the number of sides is the same. Examples of two-dimensional nets and their duals are given in Figure 5.1 and have been taken from Wannier's paper. In cases (b) and (c) the dual net is topologically identical with the original; such nets may be referred to as *self-dual.*

Kramers and Wannier (1941) had shown how it is possible to develop a low-temperature expansion for the p.f. of the Ising model starting from the ground state and considering deviations corresponding to 1, 2, 3 . . . overturned spins. For an Ising net of N spins and L links in zero magnetic field the

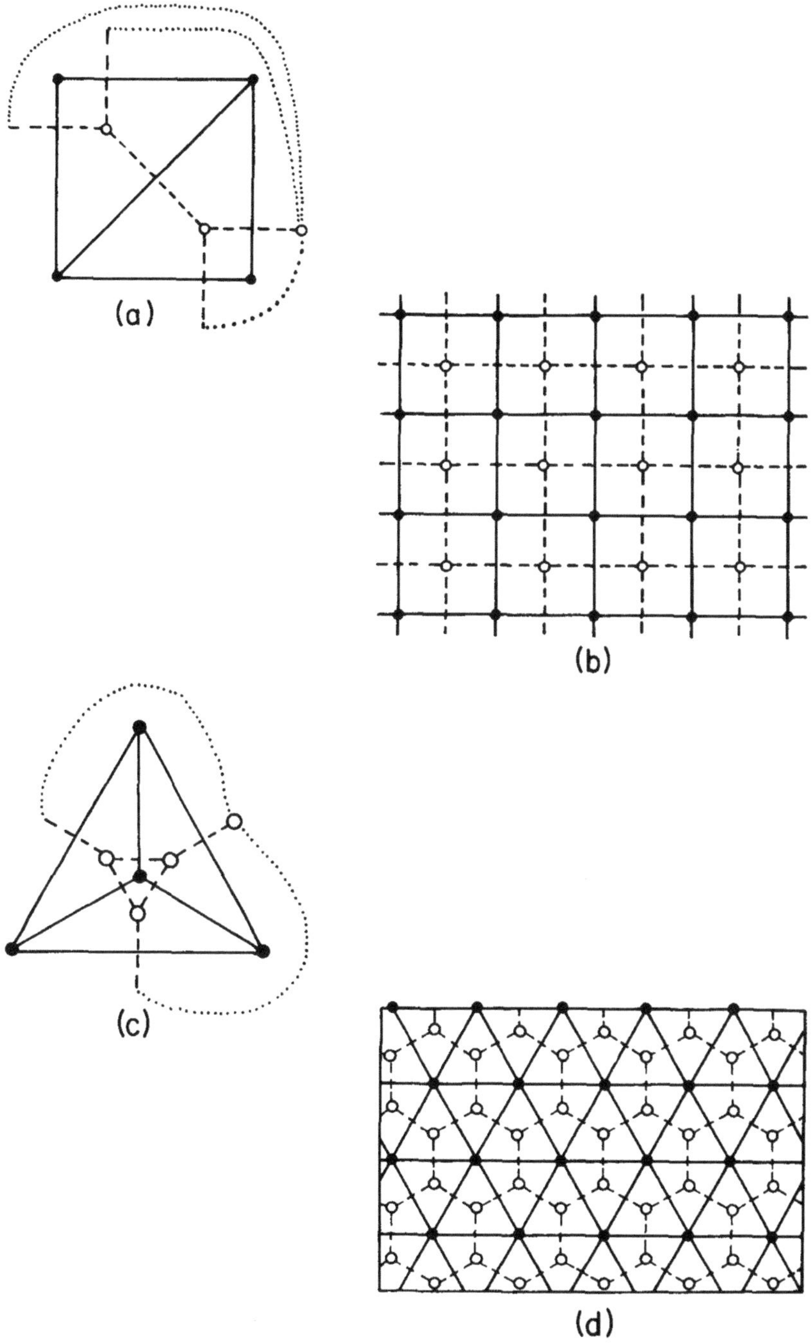

Figure 5.1 (a)–(d) Two-dimensional nets and their duals (after Wannier).

expansion can be written

$$Z_N = 2 \exp(\beta LJ)\left(1 + \sum_{r=1}^{r_{max}} v(r) \exp[-(2r\beta J)]\right). \tag{5.1}$$

The factor 2 arises because of the degeneracy of the lowest energy state, and $v(r)$ is the number of configurations corresponding to r broken links.

At high temperatures Onsager drew attention to the following alternative development. From (4.72) the p.f. in zero field can be written

$$Z_N = \exp(K\sum \sigma_i \sigma_j) \quad (K = J/kT) \tag{5.2}$$

where the sum must be taken over 2^N possible values of the σ_i for all the lattice points, and i, j run over all links of the net. The $\sigma_i \sigma_j$ satisfy the relation

$$(\sigma_i \sigma_j)^2 = (\sigma_i \sigma_j)^4 = \ldots = 1, \quad (\sigma_i \sigma_j) = (\sigma_i \sigma_j)^3 = (\sigma_i \sigma_j)^5 = \ldots$$

and we can therefore write

$$\exp(K\sigma_i \sigma_j) = \cosh K + \sigma_i \sigma_j \sinh K = \cosh K(1 + w\sigma_i \sigma_j) \quad (w = \tanh K). \tag{5.3}$$

Hence we can expand (5.2) as follows:

$$\sum_{\sigma_i, \sigma_j} \prod_{ij} (\cosh K + \sigma_i \sigma_j \sinh K) = \cosh^L K \sum_{\sigma_i, \sigma_j} \prod_{ij} (1 + w\sigma_i \sigma_j)$$
$$= \cosh^L K \sum_{\sigma_i \sigma_j} [1 + w(\sigma_i \sigma_j) + w^2(\sigma_i \sigma_j)(\sigma_k \sigma_l) + \cdots]. \tag{5.4}$$

This can now be interpreted topologically. The first term corresponds to any one of the L links of the net; the second term corresponds to any pair of non-identical links and there will be $L(L-1)/2!$ such pairs; similarly for higher terms. When we sum over the plus or minus values of the σ_i, any term with an odd σ_i left in the summation will give zero contribution. Thus the only non-zero contributions arise from closed polygons each vertex of which is the meeting point of 2, 4, 6 ... links. Typical polygons are shown in Figure 5.2.

Any such polygon gives a contribution 2 for each vertex when we sum over the appropriate σ_i; the σ_i at all remaining lattice points likewise give a factor 2

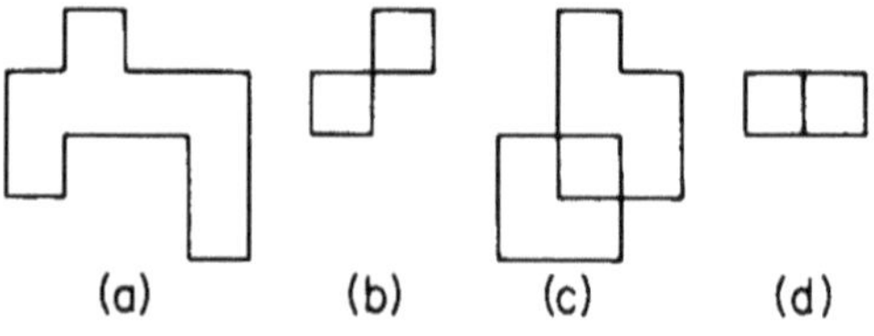

Figure 5.2 Typical polygons in the zero field expansions: (a), (b), (c) contribute, (d) does not contribute.

and we obtain a total factor of 2^N. We can therefore write the p.f. in the form

$$Z_N = 2^N(\cosh K)^L\left(1 + \sum_{r=1}^{L} p(r)\tanh^r K\right) \tag{5.5}$$

where $p(r)$ is the number of independent polygons defined above which can be constructed from links of the net.

Onsager noted that the number of configurations $v(r)$ with r broken links in (5.1) is equal to the number of closed polygons $p(r)$ for the dual lattice (Figure 5.3). Thus if we use an asterisk to denote parameters of the dual lattice, $v(r) = p^*(r)$. From (5.5) we have

$$Z_{N^*} = 2^{N^*}(\cosh K)^{L^*}\left(1 + \sum_{r=1}^{L^*} p^*(r)\tanh^r K^*\right). \tag{5.6}$$

Hence if we relate K^* to K so that

$$\tanh K^* = \exp(-2K) \tag{5.7}$$

we obtain

$$\frac{Z_{N^*}(K^*)}{2^{N^*}(\cosh K^*)^{L^*}} = \frac{Z_N(K)}{2\exp(LK)}. \tag{5.8}$$

Using the topological relations $L = L^*$, $N + N^* = L + 2$, (5.8) can be put in the more symmetrical form

$$\frac{Z_N(K)}{2^{N/2}(\cosh 2K)^{L/2}} = \frac{Z_{N^*}(K^*)}{2^{N^*/2}(\cosh 2K^*)^{L/2}}. \tag{5.9}$$

Relation (5.9) means that if we know the p.f. of any net we can immediately write down the corresponding p.f. for the dual net. The relation (5.7) can be put into the more symmetric form

$$\sinh 2K^* \sinh 2K = 1. \tag{5.10}$$

The transformation (4.94) of Kramers and Wannier is derived from (5.9) when the dual lattice is identical with the original lattice (with the exception of

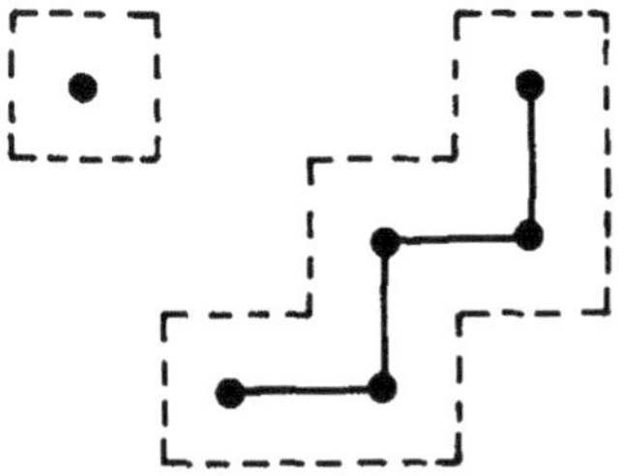

Figure 5.3 Equality of broken links on original lattice with polygons on dual lattice.

end effects) as is the case for the SQ lattice (Figure 5.1b). We find that

$$Z_{N^*}(K^*) = \frac{2^{N/2}}{2^{N^*/2}} \left(\frac{\cosh 2K^*}{\cosh 2K}\right)^{L/2} Z_N(K) \tag{5.11}$$

and the formula follows readily on substituting $N^* \simeq N \simeq L/2$, taking the Nth root, and using (5.10).

5.1.3 Star–Triangle Transformation

The dual of a triangular lattice is a honeycomb lattice (Figure 5.1d), and hence the relation (5.9) cannot be used directly to determine its Curie point. However, Onsager introduced the *star–triangle transformation* to relate the p.f.s for the triangular and honeycomb lattices, and hence to determine the Curie points of both. The star–triangle transformation replaces a star consisting of three spins interacting with a central spin by a triangle of spins interacting directly with one another (Figure 5.4); it can readily be established using (5.3). Let us sum over σ_0 in the p.f. for a star:

$$\begin{aligned}
&\sum_{\sigma_0=\pm 1} \exp[K(\sigma_0\sigma_1 + \sigma_0\sigma_2 + \sigma_0\sigma_3)] \\
&\quad = \sum_{\sigma_0=\pm 1} \cosh^3 K(1 + \sigma_0\sigma_1 \tanh K)(1 + \sigma_0\sigma_2 \tanh K)(1 + \sigma_0\sigma_3 \tanh K) \\
&\quad = 2\cosh^3 K + 2\sinh^2 K \cosh K(\sigma_1\sigma_2 + \sigma_2\sigma_3 + \sigma_3\sigma_1).
\end{aligned} \tag{5.12}$$

However the p.f. for a triangle of spins with interaction K^+ is given by

$$\begin{aligned}
&\exp[K^+(\sigma_1\sigma_2 + \sigma_2\sigma_3 + \sigma_3\sigma_1)] \\
&\quad = \cosh^3 K^+(1 + \sigma_1\sigma_2 \tanh K^+)(1 + \sigma_2\sigma_3 \tanh K^+)(1 + \sigma_3\sigma_1 \tanh K^+) \\
&\quad = (\cosh^3 K^+ + \sinh^3 K^+) + (\cosh^2 K^+ \sinh K^+ + \cosh K^+ \sinh^2 K^+) \\
&\qquad \times (\sigma_1\sigma_2 + \sigma_2\sigma_3 + \sigma_3\sigma_1).
\end{aligned} \tag{5.13}$$

Comparing (5.12) and (5.13) we find the identity

$$f \exp[K^+(\sigma_1\sigma_2 + \sigma_2\sigma_3 + \sigma_3\sigma_1)] = \cosh K(\sigma_1 + \sigma_2 + \sigma_3). \tag{5.14}$$

If we try to determine f, K^+ directly from (5.12) and (5.13) we will become involved in complicated algebraic manipulations. But since (5.14) is an identity

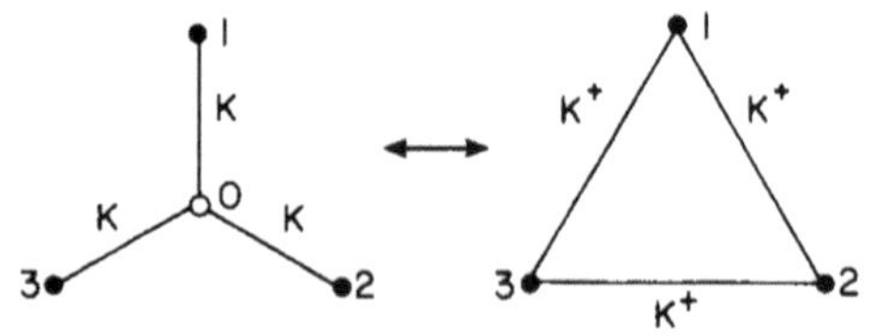

Figure 5.4 The star–triangle transformation.

we can put $\sigma_1 = \sigma_2 = \sigma_3 = 1$ to obtain

$$\cosh 3K = f \exp(3K^+) \tag{5.15a}$$

and $\sigma_1 = \sigma_2 = 1, \sigma_3 = -1$ to obtain

$$\cosh K = f \exp(-K^+). \tag{5.15b}$$

From (5.15a) and (5.15b) we find that

$$\exp(4K^+) = 2 \cosh 2K - 1 \tag{5.16a}$$

and

$$f = \exp(K^+) \cosh K = (\cosh 3K)^{1/4}(\cosh K)^{3/4}. \tag{5.16b}$$

If we now apply the star–triangle transformation to the lattice points represented by open circles in Figure 5.5, we easily deduce that

$$Z_{2N}^{\mathrm{H}}(K) = f^N Z_N^{\mathrm{T}}(K^+) \tag{5.17}$$

where the superscripts H and T refer to the honeycomb and triangular lattices.

Now from (5.9) we have the duality relationship

$$\frac{Z_N^{\mathrm{T}}(K^*)}{2^{N/2}(\cosh 2K^*)^{3N/2}} = \frac{Z_{2N}^{\mathrm{H}}(K)}{2^N(\cosh 2K)^{3N/2}}. \tag{5.18}$$

Combining (5.17) and (5.18) we obtain a relation between $Z_N^{\mathrm{T}}(K^*)$ and $Z_N^{\mathrm{T}}(K^+)$ to which the previous argument can be applied. From (5.10) and (5.16a) the reciprocal relation between K^+ and K^* is

$$[\exp(4K^+ - 1)][\exp(4K^* - 1)] = 4. \tag{5.19}$$

The Curie point of the triangular lattice is given by $K^+ = K^*$, i.e.

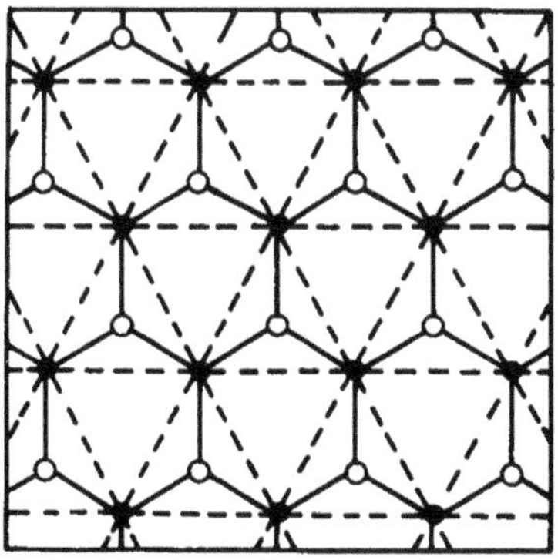

Figure 5.5 The honeycomb lattice (application of the star–triangle transformation to the spins represented by open circles and their three neighbours leads to the triangular lattice – broken lines).

$$\exp(2K^+) = 3. \tag{5.20}$$

For the honeycomb lattice (5.17) and (5.18) must be used to give a relationship between $Z^{\mathrm{H}}_{2N}(K)$ and $Z^{\mathrm{H}}_{2N}(K^{+*})$ where K^+ and K^{+*} are related by (5.10). It can readily be shown that

$$(\cosh 2K^{+*} - 1)(\cosh 2K - 1) = 1 \tag{5.21}$$

and the Curie point of the honeycomb lattice is given by

$$\cosh 2K = 2. \tag{5.22}$$

5.1.4 *Solution for the SQ and Rectangular Lattices*

At the Batelle Colloquium in 1970, Onsager told how the use of the star–triangle transformation led him to conclude that the transfer matrix **V** commutes with a linear combination of a pair of simpler matrices, $\mathbf{B} + h\mathbf{A}$, and how he then tried to find the eigenvalues of the latter for finite strips with $n = 1, 2, 3, 4, 5$ sites wide. This involved no more than quartic equations, the first of which turned out to be a biquadratic which was readily soluble. He continued the calculation for $n = 6, 7, 8$ and was able to conjecture the solution for general n. The conjecture provided the basis for the New York announcement.

The proof for general n was effected by forming the commutator $[\mathbf{AB}]$ and then the commutators of this with **A** and **B** and continuing to form commutators of commutators until a complete set of operators was obtained with the property that the commutator of any two operators is a linear combination of members of the set. The paper in 1944 was confined to the principal eigenvector; when in 1945 he was able to derive the full spectrum of the transfer matrix (as related in §5.1.1), Bruria Kaufman developed a simpler and more elegant approach.

Onsager observed that it was no more difficult to deal with the rectangular lattice having differing interactions J, J' in the principal lattice directions. The duality transformation can be applied to such an asymmetric lattice to give the two relations

$$\begin{aligned} \sinh 2K^* \sinh 2K' &= 1 \\ \sinh 2K'^* \sinh 2K &= 1. \end{aligned} \tag{5.23}$$

Relations (5.23) can no longer be solved for a given set of interaction constants to provide a unique relationship between T^* and T. However, if we interpret the transformation from K to K^* variables as converting order into disorder, the Curie points correspond to the meeting points of the ordered and disordered regions on the (K,K') plane; they will therefore be determined by

$$\sinh 2K \sinh 2K' = 1. \tag{5.24}$$

The basic properties of Onsager's solution may be summarized as follows.

5.1.4.1 *Mathematical properties of the solution*

(i) *Largest eigenvalue* The largest eigenvalue λ_{max} is given by the formula

$$\ln \lambda_{max} - \tfrac{1}{2}n \ln(2 \sinh 2K) = \tfrac{1}{2}(\gamma_1 + \gamma_3 + \cdots + \gamma_{2n-1}). \qquad (5.25)$$

The γ_r are given by

$$\begin{aligned} \cosh \gamma_r &= \cosh 2K^* \cosh 2K' - \sinh 2K^* \sinh 2K' \cos r\pi/n \\ &= \coth 2K \cosh 2K' - \operatorname{cosech} 2K \sinh 2K' \cos r\pi/n \qquad (5.26) \\ &= \coth 2K \cosh 2K - \cos r\pi/n \quad (K = K'). \end{aligned}$$

It will be seen that the γ_r are invariant under the duality transformation $K' \to K^*$, and hence so is the sum in (5.25); they are shown graphically in Figure 5.6 ($K = K'$).

In the limit when the size of the crystal becomes large the series can be replaced by an integral, and the partition function per spin is given by

$$\ln \lambda - \tfrac{1}{2} \ln(2 \sinh K) = \frac{1}{2\pi} \int_0^{\pi} \gamma(\omega)\, d\omega. \qquad (5.27)$$

This can be transformed into a double integral which is symmetric in K and K' (using the identity $\int_0^{2\pi} \ln(2 \cosh x - 2 \cos \omega)\, d\omega = 2\pi x$):

$$\ln(\lambda/2) = \tfrac{1}{2}\pi^{-2} \int_0^{\pi} \int_0^{\pi} \ln(\cosh 2K \cosh 2K' - \sinh 2K \cos \omega - \sinh 2K' \cos \omega')\, d\omega\, d\omega'. \qquad (5.28)$$

The various thermodynamic functions can be derived in the standard manner from λ, and many of them are expressible in terms of elliptic integrals.

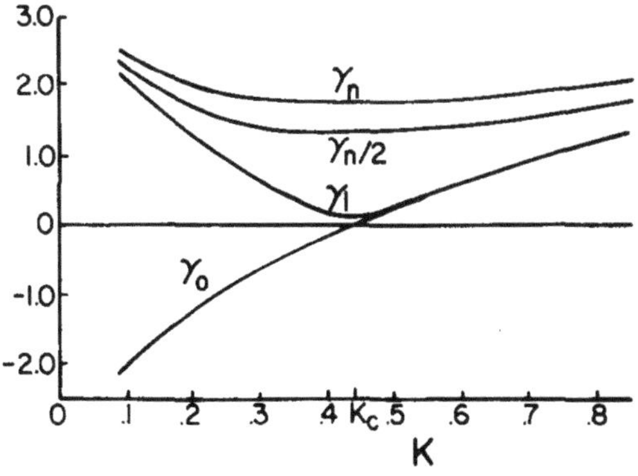

Figure 5.6 Dependence of γ_r on temperature. For $T > T_c$ ($K < K_c$), $\gamma_1 \sim -\gamma_0$. For $T < T_c$ ($K > K_c$), $\gamma_1 \sim \gamma_0$.

(ii) *The distribution of eigenvalues* Onsager and Kaufman obtained the remaining eigenvalues of **V** and these enable the ordering properties to be understood. The eigenvalues fall into two sets (Kaufman 1949)

$$\begin{aligned} \ln \lambda_i^+ + \tfrac{1}{2}n \ln(2 \sinh 2K) &= \tfrac{1}{2}(\pm\gamma_1 \pm \gamma_3 \pm \cdots \pm \gamma_{2n-1}) \\ \ln \lambda_i^- - \tfrac{1}{2}n \ln(2 \sinh 2K) &= \tfrac{1}{2}(\pm\gamma_2 \pm \gamma_4 \pm \cdots \pm \gamma_{2n}). \end{aligned} \tag{5.29}$$

Here any combination giving an even number of minus signs corresponds to an eigenvalue; there are thus 2^n eigenvalues as expected.

It will be seen from Figure 5.6 that the γ_r increase steadily from γ_1 to γ_n $(=K' + K^*)$, and for $r > n$ they satisfy the relation $\gamma_r = \gamma_{2n-r}$; for large n they form a closely spaced spectrum between $2[K' - K^*]$ and $2(K' + K^*)$. γ_0 is exceptional, however, since it changes sign at the critical point $K' = K^*$.

For large n, $\gamma_{2r+1} \simeq \gamma_{2r}$ $(r \neq 0)$ and for $K' > K_c$, i.e. $T < T_c$, $\gamma_1 \simeq \gamma_0$; thus the two sums in (5.29) are different approximations to $\frac{1}{2}\int_0^{2\pi} \gamma(\omega)\, d\omega$, and the largest eigenvalues λ_1^+ and λ_1^- are very nearly equal. (The difference is exponentially small, all correction terms in the Euler–Maclaurin formula vanishing; see Goodwin (1949).) The next eigenvalue differs from this value by a factor of approximately $\exp[4(K' - K^*)]$. However, when $K' < K_c$, i.e. $T > T_c$, $\gamma_1 \simeq -\gamma_0$. Thus the largest eigenvalue λ_1^+ is greater than λ_1^- by a factor of approximately $\exp[2(K^* - K')]$.

We may summarize the distribution of the eigenvalues as follows:

$$T < T_c, \quad \lambda_1^+ = \lambda_1^- > \text{continuum of limit } \lambda_1^+ \exp[-4(K' - K^*)]$$

$$T = T_c, \quad \lambda_1^+ = \lambda_1^- = \text{limit of continuum}$$

$$T > T_c, \quad \lambda_1^+ > \lambda_1^- = \text{limit of continuum } (\lambda_1^- = \lambda_1^+ \exp[-2(K^* - K')]).$$

This is illustrated diagrammatically in Figure 5.7.

It can be shown quite generally that long-range order is present in an assembly if and only if the largest eigenvalue is degenerate; if long-range order is absent the correlation between lattice sites n rows apart is $\sim(\lambda_1^-/\lambda_1^+)^n$ and hence the range of order is given roughly by the inverse of $\ln(\lambda_1^+/\lambda_1^-)$ (Domb 1960, pp. 173–4).

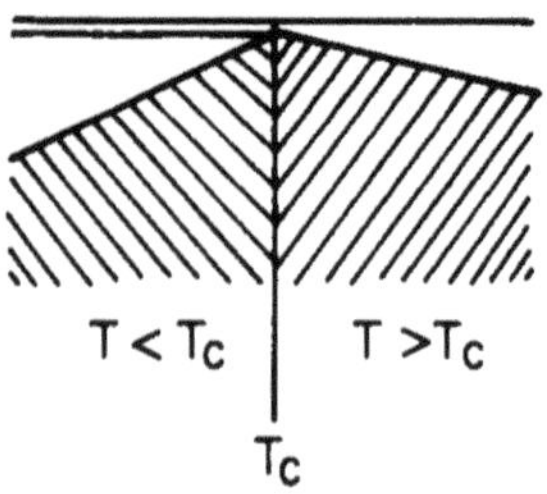

Figure 5.7 Schematic representation of the distribution of eigenvalues (shaded portion corresponds to a continuum as $n \to \infty$).

We therefore see that for $T < T_c$ there is long-range order (but this occurs only in the limit as $n \to \infty$). For $T > T_c$ there is no long-range order, and the mean distance to which order persists is given roughly by $l/[2(K^* - K')]$.

(iii) *Change in sign of the interaction* If we replace J' by $-J'$ the interaction becomes antiferromagnetic in one direction; the solution (5.29) remains valid but a detailed analysis shows a difference between odd and even values of n. We may now write

$$\cosh \gamma_r = \cosh 2K^* \cosh 2K' - \sinh 2K^* \sinh 2K' \cos(\pi - r\pi/n) \ldots \quad (5.30)$$

so that each γ_r is replaced by γ_{n-r}. When n is even λ_1^+ and λ_1^- are unchanged as we should expect. When n is odd, however, there is a misfit seam: odd γ are transformed into even γ, except for γ_n $(=K' + K^*)$ which is transformed into $-\gamma_0$ $(= -K' + K^*)$. Thus λ_1^+ is transformed into λ_1^- with the exception of the change of sign of γ_0. We may, therefore, use the discussion of (ii) to conclude that when $T > T_c$ the partition function is unchanged, but when $T < T_c$ there is an additional factor

$$\exp[-2(K' - K^*)]. \quad (5.30')$$

This will be interpreted as a *boundary tension*. However, the largest eigenvalue becomes multiply degenerate for $T < T_c$, as we should expect from the degeneracy in the position of the misfit seam.

5.1.4.2 *Physical properties of the solution*

(i) *Specific heat* The specific heat manifests a striking singularity at the Curie point. It is logarithmically infinite on both sides of the Curie temperature, and is drawn in Figure 5.8 for the symmetric net. For temperatures near to T_c the following approximation may be used:

$$\begin{aligned} C_H/Nk &\simeq (2/\pi)(\ln \cot \pi/8)^2(K_1 - 1 - \pi/4) \quad [K_1 = \ln(2^{1/2}/|K - K_c|)] \\ &\simeq -0.4945 \ln|T - T_c|. \end{aligned} \quad (5.31)$$

It should be noted that the specific heat curve differed substantially from any curves which had been observed experimentally, particularly in regard to the magnitude of the 'tail' above the Curie temperature. This can be seen more clearly from the critical value of the entropy, which shows that the entropy change above T_c $(=0.387)$ is greater than that below T_c $(=0.306)$.

For an asymmetric net the logarithmic nature of the specific heat anomaly is unchanged. As one of the interactions J' becomes small the Curie temperature moves to zero, but this movement is very slow; if the interaction J' is J/v then the reduction in T_c is of order $\ln v$ for large v. The specific heat curve for an asymmetric net in which $J' = J/100$ was calculated by Onsager, and is

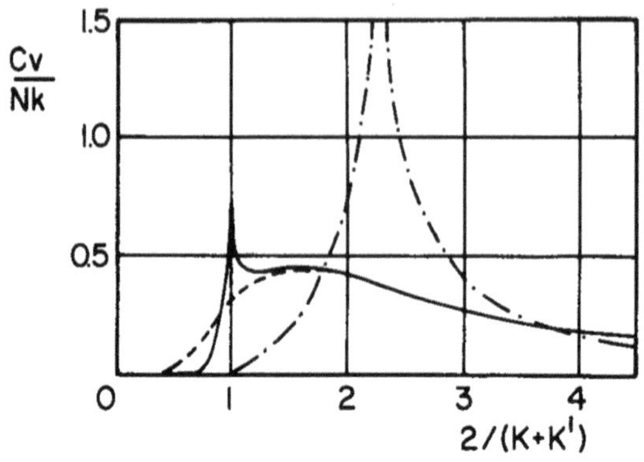

Figure 5.8 Specific heat of a rectangular lattice: chain curve, $J' = J$ (symmetric); solid curve, $J' = J/100$ (asymmetric); dashed curve, $J' = 0$ (linear chain) (after Onsager).

also shown in Figure 5.8 for comparison with the symmetric net, and with the specific heat of a one-dimensional chain ($J' = 0$). The abscissa is $2kT/(J + J')$ so that the areas under all the curves are the same. It will be seen that the large asymmetry reduces the Curie temperature by a factor of 2 only. Also there is an essential difference below the Curie temperature between a two-dimensional net with one small interaction, and a one-dimensional chain; as pointed out in §4.8.1 the latter can be regarded as an assembly whose Curie point has moved to $T = 0$.

(ii) *Other thermodynamic quantities* The internal energy E has a vertical tangent on both sides of the Curie point and is shown in Figure 5.9 (for the symmetric net). Near the Curie point it is of the form

$$A_0 + B_0(T - T_c) \ln|T - T_c|. \tag{5.32}$$

The free energy F involves an integral of the internal energy and therefore has infinite curvature at the Curie point. The entropy can readily be obtained from the relation $S = (E/T) - F$. Its behaviour at the Curie point is similar to that of the energy.

(iii) *The boundary tension* We have seen that when the sign of the interaction energy J' is changed, if n is odd there is a misfit seam which divides regions of opposite order. The partition function (per spin) of such a boundary is

$$\lambda_b' = \exp[-2(K' - K^*)] = z'(1 + z)/(1 - z). \tag{5.33}$$

The corresponding boundary free energy is

$$\sigma' = 2J' - kT \ln[\coth(J/kT)]. \tag{5.34}$$

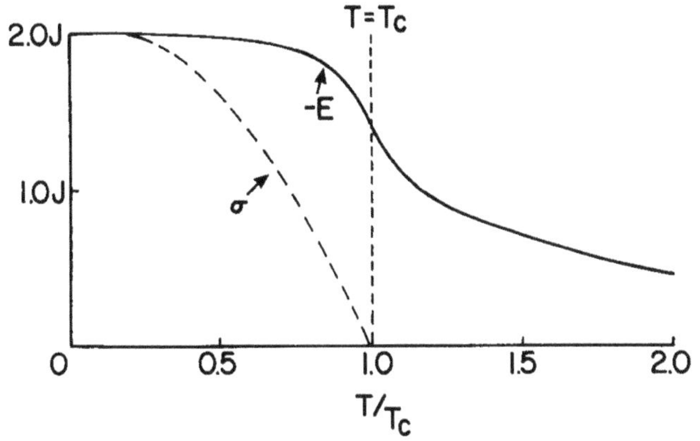

Figure 5.9 Energy and boundary tension of a quadratic lattice. E (solid curve) has infinite slope at T_c; σ (dashed curve) is zero for $T \geq T_c$.

It vanishes at the critical point and is also shown in Figure 5.9 for a symmetric net ($J = J'$). It is thus significant that the critical points are identical for the boundary and for the bulk substance. Corresponding boundary functions λ_b, σ in the perpendicular direction are obtained by interchanging J and J'.

(iv) *The specific heat of a finite crystal* The specific heat of a finite crystal of n rows can be computed from (5.25). It is finite at all temperatures, since the partition function is an analytic function of the temperature, but a maximum occurs near the transition point of the infinite crystal; this maximum increases and sharpens with increasing size of the crystal. Onsager showed that for large n the maximum specific heat is given approximately by

$$C_H/Nk \simeq 0.4945 \ln n + 0.1879 \tag{5.35}$$

and it occurs at a temperature differing by $O(n^{-2} \ln n)$ from T_c for the infinite lattice. The formula extrapolated by Kramers and Wannier (1941) from calculations for finite crystals with $n = 1, 2, 3, 4, 5, 6$ is quite close to (5.35). Note that the amplitudes in (5.31) and (5.35) are identical. We shall return to this point in §6.6.2.

5.1.5 *Correlations*

It can be shown generally that a knowledge of the complete spectrum of eigenvalues and eigenvectors of the transfer matrix enables the correlation between any two lattice sites to be calculated (Domb 1960, pp. 173–4). Kaufman and Onsager (1949) applied their general theory of the diagonalization of **W** to the evaluation of these averages. The details of the algebraic manipulations are

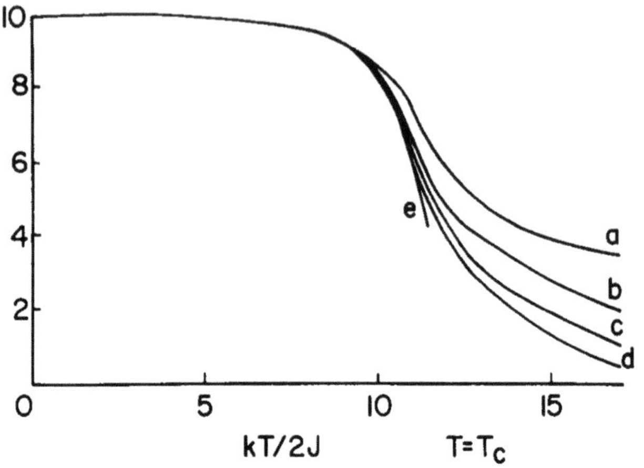

Figure 5.10 Correlations for small distances as functions of temperature: a, $\langle\sigma_1\sigma_2\rangle$; b, $\langle\sigma_{11}\sigma_{22}\rangle$; c, $\langle\sigma_1\sigma_3\rangle$; d, $\langle\sigma_{11}\sigma_{23}\rangle$; e, $\langle\sigma_1\sigma_4\rangle$. All have infinite slope at T_c (after Kaufman and Onsager 1949).

quite involved, and the resulting correlations are expressed as determinants of functions which can be reduced to elliptic integrals in the symmetric case $K = K'$. The nearest-neighbour correlation $\langle\sigma_1\sigma_2\rangle$ is equivalent to the energy; and as the distance apart k becomes very large the correlation $\langle\sigma_1\sigma_k\rangle$ approaches the long-range order θ_∞, which was evaluated exactly and will be discussed shortly. All the intermediate correlations lie between these two bounding curves. Kaufman and Onsager evaluated exactly the correlations $\langle\sigma_1\sigma_2\rangle\langle\sigma_{11}\sigma_{22}\rangle\langle\sigma_1\sigma_3\rangle\langle\sigma_{11}\sigma_{23}\rangle$ and these are shown graphically in Figure 5.10. At the Curie point it seems that all the correlation functions have a singularity similar to the energy with an infinite slope.

For the correlations $\langle\sigma_1\sigma_k\rangle$ along a given row a general formula can be obtained for all k in the form

$$(-1)^k\langle\sigma_1\sigma_{k+1}\rangle = \cosh^2 K^*\Delta_k + \sinh^2 K^*\Delta_{-k}\cdots \tag{5.36}$$

where Δ_k and Δ_{-k} are Toeplitz determinants of order k whose elements can be reduced to elliptic integrals. At the Curie point the following approximations are reasonable:*

$$\begin{aligned}|\Delta_k| &= \frac{2}{\pi}\prod_{s=1}^{k-1}\frac{\Gamma(s+1)\Gamma(s+1)}{\Gamma(s+\frac{1}{2})\Gamma(s+\frac{3}{2})}\\ |\Delta_{-k}| &= \frac{2}{\pi}\prod_{s=1}^{k-1}\frac{\Gamma(s+1)\Gamma(s+1)}{\Gamma(s+\frac{3}{2})\Gamma(s+\frac{5}{2})}.\end{aligned} \tag{5.37}$$

* In the paper by Kaufman and Onsager the products are taken to $s = k$ but Professor M. E. Fisher has pointed out that this is incorrect.

Hence for large k the correlations at the critical point $\simeq A/k^{1/4}$ and tend to zero very slowly as $k \to \infty$.

Onsager also derived an expression for the order along a main diagonal for the general anisotropic case which is much simpler than (5.36). In fact if a_r are defined by

$$\left(\frac{\sinh 2K \sinh 2K' - \exp(\mathrm{i}\omega)}{\sinh 2K \sinh 2K' - \exp(-\mathrm{i}\omega)}\right) \equiv \sum_{r=-\infty}^{\infty} a_r \exp(\mathrm{i}r\omega) \tag{5.38}$$

so that the a_r are the Fourier coefficients of this function, the order along a main diagonal is given by a single Toeplitz determinant

$$D_k = \begin{vmatrix} a_0 & a_1 & a_2 & \cdots & a_{k-1} \\ a_1 & a_0 & a_1 & \cdots & a_{k-2} \\ a_{-2} & a_{-1} & a_0 & \cdots & a_{k-3} \\ \vdots & \vdots & & & \vdots \\ a_{-(k-1)} & a_{-(k-2)} & \cdots & \cdots & a_0 \end{vmatrix} . \tag{5.39}$$

For this determinant the first equation of (5.37) is exact so that the correlations at the critical point along a diagonal $\simeq 0.645/k^{1/4}$ and the asymptotic form is approached very rapidly.

In the previous section we derived an estimate of the mean range over which order persists above the Curie temperature in the form $1/[2(K^* - K')]$; this is valid in the direction of interaction J, and the corresponding expression in direction J' is $1/[2(K^* - K)]$. In intermediate directions of angle ϕ Onsager in his first paper conjectured the implicit formula

$$\cosh 2K \cosh 2K' - \sinh 2K \cosh(\beta \sin \phi) - \sinh 2K' \cosh(\beta \cos \phi) = 0$$

for the mean range $1/\beta$ of the order. This would imply that, with $K = K'$, for temperatures near the critical point the order spreads out isotropically, whereas for high temperatures it spreads out along the links of the lattice (Figure 5.11). The exact computations available so far confirm this general behaviour.

5.1.6 *Solutions for Other Two-dimensional Lattices*

We have seen in §5.1.3 that the star–triangle transformation enables the Curie point to be located for the triangular and honeycomb lattices. Shortly after the publication of the work of Onsager and Kaufman several papers appeared independently giving the exact solution for the triangular lattice (Houtappel 1950, Husimi and Syozi 1950, Newell 1950, Temperley 1950, Wannier 1950). The solution for the honeycomb lattice can then be derived from (5.17).

The analytic form of the solution is individual to each lattice, but near T_c the pattern of critical behaviour is the same for all of them. The amplitude of

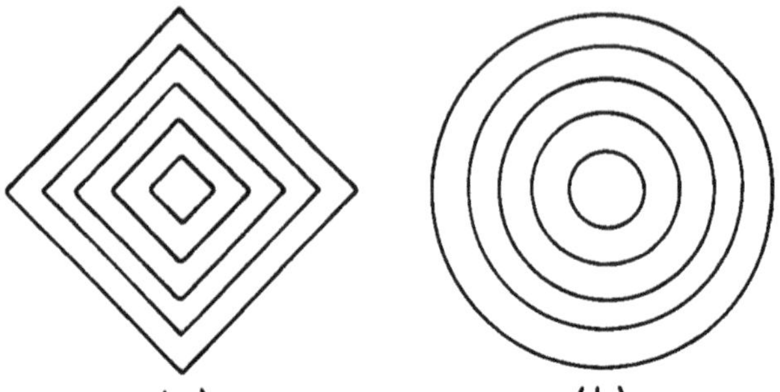

Figure 5.11 Anisotropy in the propagation of order. At high temperatures order spreads out along the links of the lattice with little correlation, and the contours of constant order are as shown in (a). Near the Curie point the correlation becomes very marked, and the contours are circles, as in (b).

the logarithmic singularity in the specific heat is insensitive to lattice structure; the coefficient 0.4945 in (5.31) is replaced by 0.4991 for the triangular lattice, and 0.4781 for the honeycomb lattice. Solutions for a number of additional two-dimensional lattices were soon forthcoming (see e.g. Domb 1960, pp. 211–20), and they all conformed to the same pattern of critical behaviour. This provided the first clear evidence that dimension rather than lattice structure plays a decisive role in the determination of critical behaviour.

5.1.6.1 *The long-range order*

There are two independent approaches to the question of long-range order. The first considers the limit of the correlation function $\langle\sigma_1\sigma_k\rangle$ as $k \to \infty$. The second evaluates the spontaneous magnetization by a small perturbation of the no-field solution for $T < T_c$; the spontaneous magnetization is equal to θ_∞, the long-range parameter defined in §4.8.4 (see Domb 1960, p. 227). As we mentioned in §5.1.1 Onsager announced the exact formula for θ_∞ at the Florence Conference in 1949 in the following form:

$$\theta_\infty = (1 - q^2)^{1/8} \quad (\sinh 2K \sinh 2K' = q^{-1}). \tag{5.40}$$

This differs only slightly from unity except in the immediate neighbourhood of the Curie point; for the symmetric lattice its value near T_c is

$$1.222\,41(1 - T/T_c)^{1/8}. \tag{5.41}$$

The spontaneous magnetization is represented graphically in Figure 5.12.

Onsager's method of calculation (as mentioned in §5.1.1) was to consider the limiting form of determinant (5.39) as $k \to \infty$. The method of Yang (1952) was to use perturbation theory. In principle if an eigenvector of a matrix is

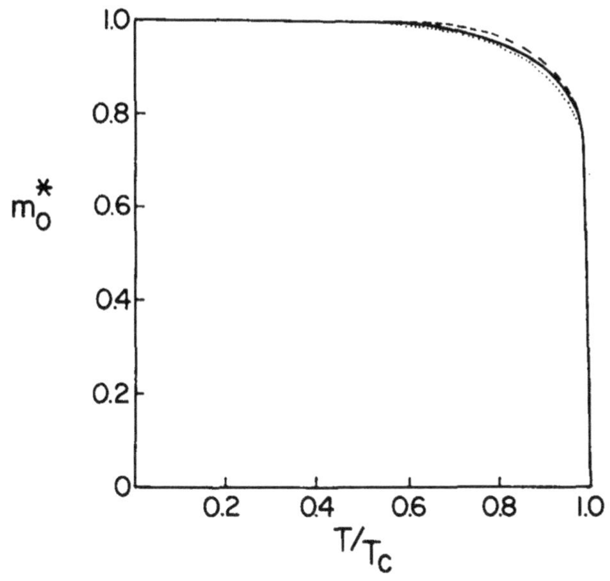

Figure 5.12 The spontaneous magnetizations of different lattices; dashed curve, honeycomb; solid curve, quadratic; dotted curve, triangular.

known at any point the first derivative of the eigenvalue can be determined. However, the details of the calculation are quite intricate.

5.1.7 The Challenge of Onsager's Solution

Onsager's exact results indicated clearly that the closed-form approximations of the classical theory give incorrect results in the critical region for a two-dimensional model, and there was no valid reason for thinking that they would prove more correct for three dimensions. Since the free energy has a singularity at T_c, no Taylor expansion is possible about this point, and hence the Landau theory which assumes such an expansion is also invalid.

However, in addition, the thermodynamic functions arising from Onsager's solution differed radically from known experimental results. The spontaneous magnetization comes to zero with an eighth-power law which was much steeper than anything which had been observed in magnetic materials; the specific heat possesses a large tail which did not accord with measurements of order–disorder and λ-point transitions; and the logarithmic singularity symmetric on both sides of T_c seemed strangely unreal. Above T_c the *coherence length* was of order $(T - T_c)^{-1}$ whereas classical theory suggested $(T - T_c)^{-1/2}$ (equation (4.50)). At T_c the $r^{-1/4}$ decay of correlations had no basis in the classical treatment. Was the disagreement with experiment due to the dimensionality of the model, the neglect of forces other than nearest neighbour, or perhaps the inadequacy of the Ising interaction? One must remember that in his original paper Ising had erroneously concluded that the model would not give rise to a spontaneous magnetization even in three dimensions; and Ising's failure to obtain a spontaneous magnetization had been quoted by Heisenberg in his classic paper (1928) which introduced the Heisenberg interaction in ferromagnets (see Brush 1967). For many years the Ising interaction was not considered seriously by research workers in ferromagnetism.

For some time after 1944 hopes ran high that a solution of the three-dimensional Ising model might soon be forthcoming. Thus Wannier ended his 1945 review with the following sentence: 'It is to be hoped that a three-dimensional calculation will, before long, furnish the answer to these questions.' Indeed, there were several claims to have solved the problem which turned out to be unjustified, the most notable being that of Maddox which was presented at the 1952 Paris Conference on Phase Transitions (see Domb 1985). But as a variety of different methods were advanced to re-derive the Onsager solution, none of which could be extended to three dimensions, the realization spread that a solution would not readily be forthcoming.

It was therefore of great importance to look for alternative methods which could provide reliable information on three-dimensional models, and on alternative forms of interaction. The most useful source of such information in the initial period following Onsager was the development of series expansions at high and low temperatures. Further important information was extracted from Onsager's solution and other exact solutions were discovered corresponding to unrealistic interactions but nevertheless providing insight into the relationship between critical behaviour and intermolecular forces. The cluster integral theory of fluids, which had been one of the triumphant theoretical developments of the late 1930s, had seemingly come up against a barrier in its attempt to explain critical behaviour. Onsager's work aroused interest in the far more primitive lattice gas model which is isomorphic with the Ising model. And finally the exact results for the two-dimensional Ising model, and the new theoretical conjectures which began to emerge for other models, provided a challenge to experimentalists to look for systems in nature which might be expected to conform to the theoretical models, and to provide accurate measurements of their critical behaviour. In the rest of this chapter we shall consider these topics in more detail.

5.2 Series Expansions

Several terms of low- and high-temperature series expansions for the Ising model on the SQ lattice were derived by Kramers and Wannier (1941), and were used to discriminate between various closed-form methods of approximation. The low-temperature series were patterned after the spin wave excitations of Bloch (1930), and the high-temperature expansions were an adaptation of an expansion for the Heisenberg model by Opechowski (1937).

The idea of generating lengthy series expansions and correlating the coefficients with critical behaviour was first advanced by the present writer in his doctoral thesis (Cambridge 1949, see Domb 1949a,b). Onsager's solution applies only in zero field, and the first target was to explore the properties of the model in a non-zero field. A series was developed for the spontaneous magnetization, and an attempt made to assess its critical behaviour (Onsager's formula was not yet known).

The method was developed with the help of Rushbrooke, Wakefield, Sykes and Fisher, and other members of the King's College group, and applied to a number of problems which will be outlined shortly. A few general remarks are necessary, however, to clarify the basis of the method. In normal statistical work it is hazardous to attempt to extrapolate asymptotic behaviour from a finite number of terms. But in the case of a ferromagnet one could anticipate on physical grounds a pattern of expected critical behaviour, and use the coefficients of the series to obtain the best fit to parameters like critical temperature and critical exponents in the assumed asymptotic form for the coefficients. Moreover, the methods of analysis and fitting could be tested on Onsager's exact solution to see how well they worked, and to get an idea of the error to be expected. The method proved capable of furnishing important reliable information, and as numerical techniques improved its use became widespread.

The problems connected with the generation and analysis of series expansions are sophisticated and are described in detail in the various chapters of Volume 3 of *Phase Transitions and Critical Phenomena* (Domb and Green 1974) (DG **3**). Our aim in this section is to provide a brief account of the most important features.

5.2.1 *Embedding Method*

5.2.1.1 *Simple Ising model $I(\frac{1}{2})$*

It was conjectured by Domb (1949b) and proved rigorously by Yang and Lee (1952) that for the simple Ising model there are no singularities except when $H = 0$. We have seen that at sufficiently low temperatures there is a discontinuity in magnetization M as $H \to 0$. Figure 5.13 illustrates the distribution of singularities in the (H,T) plane, the function being analytic except on the heavy line on the H axis.

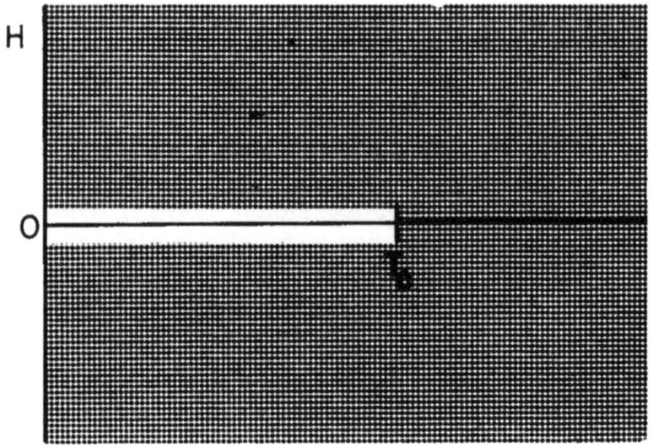

Figure 5.13 The partition function is analytic everywhere in the shaded region.

There are basically three different types of series expansion which are illustrated in Figure 5.14.

(i) *Density expansions* For this type of expansion we start with the state of complete alignment of the spins in a very large magnetic field in direction 2, and allow the field to decrease, giving rise at any particular temperature to groups 1, 2, 3 ... overturned spins. There are ${}^{N}C_1$ excited states corresponding to one overturned spin, ${}^{N}C_2$ corresponding to two overturned spins, ... ${}^{N}C_r$ corresponding to r overturned spins. The ${}^{N}C_r$ spins can be divided into different energy groups: the highest energy deviation from the ground state, $(2r\mu_0 H + 2qrJ)$ (q being the coordination number of the lattice), arises from r spins no two of which are nearest neighbours, the next from r spins with one nearest-neighbour link, the next from r spins with two nearest-neighbour links, and so on. Each nearest-neighbour link decreases the energy by $2J$; as an example the configurations which correspond to $r = 3$ are shown diagrammatically in Figure 5.15. We may thus expand $\Lambda_N(y,z)$ (equations (4.78), (4.80))

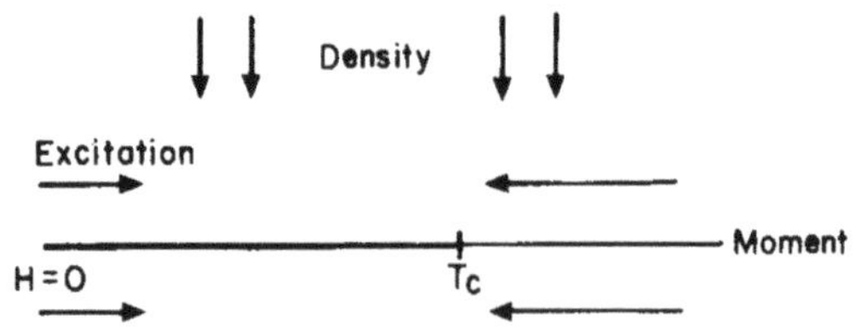

Figure 5.14 Types of series expansion for a ferromagnet.

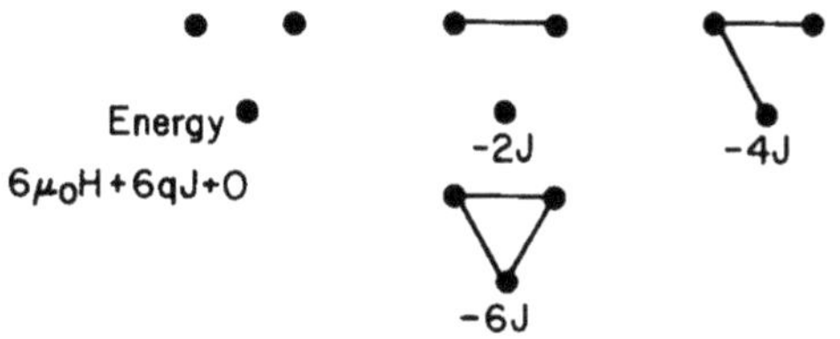

Figure 5.15 Low-temperature configurations with three spins.

in the form

$$\Lambda_N(y,z) = 1 + yF_1(N,z) + y^2F_2(N,z) + y + \cdots + y^rF_r(N,z) + \cdots \qquad (y < 1) \qquad (5.42)$$

where there are ${}^NC_r$ terms contributing to $F_r(N,z)$. For the SQ lattice with a cyclic boundary condition

$$F_1(N,z) = Nz^4, \quad F_2(N,z) = 2Nz^6 + \frac{N(N-5)}{22}z^8$$
$$F_3(N,z) = 6Nz^8 + 2N(N-8)z^{10} + \frac{N^3 - 52N^2 + 62N}{6}z^{12}. \qquad (5.43)$$

So far the expansion has been exact. For large N we should expect on physical grounds that

$$\Lambda_N(y,z) \sim [\Lambda(y,z)]^N. \qquad (5.44)$$

There might be a multiplying factor whose effect would become negligible for large N. Hence, if we assume that $\Lambda(y,z)$ can be expanded as a power series in y of the form

$$\Lambda(y,z) = 1 + yf_1(z) + y^2f_2(z) + \cdots + y^rf_r(z) + \cdots \quad (y < 1) \qquad (5.45)$$

we can equate the successive powers of N, and deduce the values of the $f_r(z)$ from the $F_r(N,z)$. In fact for a lattice satisfying cyclic boundary conditions, there are no surface effects, and it will be found that the multiplying factor is unity. We then have an identity

$$\begin{aligned} 1 + yF_1(N,z) + y^2F_2(N,z) + \cdots + y^rF_r(N,z) + \cdots \\ = 1 + {}^NC_1[yf_1(z) + y^2f_2(z) + \cdots] \\ + {}^NC_2[yf_1(z) + y^2f_2(z) + \cdots]^2 + \cdots \\ + {}^NC_s[yf_1(z) + y^2f_2(z) + \cdots]^s + \cdots. \end{aligned} \qquad (5.46)$$

Putting $N = 1$ formally we find the simple relation

$$f_r(z) = F_r(1,z). \qquad (5.47)$$

The derivation of this result depends on the identity of both sides of (5.46), and to convince oneself of its correctness it is advisable to write out fully the first few terms of each side of (5.46) in order to verify that it is indeed an identity. For the SQ lattice from (5.43) we have

$$f_1(z) = z^4, \quad f_2(z) = 2z^6 - 2z^8, \quad f_3(z) = 6z^8 - 14z^{10} + 8z^{12}. \tag{5.48}$$

It is usually more convenient to use the expansion for $\ln \Lambda(y,z)$ in the form

$$\ln \Lambda(y,z) = yg_1(z) + y^2 g_2(z) + \cdots + y^r g_r(z) + \cdots. \tag{5.49}$$

By equating $\exp[N \ln \Lambda(y,z)]$ with $\Lambda_N(y,z)$ in (5.42) it is easy to show formally that $g_r(z)$ is given by the coefficient of N in $F_r(N,z)$ so that

$$g_1(z) = z^4, \quad g_2(z) = 2z^6 - \tfrac{5}{2}z^8, \quad g_3(z) = 6z^8 - 16z^{10} + \tfrac{10}{3}z^{12}. \tag{5.50}$$

But more fundamentally if we take the logarithm of (5.42) we find that all powers of N other than the first disappear. In configurational terms the contribution of all graphs involving disjoint spins have been incorporated into the connected graphs, so that the expansion can be written in terms of connected graphs only. Such an expansion is usually termed a *linked cluster expansion*; its origin lies in general theorems of graph theory (DG **3**, Ch. 1), and it has wide applications.

So far states in which more than half of the spins point in direction 1 have been ignored, and in the presence of a magnetic field, however small, this is certainly valid for large N. When the field is zero these states become significant; they can easily be taken into account because of the symmetry between directions 1 and 2, and we can write instead of (5.42)

$$\begin{aligned}\Lambda_N(y,z) = (1 + y^N) + (y + y^{N-1})F_1(N,z) + (y^2 + y^{N-2})F_2(N,z) + \cdots \\ + (y^4 + y^{N-r})F_r(N,z) + \cdots \quad \text{(for all } y\text{)}.\end{aligned} \tag{5.51}$$

When y is less than 1 the second term in each bracket of (5.51) is negligible and we recover (5.42).

When $y = 1$, both terms in each bracket are equal, and the expression (5.42) must be multiplied by 2. On taking the Nth root, the expression (5.45) for $\Lambda(y,z)$ must be multiplied by $2^{1/N}$, and this differs negligibly from unity for large N. Hence $\Lambda(y,z)$ is continuous at $y = 1$.

The magnetization is given by

$$M = kT \frac{\partial}{\partial H} \ln Z_N = \mu_0\left(N - 2y \frac{\partial}{\partial y} \ln \Lambda_N(y,z)\right). \tag{5.52}$$

When $y = 1$ it will be seen from (5.51) that $\partial\Lambda_N/\partial y$ is equal to $N\Lambda_N$, and hence that $M = 0$. When $y < 1$, however, if we use the expansion (5.49) the magnetization is equal to

$$N\mu_0\{1 - 2[yg_1(z) + 2y^2 g_2(z) + \cdots + ry^r g_r(z) + \cdots]\} \tag{5.53}$$

and, as long as the expansion exists at $y = 1$ for sufficiently small z, this tends to a non-zero value as $y \to 1$. Hence there is a spontaneous magnetization for sufficiently small z.

The existence of a spontaneous magnetization therefore depends on the possibility of a valid expansion of type (5.49) for $y = 1$ near $z = 0$. For a one-dimensional chain every $f_r(z)$ starts with z^2 since any number of spins can be overturned with only two adjacent 1–2 links; hence the coefficient of z^2 in the expansions (5.45) and (5.49) becomes infinite when $y = 1$ and there is no spontaneous magnetization. Similarly for any strip of finite width, since a finite number of overturned spins can divide the chain into two parts, one of the coefficients in the expansions would become infinite. It is only when the model extends to infinity in two or more dimensions that high powers of y necessitate correspondingly high powers of z, the expansions (5.45) and (5.49) are valid when $y = 1$, and there is a spontaneous magnetization.

(ii) *Low-temperature excitation expansions* These are expansions in z whose coefficients are functions of y, and are simply a rearrangement of the density series. For example, for the SQ lattice

$$\ln \Lambda(y,z) = z^4 y + 2z^6 y^2 + z^8(-\tfrac{5}{2}y^2 + 6y^3 + y^4) + \cdots. \tag{5.54}$$

From the discussion in the previous paragraph it is clear that such an expansion is possible for a system which extends to infinity in two or more dimensions. For a finite strip it is possible only if $y \neq 1$.

(iii) *High-temperature expansions* By analogy with the transformation (5.3) for the interaction variable we can write

$$\exp(\beta\mu_0 H\sigma_i) = \cosh \beta\mu_0 H + \sigma_i \sin \beta\mu_0 H = \cosh \beta\mu_0 H(1 + \tau\sigma_i)$$
$$(\tau = \tanh \beta\mu_0 H). \tag{5.55}$$

Hence we can write the p.f. for the simple Ising model in the form

$$\exp\left(K\sigma \sum_{i,j} \sigma_i\sigma_j\right) \exp\left(\beta\mu_0 H \sum_i \sigma_i\right)$$
$$= \cosh^L K \cosh^N(\beta\mu_0 H) \sum \prod_{i,j} (1 + w\sigma_i\sigma_j)(1 + \tau\sigma_i). \tag{5.56}$$

As in §5.1.2 the first product in (5.56) can be expanded as a power series in w, the first term corresponding to ${}^L C_1$ links of the lattice, the second term to ${}^L C_2$ pairs of links on the lattice, ..., the rth term to ${}^L C_r$ sets of r links on the lattice. The second product can be expanded as

$$\left[1 + \tau \sum \sigma_i + \tau^2 \sum_{i,j} \sigma_i\sigma_j + \cdots\right]. \tag{5.57}$$

When we now group the σ_i from the two expansions together, remembering that any free σ_i sums to zero, we find that any graph with l links and s odd vertices contributes $2^N w^l \tau^{2s}$ (the number of odd vertices on a graph is always even). Hence we can write

$$Z_N = 2^N \cosh^N K \cosh^L \beta\mu_0 H \times [1 + \Psi_1(N,\tau)w + \Psi_2(N,\tau)w^2 + \cdots + \Psi_r(N,r)w^r + \cdots]. \quad (5.58)$$

The $\Psi_r(N,\tau)$ play exactly the same role as the $F_r(N,z)$ in the density expansions, and the discussion proceeds on identical lines. We may note that in zero field only even-vertex graphs enter, for the coefficient of τ^2 (the initial magnetic susceptibility χ_0) only graphs with two odd vertices enter, and for the coefficient of τ^{2s}, which is related to the $(2s)$th derivative of the p.f. with respect to H, only graphs with $2s$ odd vertices enter. If we move over to $\ln Z_N$ we can again obtain an expansion involving only connected graphs on the lattice.

5.2.1.2 *More general models*

The more general models which attracted attention were the spin s Ising model, and the quantum and classical Heisenberg models, which were defined in §3.4.1. For the spin s Ising model density and low-temperature expansions were derived; each overturned spin can have a number of different excited states, and the different interactions between the various excited states must be taken into account so that there are considerable complications. For the Heisenberg model the density and low-temperature expansions are related to the interaction of spin waves, a difficult problem first tackled effectively by Dyson (1956); but the number of terms he calculated are insufficient to throw any light on critical behaviour.

However, high-temperature expansions can be derived and used for all these models. Quite generally we can write

$$Z_N = \langle \exp(-\beta H) \rangle = 1 - \beta\langle H \rangle + \frac{\beta^2}{2!}\langle H^2 \rangle + \cdots + \frac{(-\beta)^r}{r!}\langle H^r \rangle + \cdots \quad (5.59)$$

where $\langle\ \rangle$ now denote the trace for a quantum interaction or sum over all configurations for a classical interaction. This has the form of a moment expansion in statistics, and if we expand $\ln Z_N$ we will obtain a cumulant expansion:

$$\ln Z_N = -\beta\langle H \rangle + \frac{\beta^2}{2!}(\langle H^2 \rangle - \langle H \rangle^2) - \frac{\beta^3}{3!} \times (\langle H^3 \rangle - 3\langle H^2 \rangle\langle H \rangle + 2\langle H \rangle^3) + \cdots. \quad (5.60)$$

When a topological interpretation is given of the terms in (5.59) and (5.60), the graphs which enter can have multiple bonds, and for quantum models with non-commuting operator interactions different permutations of these operators give rise to different traces and must be treated separately. A great deal of ingenuity was introduced to facilitate such calculations which is described in DG **3** (Chs 1, 5 and 6). For the spin s Ising model with classical interactions the calculations are much simpler.

Basically there are two problems to be tackled in deriving series expansions of any length:

1 *The embedding problem.* Determination of the number of different ways in which the graphs which enter at order r can be fitted onto the lattice.
2 *The weighting problem.* Determination of the weight associated with a particular graph.

The transformation to connected graphs significantly complicates the weighting problem, but is usually worth while. Initially geometrical and analytical methods were used to calculate the embeddings, but as the series lengthened computers were brought in, and proved more effective (DG **3**, Ch. 2). The limiting factor for most models is the very rapid growth of the number of connected graphs which have to be considered.

Important simplifications for the simple Ising model are (i) the lack of multiple bonds, (ii) the possibility of transforming to expansions in which only *multiply connected graphs* enter whose number is significantly smaller than that of connected graphs (DG **3**, Ch 1). For this reason it has been possible to derive series expansions of much greater length than for any other model, and the simple Ising model has served as a pioneer in the exploration of critical behaviour by series expansion techniques.

5.2.1.3 *Correlations and long-range order*

The methods of the previous section can be applied to the development of the correlation between a pair of sites in the lattice as a series expansion. We illustrate by considering the simple Ising model. Using the definition of $\langle \sigma_0 \sigma_{\mathbf{r}} \rangle$ given in (4.104) we can develop a high-temperature expansion as in (5.56) and derive the relation

$$Z_N \langle \sigma_0 \sigma_{\mathbf{r}} \rangle = \cosh^L K \cosh^N(\beta \mu_0 H) \sum \prod_{i,j} \sigma_0 \sigma_{\mathbf{r}} (1 + w\sigma_i \sigma_j)(1 + \tau \sigma_i) \tag{5.61}$$

and the two brackets on the right-hand side can be expanded and the terms interpreted topologically. When $H = 0$ it is easy to see that the only non-zero diagrams are those with odd vertices at (0,**r**) and even vertices elsewhere. The series starts with a term w^l where l is the length of the shortest lattice path or *self-avoiding walk* (SAW) connecting the site (0,**r**).

The primitive expansion will involve disjoint graphs, but division of the RHS by Z_N, expanded diagrammatically as in (5.58), will eliminate all disjoint graphs. As in the case of the free energy, it is possible to calculate the correlations by connected graphs only.

We can similarly derive an expansion for the correlations analogous to the density expansion (5.42) in the presence of a magnetic field in direction 2. Using the notation of §4.8.4, we write

$$\Lambda_N(y,z)\alpha_{11}(\mathbf{r}) = y^2F_2(N,z,\mathbf{r}) + y^3F_3(N,z,\mathbf{r}) + \cdots. \tag{5.62}$$

Here $\alpha_{11}(\mathbf{r})$ is the mean fraction of one-to-one spins at $(0,\mathbf{r})$, and $F_s(N,z,\mathbf{r})$ is the subgroup of $F_s(N,z)$ in which the spins at $(0,\mathbf{r})$ are overturned. This can be converted into a connected graph cumulant expansion from which we derive

$$\alpha_{11}(\mathbf{r}) - \alpha_1^2 \tag{5.63}$$

in terms of clusters of overturned spins connecting the sites $(0,\mathbf{r})$. The lowest term will be determined by a chain of SAW spins connecting $(0,\mathbf{r})$ and will contain a factor y^{l+1}. As the distance r becomes large (5.63) will become negligibly small; there is no long-range order in the assembly.

But in the absence of a magnetic field both of the equivalent lowest energy states must be taken into account. If the lowest energy state at absolute zero consists of all spins pointing in direction 2, denote the mean fraction of spins which have overturned in direction 1 at any temperature by $\eta(z)$; the mean fraction pointing in direction 2 is $1 - \eta(z)$. Thus since both lowest energy states in directions 1 and 2 are equally probable

$$\alpha_{11}(\infty) = \tfrac{1}{2}[1 - \eta(z)]^2 + \tfrac{1}{2}\eta(z)^2 = \tfrac{1}{2}[1 - 2\eta(z) + 2\eta(z)^2] \tag{5.64}$$

α_1 is equal to $\frac{1}{2}$, and hence using the definition of order given in (4.116)

$$\theta(\infty) = 4\alpha_{11}(\infty) - 1 = [1 - 2\eta(z)]^2. \tag{5.65}$$

From (5.53) the spontaneous magnetization is equal to $[1 - 2\eta(z)]$. Hence we see that the parameter t introduced by Bragg and Williams to characterize long-range order is identical with the spontaneous magnetization. A similar discussion applies to an antiferromagnet.

We see physically that long-range order cannot be present when there is a single unique lowest energy state. It arises from the interference of two (or more) lowest energy states.

5.2.2 *Linked Cluster Method* (for a detailed exposition see DG **3**, Ch. 3)

The linked cluster method made its first appearance many years after the embedding method. Its formal development was due to Brout (1959), Horwitz and Callen (1961) and Englert (1963). In principle it is extremely comprehensive, and is able to use the same formalism for a variety of different models, for interactions extending to all ranges, and for pair and multiple correlation functions. In practice the number of diagrams grows extremely rapidly, and renormalization techniques taken from quantum field theory were introduced to enable this number to be reduced to manageable proportions. The method found practical application for series expansions in the hands of Wortis and his collaborators (e.g. Jasnow and Wortis 1968, Moore, Jasnow and Wortis 1969, Ferer, Moore and Wortis 1969). They exploited the renormalization techniques to sum large classes of diagrams of simple topology, and the remaining diagrams of complex topology were dealt with individually. Our treatment is based on the article by Wortis.

We illustrate the method by considering the Ising model of spin s (see §3.4.1) whose Hamiltonian can be written in the form

$$\mathscr{H} = -\frac{1}{s^2}\sum_{\langle ij\rangle} J_{ij}s_{zi}s_{zj} - \frac{m}{s}\sum_i H_i s_{zi}. \tag{5.66}$$

A very general model is being considered in which interactions J_{ij} are present between any two sites of the lattice, and the field H_i varies from site to site of the lattice. It is convenient to rewrite the Hamiltonian as follows:

$$-\beta\mathscr{H} = \tfrac{1}{2}\sum_{1,2} v(12)\mu(1)\mu(2) + \sum_1 h(1)\mu(1) \tag{5.67}$$

where

$$v(ij) = \beta J_{ij}, \quad v(ii) = 0, \quad h(i) = \beta\mu_0 H_i, \quad \mu(i) = s_{zi}/s$$

and the sum is now over all sites 1,2 of the lattice. The p.f. is given by

$$Z_N = \sum_t \exp(-\beta\mathscr{H}) \tag{5.68}$$

where the sum is taken over all the states t of all the spins. For convenience write

$$W = \ln Z_N = W[h,v] \tag{5.69}$$

where h represents all the different $h(i)$ and v all the different $v(ij)$.

The aim of the method is to expand W as a Taylor series in all the different variables. By keeping the H_i distinct it is possible to determine the correlation between any pair of sites from this expansion. The coefficients in this expansion are the derivatives of W w.r.t. $h(i)$, $v(ij)$, and following Wortis (DG 3) we will use the functional differentiation notation

$$\frac{\delta}{\delta h(i)} \quad \frac{\delta}{\delta v(ij)} \tag{5.70}$$

to remind us of the multivariable nature of the expansion. Finally any derivative w.r.t. $v(ij)$ can be simply expressed in terms of the derivatives w.r.t. $h(i)h(j)$, so that all the Taylor coefficients in a $v(ij)$ expansion can be expressed in terms of derivatives w.r.t. $h(i)h(j)$ at $v(ij) = 0$, which corresponds to a simple uncoupled system.

Using the standard notation

$$\langle X \rangle = \sum X \exp(-\beta\mathscr{H}/Z_N) \tag{5.71}$$

we find that

$$\frac{\delta W}{\delta h(1)} = \langle \mu(1) \rangle = K_1(1) \tag{5.72}$$

$$\frac{\delta^2 W}{\delta h(1)\delta h(2)} = \langle \mu(1)\mu(2) \rangle - \langle \mu(1) \rangle \langle \mu(2) \rangle = K_2(12) \tag{5.73}$$

and more generally

$$\frac{\delta^n W}{\delta h(1)\delta h(2) \ldots \delta h(n)} = K_n(12 \ldots n) \tag{5.74}$$

is the n-point cumulant (we use the letter K for cumulant as is customary in statistics). $K_1(1)$ represents the magnetization, and $K_2(12)$ the pair correlation.

We wish to develop a power series for $W[h,v]$ in the $v(12)$ which will be of the form

$$\begin{aligned} W[h,v] = W[h,0] &+ \sum v(12)\left[\frac{\delta W}{\delta v(12)}\right]_0 \\ &+ \frac{1}{2!}\sum v(12)v(34)\left[\frac{\delta^2 W}{\delta v(12)\delta v(34)}\right]_0 + \cdots. \end{aligned} \tag{5.75}$$

Clearly, as in (3.19),

$$W[h,0] = \sum_1 \ln \frac{\sinh[(2s+1)h(1)/2s]}{\sinh[h(1)/2s]} = \sum K_0^0[h(1)] \tag{5.76}$$

corresponding to uncoupled spins. But also from (5.67)

$$\frac{\delta W}{\delta v(12)} = \langle \mu(1)\mu(2) \rangle = \frac{\delta^2 W}{\delta h(1)\delta h(2)} + \frac{\delta W}{\delta h(1)}\frac{\delta W}{\delta h(2)} = K_2(12) + K_1(1)K_1(2). \tag{5.77}$$

At $v = 0$ the lattice sites are entirely uncorrelated, so that

$$K_n(1, 2, \ldots, n) = \delta(1, 2, \ldots, n)K_n^0[h(1)] \tag{5.78}$$

where $\delta(1, 2, \ldots, n) = 1$ when all lattice points $1, \ldots, n$ are identical, and vanish otherwise, and K_n^0 is the nth derivative of K_0^0. Thus the coefficient of $v(12)$ in the expansion (5.75) is $K_1^0(1)K_1^0(2)$, since 1 and 2 are different.

To determine the next term in the series we must evaluate $\delta^2 W/\delta v(12)\delta v(34)$; this is straightforward if we differentiate (5.77) w.r.t. $v(34)$ as follows:

$$\frac{\delta W}{\delta v(12)\delta v(34)} = \frac{\delta^2}{\delta h(1)\delta h(2)}\frac{\delta W}{\delta v(34)} + \frac{\delta W}{\delta h(2)}\frac{\delta}{\delta h(1)}\frac{\delta W}{\delta v(34)} + \frac{\delta W}{\delta h(1)}\frac{\delta}{\delta h(2)}\frac{\delta W}{\delta v(34)} \tag{5.79}$$

and substitute (5.77) for $\delta W/\delta v(34)$. The algebra is tedious but straightforward, and we find up to the second-order terms

$$W[h,v] = \sum_{1} K_0^0(1) + \tfrac{1}{2}\sum_{1,2} K_1^0(1)v(12)K_1^0(2) + \tfrac{1}{4}\sum_{1,2} K_2^0(1)v^2(12)K_2^0(2)$$
$$+ \tfrac{1}{2}\sum_{1,2,3} K_1^0(1)v(12)K_2^0(2)v(23)K_1^0(3). \qquad (5.80)$$

The procedure can be completely systematized in graph theoretical terms. We quote the following rules given by Wortis.

5.2.2.1 *Calculation of free energy*

1 Enumerate all different connected graphs (including multiply bonded graphs). (For a survey of graph theory see DG **3**, Ch. 1.)
2 Assign a dummy label to each vertex.
3 For each edge joining vertices i and j write a factor $v(ij)$.
4 For each l-valent vertex i write a factor $K_l^0(i)$.
5 Divide by the symmetry number of the graph.
6 Sum each vertex label freely over the lattice.

Following these rules we write (up to third order in v)

$W[h,v] =$ 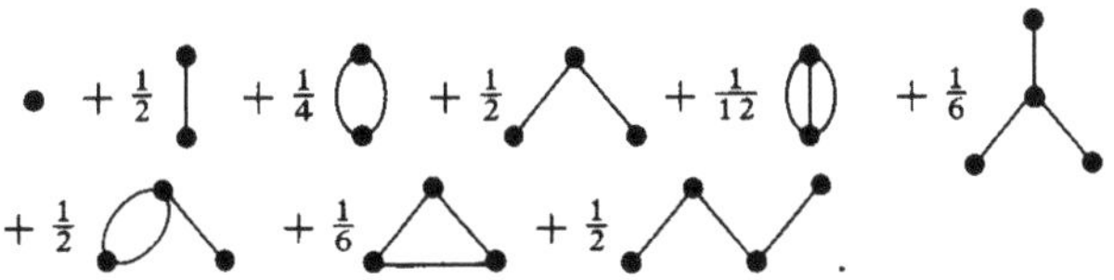

(5.81)

Equation (5.81) applies to arbitrary inhomogeneous fields and interactions. If we particularize to a homogeneous field h and interactions $v(\mathbf{r})$ depending on distance $\mathbf{r}$, W becomes extensive and we can write

$$\frac{W}{N} = K_0^0(h) + \tfrac{1}{2}[K_1^0(h)]^2 \sum_{\mathbf{r}} v(\mathbf{r})$$
$$+ \tfrac{1}{4}[K_2^0(h)]^2 \sum_{\mathbf{r}} v^2(\mathbf{r}) + \tfrac{1}{2}[K_1^0(h)]^2 K_2^0(h)\left(\sum_{\mathbf{r}} v(\mathbf{r})\right)^2 + O(v^3). \qquad (5.82)$$

Finally if we restrict the interactions to nearest neighbours, we obtain

$$\frac{W}{N} = K_0^0(h) + \frac{q}{2}[K_1^0(h)]^2 v$$
$$+ \left\{\frac{q}{4}[K_2^0(h)]^2 + \frac{q^2}{2}[K_1^0(h)]^2 K_2^0(h)\right\} v^2 + O(v^3) + \cdots \qquad (5.83)$$

where q is the coordination number of the lattice. The number of terms corresponding to

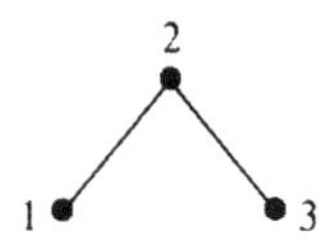

is q^2 since the vertices are free, and 1 and 3 are allowed to coincide on the same lattice point. For any graph we need the *free multiplicity* instead of the number of embeddings of the previous section in which 1 and 3 must be separate, and this is much easier to calculate.

The power of the general expansion (5.81) becomes clear when we come to calculate the correlations, simple and multiple. Let us first consider the expansion for the magnetization

$$K_1 = \delta W/\delta h(1).$$

We can calculate this very simply by fixing vertex 1 and differentiating w.r.t. $h(1)$, remembering that the derivative of $K_n^0(h)$ is $K_{n+1}^0(h)$. From (5.80) we find

$$\begin{aligned} K_1(1) = {} & K_1^0(1) + \sum_{2} K_2^0(1)v(12)K_1^0(2) \\ & + \tfrac{1}{2}\sum_{2} K_3^0(1)v^2(12)K_2^0(2) + \sum_{2,3} K_2^0(1)v(12)K_2^0(2)v(23)K_1^0(3) \\ & + \tfrac{1}{2}\sum_{2,3} K_1^0(2)v(21)K_3^0(1)v(13)K_1^0(3) + O(v^3). \end{aligned} \tag{5.84}$$

Again this can be interpreted diagrammatically, but we now need to differentiate the *rooted* vertex 1, denoted by an open circle ○, from the other vertices. With no more difficulty we can deal with the nth derivative $K_n(1)$ according to the following rules.

5.2.2.2 *Calculation of renormalized cumulants*

1 Enumerate all different 1-rooted connected graphs.
2 Assign the label 1 to the rooted vertex and a dummy label to all other vertices.
3 For each edge joining vertices i and j write a factor $v(ij)$.
4 For each l-valent non-rooted vertex write a factor $K_l^0(i)$.
  For each l-valent rooted vertex write a factor $K_{l+n}^0(1)$.
5 Sum each non-rooted vertex label over the lattice.
6 Divide by the symmetry number of the 1-rooted graph.

The graphical representation of (5.84) starts

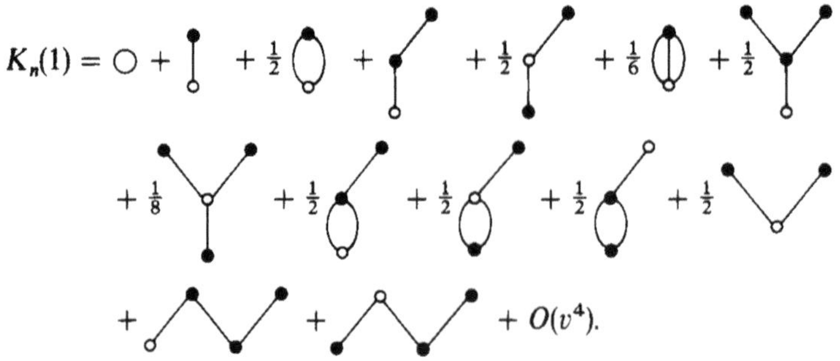

(5.85)

For the pair correlation, two vertices 1 and 2 must be fixed, and the differentiation performed w.r.t. $h(1)$ and $h(2)$. The procedure is a straightforward generalization of (5.84) and the diagrammatic rules are as follows.

5.2.2.3 *Calculation of pair correlation*

1. Enumerate all different 2-rooted connected graphs.
2. Assign the labels 1 and 2 to the rooted vertices and a dummy label to all other vertices.
3. For each edge joining vertices i and j write a factor $v(ij)$.
4. For each l-valent non-rooted vertex write a factor $K_l^0(i)$.
   For each l-valent rooted vertex 1,2 write a factor $K_{l+1}^0(1)$, $K_{l+1}^0(2)$.
5. Sum the non-rooted vertices over the entire lattice.
6. Divide by the symmetry number of the 2-rooted graph.

The diagrammatic expansion now starts as follows

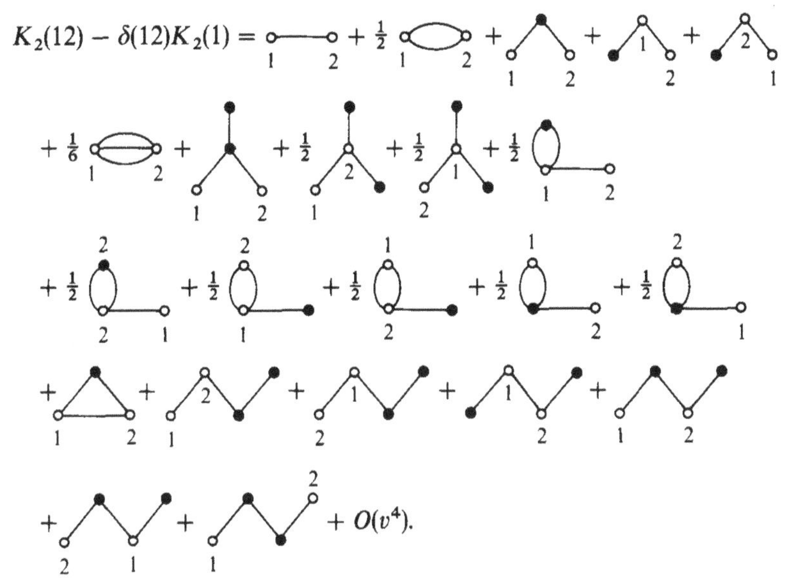

(5.86)

The elegant formalism developed above is now used to resum large classes of articulated graphs by means of vertex renormalization, i.e. by starting with skeleton graphs having no articulation points and attaching classes of graphs at the vertices in a symmetric manner. The procedure is described in detail in the article by Wortis. Many such resummations correspond to replacing the $K_n^0(h)$ in formulae like (5.82) by the $K_n(h)$ given by (5.85).

When the linked cluster technique had been fully developed it was able to provide up to 12 terms of high-temperature expansions of the free energy, susceptibility, and pair correlations for general models with classical interactions. The embedding method could go further only for the simple Ising model. In any case it proved useful in complex calculations of this kind to have them checked by two completely independent methods to ensure their correctness.

5.2.3 *Asymptotic Analysis of Coefficients* (for a more detailed account see e.g. Gaunt and Guttman (1974) and Guttman (1989))

The methods of the previous sections enabled lengthy series expansions to be developed for standard thermodynamic properties like specific heat and magnetic susceptibility of magnetic models, and in this section we shall consider how information regarding critical behaviour can be extracted from the coefficients of these expansions. The most useful series with which to start the investigation are those which converge most rapidly, and it soon became clear that high-temperature expansions behave more smoothly than their low-temperature counterparts. Also, for nearly all lattices investigated, the high-temperature coefficients were all positive, which we shall shortly find to be very advantageous.

The first calculations were undertaken for the simple Ising model. Series for the initial magnetic susceptibility were found to behave more smoothly than those for the specific heat. Calculations were undertaken for a variety of lattices: loose-packed lattices like the SC (Simple Cubic) and BCC (Body-Centred Cubic) contain odd–even oscillations for which data smoothing is desirable, whereas for close-packed lattices like the PT (Plane Triangular) and the FCC (Face-Centred Cubic) the data can be used directly. We shall start our discussion, therefore, with an analysis of the high-temperature susceptibility series for the triangular and FCC lattices since these served as pioneers in the estimation of critical exponents.

First a few observations on power series (Dienes 1931). Consider a function $f(z)$ defined by

$$f(z) = \sum_{n=0}^{\infty} a_n z^n. \tag{5.87}$$

Then if

$$\mathop{\mathrm{Lt}}_{n\to\infty} |a_n|^{1/n} \tag{5.88}$$

exists and is equal to $1/z_c$, the series converges for $|z| < z_c$. We can then write

$$|a_n| \sim \phi(n)/z_c^n \tag{5.89}$$

where

$$\mathop{\mathrm{Lt}}_{n\to\infty} [\phi(n)]^{1/n} = 1. \tag{5.90}$$

There is always a singularity on the circle, $z = z_c$. If all of the a_n are consistent in sign, then the dominant singularity lies on the positive real axis. Replacing z by $-z$, we see that if the a_n alternate regularly the dominant singularity lies on the negative real axis. More irregular alternations indicate dominant singularities in the complex plane. If a_n is real these must occur in complex pairs $(1/z_c)\exp[\pm(i\sigma)]$. For a single pair we should expect

$$a_n \sim \frac{\phi(n)}{z_c^n}\cos n\sigma. \tag{5.91}$$

If σ is a simple fraction, this gives rise to cyclic behaviour, otherwise it is more random.

If the a_n are all known *exactly* we can (in principle) continue the function analytically across the whole plane. Asymptotic values of a_n determine the behaviour near to the dominant singularity. A dominant singularity on the positive real axis corresponds to a positive temperature, and can reasonably be identified with the Curie point. Hence we see that series of terms consistent in sign are particularly useful since a numerical analysis of the a_n provides direct information about critical behaviour. The magnetic susceptibility χ_0 for the PT and FCC lattices can be written as follows.

$$\begin{aligned}\chi_0/\beta\mu_0^2 &= 1 + 6w + 30w^2 + 138w^3 + 606w^4 + 2586w^5 \\ &\quad + 10\,818w^6 + 44\,574w^7 + 181\,542w^8 \\ &\quad + 732\,678w^9 + 2\,953\,218w^{10} + 11\,687\,202w^{11} \\ &\quad + 46\,296\,210w^{12} + 182\,588\,850w^{13} + \cdots \text{ (PT)}\end{aligned} \tag{5.92}$$

$$\begin{aligned}\chi_0/\beta\mu_0^2 &= 1 + 12w + 132w^2 + 1404w^3 + 14\,652w^4 \\ &\quad + 151\,116w^5 + 1\,546\,332w^6 + 15\,734\,460w^7 \\ &\quad + 59\,425\,580w^8 + \cdots \text{ (FCC)} \quad (w = \tanh K).\end{aligned} \tag{5.93}$$

The variable w is a natural one for the simple Ising model (§5.2.1.1 (iii)), and leads to integer coefficients.

In view of (5.90), Domb and Sykes (1957a, 1961) made the simple assumption

$$\phi(n) \sim An^g \tag{5.94}$$

and to assess whether this assumption is in accord with the numerical data, they found it convenient to plot a_n/a_{n-1} as a function of $1/n$. If assumption (5.94) is valid,

$$a_n/a_{n-1} \simeq (1/w_c)(1 + g/n) \tag{5.95}$$

and hence a straight line should be obtained whose intersection with $1/n = 0$ determines w_c, and whose slope determines g.

Figure 5.16 is a *ratio plot* reproduced from Domb and Sykes (1961). It is convenient to plot a_n/qa_{n-1}, since all lattices then start from the same value at $n = 1$. For large q a mean field approach is valid (§3.4.1) and the ratio a_n/qa_{n-1} remains constant and equal to 1. We see immediately how deviations from the mean field value become more marked as the dimension decreases. But we also

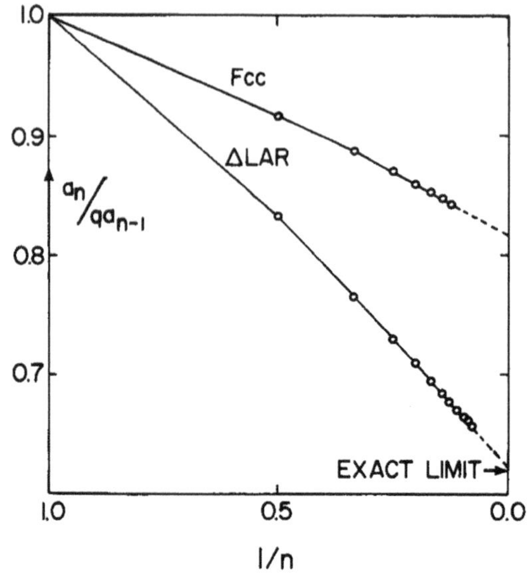

Figure 5.16 Ising model. Successive ratios in the susceptibility expansions of the triangular and FCC lattices as functions of $1/n$.

note that the ratios approach an asymptotic linear value quite rapidly, and the PT lattice provides a convenient check on the procedure since its Curie point is known exactly.

On the basis of this analysis Domb and Sykes suggested that χ_0 for both lattices can reasonably be represented near the Curie point by

$$\chi_0 \sim A(1 - w/w_c)^{-g-1} \tag{5.96}$$

where g is approximately $\frac{3}{4}$ for the PT lattice and $\frac{1}{4}$ for the FCC lattice (the mean field value of g is zero (equation (3.53)). Convergence was more rapid for the FCC than for the PT lattice, but the knowledge of the exact value of w_c in the latter case greatly improved the estimate of g. The *critical temperature* for the FCC could be estimated with great accuracy (relative error $\sim 10^{-4}$), the *critical exponent* g and critical amplitude A with somewhat less accuracy (relative error $\sim 10^{-2}$). For a branch point singularity of the form (5.96), the ratio assumption (5.95) is exact, by the binomial theorem.

A similar type of analysis was undertaken for the Ising model of general spin s (Domb and Sykes 1962), and for the classical Heisenberg model (Rushbrooke and Wood 1958, Domb and Sykes 1962). Fewer terms were available, and hence the estimates were more tentative. For the simple Ising model, as more terms became available, the form (5.96) was modified to take account of correction terms, and the ratio method was refined. Under the influence of Onsager's solution correction terms were assumed to be of

Darboux form

$$\chi_0 \sim A(w)(1 - w/w_c)^{-\gamma} + B(w) \tag{5.97}$$

where $A(w)$ and $B(w)$ are analytic at w_c (Ninham 1963). It became clear later that this assumption, although satisfactory for the simple Ising model in two dimensions, is inadequate for more general models (§7.5.7).

In contrast to high-temperature series, low-temperature series for the simple Ising model are inconsistent in sign and irregularly behaved for many common lattices. For example, the spontaneous magnetization m_0 for the SC lattice can be put in the form

$$m_0/\mu_0 = 1 - 2u^3 - u^5(12 - 14u + 90u^2 - 192u^3 + 792u^4 - 2148u^5 + 7716u^6 - 23\,262u^7 + 79\,512u^8 - 252\,054u^9 \ldots) \quad [u = z^2 = \exp(-4\beta J)]. \tag{5.98}$$

Successive terms alternate in sign, indicating a non-physical singularity at about $u \sim -3$ which corresponds to an imaginary temperature and which masks the physical singularity. For the FCC lattice the corresponding series is

$$m_0/\mu_0 = 1 - 2u^6 - u^{11}(24 - 26u + 48u^4 + 252u^5 - 720u^6 + 438u^7 + 192u^8 + 984u^9 + 1008u^{10} - 12\,924u^{11} + 19\,536u^{12} \ldots) \tag{5.99}$$

and it is difficult to detect any pattern in the coefficients. For the SQ lattice the terms are consistent in sign, and the ratio method can be applied. (In fact it was used by Domb (1949b) to estimate critical behaviour before he became aware of the exact solution of Onsager.)

Attempts were made to deal with series like (5.98) and (5.99) but they were not very successful. The breakthrough came when Baker (1961) introduced the Padé approximant into the field. Padé approximants date back to 1892, and were introduced as a method of providing an analytic continuation of a function defined by a power series. The $[L,M]$ Padé approximant to a function $F(z)$ is the ratio of a polynomial $P_L(z)$ of degree L to a polynomial $Q_M(z)$ of degree M,

$$[L,M] \equiv \frac{P_L(z)}{Q_M(z)} = \frac{p_0 + p_1 z + p_2 z^2 + \cdots + p_L z^L}{1 + q_1 z + q_2 z^2 + \cdots + q_M z^M}. \tag{5.100}$$

The coefficients $p_0, p_1, p_2, \ldots, p_L$ and $q_1, q_2, \ldots, q_M$ are chosen so that the expansion of $[L,M]$ agrees with the expansion of $F(z)$ to order $L + M$, i.e.

$$F(z) = [L,M] + O(z)^{L+M+1}. \tag{5.101}$$

It can be shown (Baker 1965) that these coefficients are unique.

The Padé approximant can represent any meromorphic function exactly, the poles corresponding to the zero of the denominator $Q_M(z)$. For functions with branch point singularities it is less successful directly, but Baker pointed out that if a branch point singularity is anticipated in $F(z)$ at a point z_c, then

$$\frac{d}{dz}[\ln F(z)] = \frac{F'(z)}{F(z)} \tag{5.102}$$

will have a simple pole at z_c. If a logarithmic singularity is anticipated in $F(z)$ then $F'(z)$ will have a simple pole at z_c.

In general the approximants with L and M nearly equal provide the most useful information. Baker, Gammel and Wills (1961) showed that the $[M,M]$ approximants are invariant under the group of Euler transformations $z = aw/(1 + bw)$, which had been used previously to remove dominant non-physical singularities. Padé approximants can also be used for series with terms consistent in sign as an alternative to the ratio method.

In principle the Padé approximant enables *all* the singularities of a function defined by a power series to be studied. It is a powerful tool which can be applied in a variety of different ways, and its introduction into critical phenomena was an event of major importance. For the high-temperature susceptibilities, Padé approximant analysis undertaken by Baker (1961) confirmed the values given previously by Domb and Sykes. For the spontaneous magnetization, using series (5.98) and (5.99) and the BCC counterpart, Baker suggested that near the critical point the spontaneous magnetization can be represented by

$$m_0 \sim (z_c - z)^{0.3} \tag{5.103}$$

for all three lattices. This is very different from the two-dimensional value of 1/8, but also significantly different from the mean field value of 1/2 (equation (3.52)).

Later more refined estimates put forward a value close to 5/16 for the exponent (Essam and Fisher 1963). (At this period there was optimism that the exponents might turn out to be simple rational fractions, in the pattern of the Onsager solution.) Essam and Sykes (1963) discovered that for the diamond lattice the terms of the low-temperature series are consistent in sign, and the ratio method can be used (the property of consistency in sign seems to depend largely on the coordination number of the lattice and not on its dimension). They obtained results similar to those of Essam and Fisher.

The ratio method and the Padé approximant served initially as the major instruments for the extraction of physical information from series expansions, and were applied to estimate critical behaviour for a variety of models and thermodynamic properties.

5.3 Assessment of Critical Behaviour

The series expansion method enabled a fairly coherent picture to be constructed of the true behaviour of standard magnetic models in the critical region. Each model and each property required separate treatment, and the calculations were lengthy. Some properties could be assessed with greater accuracy, other with less, but although error bounds given in estimates were largely subjective, there was a general awareness of the reliability of different conjectures. It was well known, for example, that the most reliable results in the field corresponded to high-temperature series for the simple Ising model $I(\frac{1}{2})$, (equation (4.72)). Undoubtedly the most important result of these investigations was the emergence of a pattern of critical behaviour as a function of lattice structure, dimension and specific model.

5.3.1 Critical Parameters

For $I(\frac{1}{2})$, the value of T_c, the critical temperature, was known exactly for many two-dimensional lattices, and very accurately ($\sim 10^{-4}$) for the common three-dimensional lattices. The high accuracy is important since estimates of critical parameters and other aspects of critical behaviour are very sensitive to the choice of T_c.

Table 5.1 (with minor modifications) is taken from a review article by Domb and Miedema (1964), and makes use, for each lattice, of the critical temperatures, T_c, and critical values of the entropy, S_c, and energy, U_c.

We have seen in §4.8.1 that the thermodynamic and magnetic properties of one-dimensional models do not show discontinuities except at $T = 0$. It is convenient to regard a one-dimensional model as having a Curie point at $T = 0$, and this is consistent with Onsager's solution for an asymmetric net in which the second interaction, J', goes to zero (§5.1.4.2(i)).

The third column of Table 5.1 gives the values of kT_c/qJ which is equal to 1 in the mean field limit (equation (3.42)). We see that as the coordination number, q, increases the values of kT_c/qJ also increase in a given dimension, but a major change occurs as the dimension changes from two to three.

Table 5.1 Critical values for the $I(\frac{1}{2})$ model

Lattice	q	kT_c/qJ	S_c/k	$(S_\infty - S_c)/k$	$(U_c - U_0)/kT_c$	$-U_c/kT_c$
Linear chain	2	0	0	0.693	0	∞
Honeycomb	3	0.506	0.265	0.428	0.227	0.760
SQ	4	0.567	0.306	0.387	0.258	0.623
Triangular	6	0.607	0.330	0.363	0.275	0.549
Diamond	4	0.676	0.510	0.183	0.417	0.323
SC	6	0.752	0.558	0.135	0.445	0.220
BCC	8	0.794	0.582	0.111	0.458	0.172
FCC	12	0.816	0.590	0.103	0.461	0.152
Mean field	∞	1.0	0.693	0	0.5	0

The critical values of energy and entropy give important information about the nature of the specific heat curve. Taking T_c as the unit of temperature, and considering the specific heat, C_m, as a function of τ $(=T/T_c)$, we have

$$\begin{aligned} S_c &= \int_0^1 \frac{C_m}{\tau}\,d\tau \\ S_\infty - S_c &= \int_1^\infty \frac{C_m}{\tau}\,d\tau. \end{aligned} \tag{5.104}$$

The tabulation in columns 4 and 5 of S_c and $S_\infty - S_c$ for various lattices enables a comparison to be made between the magnitude of the specific heat curves below and above the Curie temperature. The sum of the two terms, S_∞, is the same for all lattices and is equal to $k \ln 2$ $(=0.693k)$. These quantities are also particularly useful for comparison with experiment since they do not depend on the magnitude of the interaction J.

It is useful similarly to tabulate (columns 6 and 7)

$$\begin{aligned} (U_c - U_0)/kT_c &= \frac{1}{k}\int_0^1 C_m\,d\tau \\ -U_c/kT_c &= \frac{1}{k}\int_1^\infty C_m\,d\tau \end{aligned} \tag{5.105}$$

which represent directly the areas under the specific heat curve below and above the Curie temperature. The sum of these two terms, $-U_0/kT_c$, is no longer constant, but decreases from infinity to a limiting value of $\frac{1}{2}$ as $q \to \infty$.

It will be seen that it is dimension rather than lattice coordination number which is the dominant factor in determining these tabulated values, and that there is a major difference between the results for two- and three-dimensional lattices. The 'tail' of the specific heat curve is much smaller for the latter, and it could be stated definitely that one of the major differences between the Onsager curve and experimental observation is due to the two-dimensional character of the Onsager calculations.

Moving now to the Ising model of spin s, $I(s)$, (equation (4.73)), the most immediate effect of increasing s is to increase the total entropy change of the system from $T = 0$ to $T = \infty$ to $k \ln(2s + 1)$. Table 5.2 taken from the same review article gives corresponding data as a function of s for the FCC lattice in three dimensions. It is now convenient to tabulate in column 2,

$$\frac{3s}{s+1}\frac{kT_c}{qJ} \tag{5.106}$$

which is equal to 1 in the mean field approximation (equation (3.42)). It will be seen from columns 3 and 4 that nearly all of the increase in entropy takes places in the region *below* the Curie temperature; even when s goes to infinity the increase in entropy above the Curie temperature is not more than 30%. This pattern of behaviour is confirmed in columns 5 and 6; the area under the

Table 5.2 Critical values for the $I(s)$ model (FCC lattice)

s	$\frac{3skT_c}{(s+1)qJ}$	S_c/k	$(S_\infty - S_c)/k$	$(U_c - U_0)/kT_c$	$-U_0/kT_c$
$\frac{1}{2}$	0.816	0.590	0.103	0.461	0.152
1	0.851	0.983	0.116	0.721	0.160
2	0.864	1.486	0.123	0.990	0.167
∞	0.874	∞	0.131	1.541	0.175

specific heat curve remains finite when $s \to \infty$, and a change of s makes little difference to the area in the region $T > T_c$.

On physical grounds this result is not surprising. High-temperature properties arise from averages which are not very sensitive to the number and form of the states over which the averaging takes place. But at low temperatures the ground state and elementary excitations are very sensitive to the individual energy levels.

For the quantum mechanical Heisenberg model (equation (3.35)), despite great ingenuity in dealing with non-commuting spin operators, relatively few terms have been calculated, and the series are not too well behaved. Nevertheless it is possible to make estimates of critical parameters with modest accuracy, and hence to assess approximately the change in the specific heat curve arising from the different nature of the interaction. Table 5.3 taken again from Domb and Miedema (1964) gives the same data for the Heisenberg model of spin s for the FCC lattice. On comparison with Table 5.2 it will be seen that the tail of the specific heat curve is nearly three times as large as for the Ising model.

5.3.2 *Shape of Specific Heat Curve*

We shall see shortly that the critical exponent for the specific heat in three dimensions could not be determined initially with great accuracy because of its weakly divergent character. However, if one was not too concerned with the immediate neighbourhood of the critical point but with overall shape of the

Table 5.3 Critical values for the Heisenberg spin s model (FCC lattice)

s	$\frac{3skT_c}{(s+1)qJ}$	S_c/k	$(S_\infty - S_c)/k$	$(U_c - U_0)/kT_c$	$-U_0/kT_c$
$\frac{1}{2}$	0.679	0.428	0.265	0.297	0.439
1	0.747	0.810	0.289	0.555	0.449
2	0.774	1.305	0.304	0.833	0.459
∞	0.798	∞	0.322	1.406	0.474

specific heat, a curve could be constructed with moderate accuracy for comparison with the exact calculation for a two-dimensional model. Such a comparison is shown in Figures 5.17 and 5.18 taken from Domb (1960), and it became clear that the three-dimensional estimate was reasonably close to experimental specific heat curves which had been observed in practice.

This provided useful evidence that a model using a short-range Ising interaction might be capable of accounting for experimental results observed in practice. It also showed that closed-form approximations of mean field type are nearer to the correct curve in three dimensions than in two dimensions.

5.3.3 *Critical Exponents*

We first define a critical exponent (or index), θ, of a thermodynamic quantity Ψ by the relation

$$\theta = \operatorname*{Lt}_{\tau \to 0} \frac{\ln \Psi}{\ln \tau}, \quad \Psi \sim A\tau^{\theta}. \tag{5.107}$$

This definition ignores logarithmic factors. In a more refined analysis one might need to modify (5.107) to

$$\Psi \sim A\tau^{\theta}(\ln \tau)^{\phi}. \tag{5.108}$$

Fortunately in the practical application to critical phenomena in two and three dimensions (5.107) is usually sufficient, although in four and higher

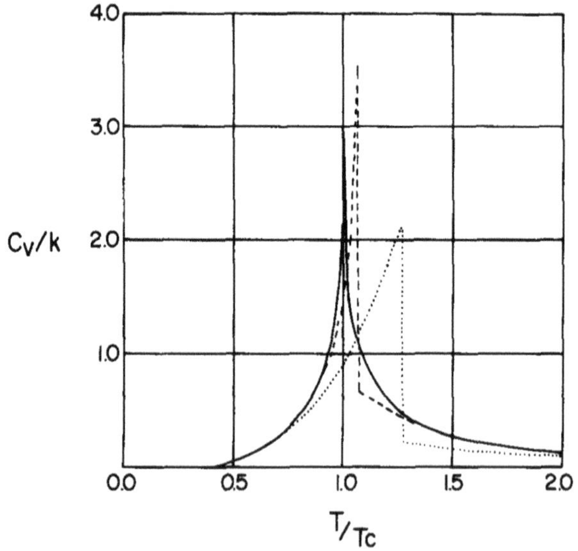

Figure 5.17 Comparison of specific heat curves given by various approximations for SQ lattice (after Onsager): solid curve, exact; dashed curve, Kramers and Wannier variation (≡Kikuchi); dotted curve, Bethe.

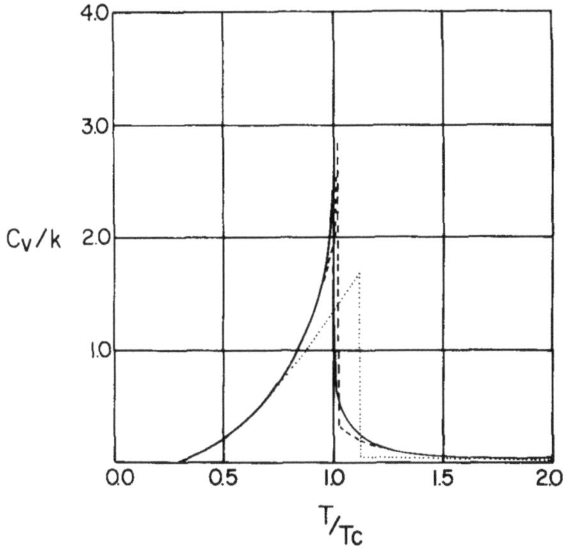

Figure 5.18 Comparison of specific heat curves given by various approximations for the FCC lattice; solid curve, based on series expansions; dashed curve, Kikuchi approximation; dotted curve, first-order Bethe approximation.

dimensions (5.108) is often needed. Also when $\theta = 0$ it is usual to specify whether the behaviour is discontinuous (e.g. classical specific heat), or logarithmic (e.g. specific heat in the Onsager solution).

The notation for magnetic critical exponents developed empirically, and the standardization provided by Fisher at the Washington Conference in 1965 received wide acceptance. Table 5.4 defines the commonly used exponents and lists their classical values. The correlation exponents ν, ν', η will be discussed in §5.5, and the gap exponents Δ, Δ' in §6.2.3.

5.3.3.1 *High-temperature susceptibility* γ

This is the oldest and best established critical exponent. Soon after Domb and Sykes (1957a) suggested the value 7/4 for the two-dimensional $I(\frac{1}{2})$ model, Fisher (1959) gave a theoretical argument to justify this value. In subsequent work by a number of authors the argument was refined and rendered more rigorous, and we shall discuss this in more detail in §5.4.1. Historically the value of 7/4 was widely accepted quite early as exact.

For the three-dimensional lattices Domb and Sykes (1957a) initially suggested $\gamma = 1.250$ for the SC and FCC lattices and 1.244 for the BCC lattice. However, they later conjectured (1961) that $\gamma = 5/4$ for *all* three-dimensional lattices. The idea that *critical exponents for a given model depend on dimension and not on lattice structure* was a first step towards the *universality hypothesis*

Table 5.4 Magnetic critical exponents

Exponent	Quantity	Definition	Curve	Classical value
$T > T_c$				
α	Specific heat	$C_m^{\sim \tau^{-\alpha}}$	$m = 0$	0 (discon.)
γ	Initial susceptibility	$\chi_0 \sim \tau^{-\gamma}$	$m = 0$	1
ν	Coherence length	$\kappa \sim \tau^{\nu}$	$m = 0$	$\frac{1}{2}$
2Δ	Derivatives of susceptibility	$\frac{d^2\chi_0}{dH^2} \sim \tau^{-\gamma-2\Delta}$	$m = 0$	3
$T < T_c$				
α'	Specific heat	$C_m^{\sim \tau'^{-\alpha'}}$	$m = 0$	0 (discon.)
β	Spontaneous magnetization	$m_0 \sim \tau'^{\beta}$	$H \to 0_+$	$\frac{1}{2}$
γ'	Initial susceptibility	$\chi_0 \sim \tau'^{-\gamma'}$	$m = m_0$	1
ν'	Coherence length	$\kappa \sim \tau'^{\nu'}$	$m = m_0$	$\frac{1}{2}$
Δ'	Derivatives of susceptibility	$\frac{d\chi_0}{dH} \sim \tau'^{-\gamma'-\Delta'}$	$m = m_0$	$\frac{3}{2}$
$T = T_c$				
δ	Field/magnetization	$H \sim m^{\delta}$		3
η	Pair correlation function	$\Gamma(r, 0) \sim r^{d-2+\eta}$	$m = 0$	0

(§6.5). We have already seen (§5.1.6) that exact solutions for different lattices in two dimensions give rise to the same exponents for specific heat and spontaneous magnetization. The value of 5/4 was widely accepted for many years until it was challenged by the ϵ expansions of the renormalization group (Chapter 7).

For the $I(s)$ model with general spin s, although fewer terms were available, Domb and Sykes (1962) suggested that γ *is independent of spin*, and this initiated a second step towards the universality hypothesis.

For the Heisenberg model the situation was less satisfactory. Using the data of Rushbrooke and Wood (1958) for the classical Heisenberg model ($s \to \infty$), Domb and Sykes (1962) suggested that $\gamma = 4/3$, and this result was supported by Marshall, Gammel and Morgan (1963) who used Padé approximants. As more terms became available it became clear that the curvature of $1/\chi_0$ had been underestimated in this early work, and that the true value was closer to 1.38 or even 1.40 (for further details see Rushbrooke, Baker and Wood (1974)).

An important question was posed by the anisotropic Heisenberg model with interaction between spins i and j equal to

$$J\sigma_{zi}\sigma_{zj} + J'(\sigma_{xi}\sigma_{xj} + \sigma_{yi}\sigma_{yj}). \tag{5.109}$$

If $J' = 0$ the model reduces to $I(\frac{1}{2})$ with exponent 1.25, whereas if J' grows to J

the model becomes classical Heisenberg with exponent 1.38. What happens when $0 < J' < J$? Fisher (1966) conjectured that the change takes place discontinuously at $J' = J$, and Dalton and Wood (1967) produced evidence to support this conjecture. A definitive investigation was undertaken by Jasnow and Wortis (1968) who dealt also with the cases $J' > J$ and $J = 0$. The latter is usually called the classical x–y model, and corresponds to classical vector spins in two *spin* dimensions. Jasnow and Wortis found that the critical exponents could be divided into three groups:

$$\begin{array}{lll} J' < J & \text{Ising-like} & \gamma \simeq 1.23 \\ J' = J & \text{Heisenberg-like} & \gamma \simeq 1.38 \\ J' > J & x\text{–}y\text{-like} & \gamma \simeq 1.32. \end{array} \tag{5.110}$$

The correlation exponents ν followed a similar pattern of behaviour. From this evidence they drew the following conclusion: *The high-temperature critical indices are ordered in a one-to-one manner with the symmetry group of the order parameter in the ground state manifold.* This constituted a third step towards the universality hypothesis. (The difference between $\gamma = 1.23$ and the generally accepted value of 1.25 was later attributed to a confluent singularity, see Saul, Wortis and Jasnow (1975)).

The Heisenberg model for two-dimensional lattices has an interesting history. The early analysis by Bloch (1930) of spin wave excitations gave a strong indication that the ordering of spins at $T = 0$ is destroyed at any finite temperature however small, and that there could be no spontaneous magnetization. Hence it was generally assumed that no phase transition occurs for this model in two dimensions. However, Rushbrooke and Wood (1958) noted in analyzing the susceptibility series for the PT lattice that they correspond to a Curie temperature about two-thirds that of the three-dimensional SC lattice.

Stanley and Kaplan (1966) raised the question more emphatically, pointing out that it was possible to have a new type of transition with an infinite susceptibility at a finite Curie temperature, and no spontaneous magnetization below the Curie temperature. They were supported in this idea by some empirical unpublished investigations by Dyson on correlations which decay very slowly (Stanley 1974). Incidentally shortly after Stanley and Kaplan's paper appeared, a rigorous proof was given by Mermin and Wagner (1966) of the absence of spontaneous magnetization for the two-dimensional Heisenberg model for $T > 0$. For the x–y model similar arguments apply at low temperatures regarding the absence of long-range order. But the evidence for the existence of a phase transition in this case was stronger than for the Heisenberg model (Stanley 1968a, Moore 1969).

In 1973 Kosterlitz and Thouless put forward the idea of a new type of ordering which they called topological long-range, related to the presence of vortices, which is relevant for the x–y model but *not* for the classical Heisenberg model. They showed that there exist metastable states corresponding to vortices which are closely bound in pairs below some Curie temperature, while above it they become free. This idea has proved very fruitful and has initiated a whole new area of research (see e.g. Nelson 1983). The transition is now considered to be well established for the x–y model. However, following a conjecture by Kosterlitz and Thouless, it is now generally accepted that there is no transition for the classical Heisenberg model in two dimensions.

5.3.3.2 *High-temperature specific heat* α

The high-temperature series for ln Z take considerably longer to settle down to their asymptotic behaviour than those for χ_0. This is particularly true for 'even' lattices like the SC and BCC where all odd coefficients vanish, but even for the FCC lattice the convergence is slow. It has already been mentioned in relation to classical theory that the specific heat diverges only weakly (§2.6).

The earliest estimate of α for $I(\frac{1}{2})$ by Domb and Sykes (1957b) from the terms then available for the FCC lattice suggested a small positive power $\leq 1/4$. The scaling relations to be discussed in the next chapter led to the conjecture $\alpha = 1/8$, and this stimulated Sykes and his collaborators to add several new terms to the expansion for the FCC lattice so as to bring the series to the region of better convergence. They seemed to provide strong evidence in support of the conjecture, and their *ratio plot* for the susceptibility and specific heat is reproduced in Figure 6.1. Subsequently they extended the calculations to the SC and BCC lattices and obtained the same value of α. This value was subsequently challenged and somewhat modified as a result of the ϵ expansion of the renormalization group.

Since even the $I(\frac{1}{2})$ model gave difficulty in relation to the exponent α, there was little direct evidence of its value for the other models. It was usual to make use of the scaling relation (6.55)

$$\alpha + 2\beta + \gamma = 2 \tag{5.111}$$

and the well-determined exponents γ and β to estimate the value of α.

5.3.3.3 *Critical isotherm* δ

The series expansion (5.53) taken at the critical temperature defines the m–H relationship along the critical isotherm. The terms are all consistent in sign, the dominant singularity corresponding to $y = 1$ or $H = 0$; hence the critical exponent is determined by the asymptotic value of the coefficients $g_r(z_c)$. It is again important to have an accurate value of z_c.

The first attempt to establish δ in this way was made by Gaunt *et al.* (1964)

and yielded the values

$$\begin{aligned} \delta &= 15.00 \pm 0.08 \quad \text{(two dimensions)} \\ \delta &= 5.20 \pm 0.15 \quad \text{(three dimensions).} \end{aligned} \qquad (5.112)$$

It was suggested that the two-dimensional value might be exactly 15 since this would satisfy relation (1.43) given by the droplet model (Essam and Fisher 1963, see §6.1)

Scaling relations to be discussed in the next chapter led to the conjecture that δ should be exactly 5 in three dimensions, and a re-examination of the series with additional terms gave results consistent with this conjecture (Gaunt and Sykes 1972). However, this appealing suggestion has also become a casualty of the refined ϵ expansion (§7.7). Estimates of $I(s)$ for $s = 1$ and $s = \frac{3}{2}$ were consistent with the assumption that δ is independent of s (Fox and Gaunt 1972).

5.3.3.4 *Spontaneous magnetization β*

For a number of two-dimensional lattices closed-form solutions were available for the spontaneous magnetization which gave $\beta = 1/8$. In three dimensions the discussion in §5.2.3 has reported the arguments which led to adoption of 5/16 for β for $I(\frac{1}{2})$. For $I(s)$ with $s = 1$ and 3/2 results obtained were consistent with the same values of β: 1/8 in two dimensions and 5/16 in three dimensions (Guttman, Domb and Fox 1971, Fox and Guttman 1973). It is interesting to see how the spontaneous magnetization changes with s for a given lattice; Figure 5.19 taken from Fox and Guttman (1973) shows that the variation is in fact quite small, and of the same order as the change with lattice structure in a given dimension.

5.3.3.5 *Low-temperature exponents γ' and α'*

We have already referred to the difficulties posed by low-temperature series, and the help given by the Padé approximant (PA) in dealing with them. However, PA analysis was less satisfactory in its application to γ', α' than in its application to β. For the SQ lattice Essam and Fisher (1963) estimated that $\gamma' \simeq 1.75 \pm 0.01$, which is sufficiently close to γ to suggest symmetry of the exponents below and above T_c. Theoretical arguments were soon forthcoming to support the value 7/4 for γ' (Kadanoff 1966).

For the SC, BCC and FCC lattices they still suggested symmetry ($\gamma' \simeq 1.25^{+0.07}_{-0.02}$) although the error bounds were considerably larger. For a heuristic model, the droplet model (§6.1.1), Essam and Fisher found that the relation

$$\alpha' + 2\beta + \gamma' = 2$$

was satisfied exactly. They pointed out that this relation was also satisfied by the two-dimensional Ising model, and suggested that if it were satisfied by the

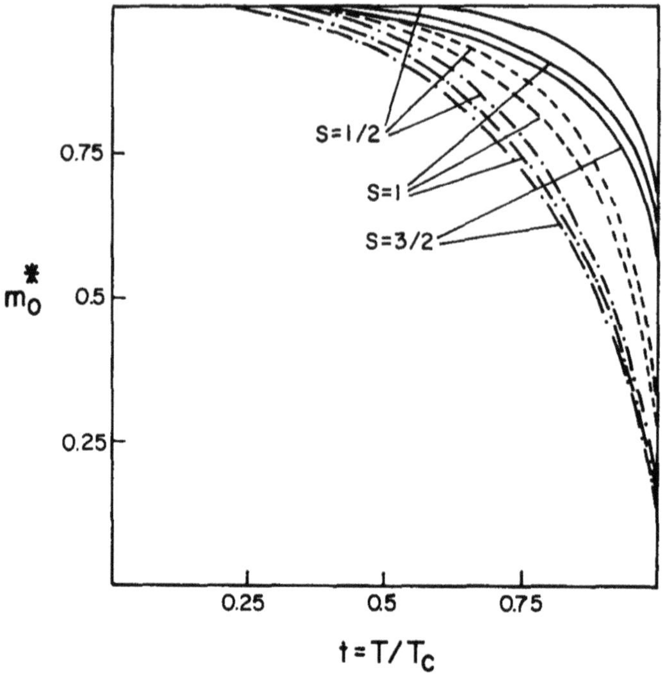

Figure 5.19 Variation of the spontaneous magnetization of the Ising model on the triangular and BCC lattices with spins and relative temperature $t = T/T_c$ (molecular field theory results are also shown); solid curves, triangular; dashed curves, BCC; chain curves, molecular field (from Fox and Guttman 1973).

three-dimensional model, it would lead to a value of α' equal to 1/8. However, direct PA analysis of the SC, BCC and FCC lattices gave $\alpha' \simeq 0$ (Baker 1963), and for the diamond lattice for which all the terms are positive Essam and Sykes (1963) showed that the ratio method also led to $\alpha' \simeq 0$. Rushbrooke (1963) then made the important observation that thermodynamics alone (§6.1.2) enables one to derive an inequality

$$\alpha' + 2\beta + \gamma' \geq 2 \tag{5.113}$$

which is violated if $\alpha' \simeq 0$, $\beta \simeq 5/16$ and $\gamma' \simeq 5/4$, the currently proposed estimates.

For some time there was turmoil as alternative suggestions were advanced for resolving the conflict. Eventually a significant consensus supported the view that the critical exponents are symmetric, but the low-temperature series converge only slowly, and that the current estimates did not represent the true exponents. However, even the addition of several more terms to the diamond and FCC lattices did not resolve the matter satisfactorily. Gaunt and Sykes (1973) pointed out that whilst the data were not inconsistent with symmetry, the ratio of convergence was too slow for precise conclusions to be drawn.

Analysis for spin s led to the conclusion that the exponents do not depend noticeably on s (Fox and Guttman 1970, 1973, Guttman, Domb and Fox 1971).

5.3.4 *Antiferromagnetism and the Néel Point*

At the end of §4.8 attention was drawn to the analogy between the order–disorder transition in a binary alloy and the Ising $I(\frac{1}{2})$ model of an antiferromagnet. In zero field there is a precise correspondence between a ferromagnet and an antiferromagnet. But once a non-zero magnetic field is introduced this correspondence disappears, and the physical behaviour of the two assemblies differs fundamentally.

This can be readily seen by a consideration of their ground states. In zero field both assemblies have two equivalent ground states. For the ferromagnet the spins are all aligned in direction 1 or in direction 2, and for the antiferromagnet there is a state of alternating order, and the equivalent state formed by switching the spins.

The energy of any configuration of the assembly is given by (4.77) as

$$-(\mu_0 H + \tfrac{1}{2}qJ)N + 2\mu_0 HN_1 + 2JN_{12}.$$

For $J > 0$ (ferromagnet) N_{12} must be minimized, and the lowest energy state for $H > 0$ corresponds to $N_1 = 0$, $N_{12} = 0$. For $J < 0$ (antiferromagnet) N_{12} must be maximized to $\frac{1}{2}qN$, and even when $H > 0$ this condition still holds until the field becomes sufficiently large to counter the coupling between the spins,

$$\mu_0 H_c + qJ = 0. \tag{5.114}$$

The difference between the two assemblies is made clear in Figure 5.20 which shows the singularity structure for an antiferromagnet, and which should be contrasted with Figure 5.13 for a ferromagnet.

Our reference to the properties of an antiferromagnet in a non-zero field will be brief, and will be concerned with the initial magnetic susceptibility $\chi_0^{(a)}$.

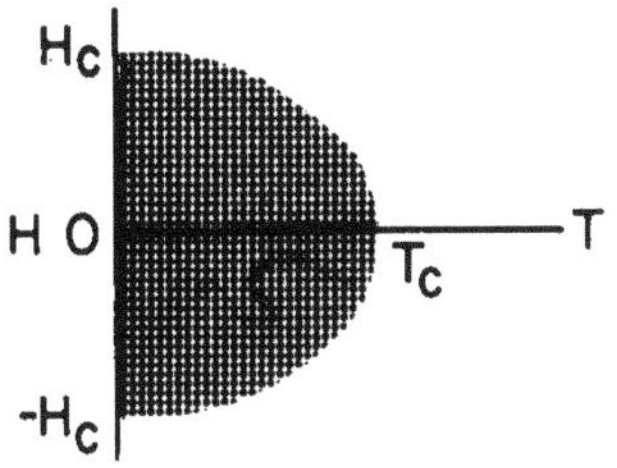

Figure 5.20 Line of singularities for an antiferromagnet for the SQ, SC, BCC and diamond lattices (the shaded region corresponds to long-range order).

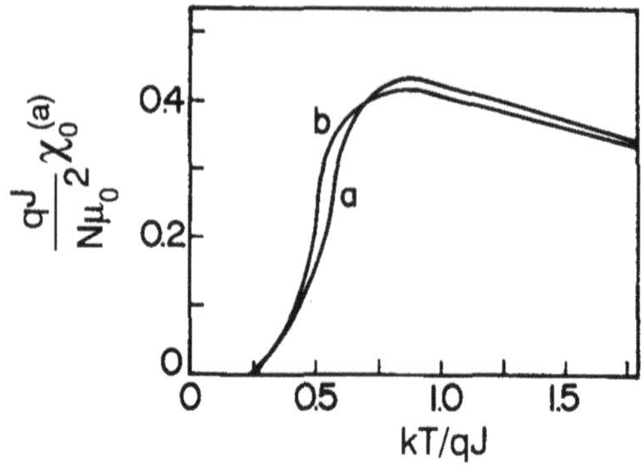

Figure 5.21 Antiferromagnetic susceptibility of (a) the SQ and (b) the honeycomb lattice (after Sykes and Fisher 1962).

Series expansions are immediately available for χ_0 if w is replaced by $-w$ in the ferromagnetic series, and they therefore correspond to the same series with alternating positive and negative signs. However, because of the alternation of signs the analysis of critical behaviour is more difficult. The first attempt at such an analysis by Brooks and Domb (1951) suggested a maximum of the susceptibility at the Néel temperature (the analogue of the Curie temperature for an antiferromagnet), accompanied by a discontinuity of slope, which is in qualitative agreement with mean field theory (Bitter 1937).

The true nature of the antiferromagnetic susceptibility in the critical region was first revealed by Sykes and Fisher (1958, 1962), Fisher (1959, 1962) and Sykes (1961). These authors demonstrated that the critical behaviour of $\chi_0^{(a)}$ parallels that of the energy U so that it has a vertical tangent at T_c. Since it goes to zero as $T \to \infty$ it is clear that it must have a maximum for some value $T_m > T_c$. Theoretical support for this behaviour provided by Fisher from a consideration of the correlations will be discussed in §5.4.1.

Detailed numerical analyses of $\chi^{(a)}$ for the SQ and honeycomb (HC) lattices in two dimensions were given by Sykes and Fisher (1962), and for the SC and BCC lattices by Fisher and Sykes (1962). The form of the $\chi^{(a)}$ for the two-dimensional lattices is reproduced in Figure 5.21. The temperature of the maximum, T_m, was found to be significantly larger then T_c in two dimensions ($T_m \sim 1.5T_c$) but much less in three dimensions ($T_m \sim 1.08T_c$). Again it became clear that deviation from mean field behaviour decreases with increasing dimension. However, a number of features of the 1962 analyses were unsatisfactory and have been modified in a later treatment (Sykes *et al.* 1972).

5.4 Other Exact Results

The crucial role which Onsager's exact solution played in the development of the modern theory of critical phenomena is manifest from the discussion in the earlier sections of the chapter. A few additional exact results became available subsequently and they also helped significantly in the construction of a picture of critical behaviour, and its dependence on dimension, symmetry and range of intermolecular forces.

We have seen in §4.8.1 that the largest eigenvalue of the transfer matrix determines the free energy. It can readily be shown (see e.g. Domb 1949b, 1960, p. 171) that the *first* derivative of this eigenvalue is determined by the components of the largest *eigenvector*. Since Onsager's original solution for the two-dimensional Ising model $I(\frac{1}{2})$ evaluated both the largest eigenvalue and largest eigenvector, it is not surprising that exact solutions for the free energy and its first derivative (the magnetization) soon became available.

The susceptibility is given by the second derivative of the free energy, and this requires a knowledge of *all* of the eigenvalues of the transfer matrix. But the work of Kaufman and Onsager (1949) had tackled this problem, and therefore in principle an exact calculation of the initial susceptibility should have been possible. A major step towards a theoretical assessment of its critical behaviour was taken by Fisher (1959) but the path to a closed-form solution proved difficult and tenuous, and was achieved ultimately only by a *tour de force* calculation in 1976 by Wu *et al.* The detailed history will be discussed in §5.4.1.

In 1952 Berlin and Kac discussed two new models for which they were able to calculate an exact solution in three dimensions in a non-zero field. The Gaussian model consists of independent spins with a nearest-neighbour Ising interaction, the magnitude of each spin being governed by a probability distribution

$$(2\pi)^{-1/2} \exp[-(\sigma^2/2)]\, d\sigma. \tag{5.115}$$

The spherical model allows the spins to take any value subject to the overall restriction

$$\sigma_1^2 + \sigma_2^2 + \cdots + \sigma_N^2 = N. \tag{5.116}$$

Neither model is realized by any known physical system, and the solutions exhibit certain marked non-physical features.

For many years the models were regarded as of mathematical interest only. The discovery by Stanley (1968b) that the spherical model can be regarded as the limit of a classical vector spin model in which the *spin* dimension becomes infinite secured an important place for the model in the general pattern for critical behaviour. The Gaussian model found its place as a first step in Wilson's (1971) application of the renormalization group. The properties of these models will be discussed in §5.4.2.

We have seen that the mean field type of approximation was discredited in two dimensions by Onsager's exact calculation for $I(\frac{1}{2})$ and subsequent work showed that it is also inadequate in three dimensions. Might there nevertheless be models for which the mean field solution is completely correct? On physical grounds one would expect the approximation to improve as the number of interacting neighbours increases, so that it might be correct for sufficiently long-range forces. Kac and Uhlenbeck and their collaborators showed that this is indeed true for long-range interactions of the form in $\mathrm{Lt}_{\lambda\to 0} J\lambda^d \exp(-\lambda J)$ in d dimensions.

Thus the mean field solution also has a legitimate place in the critical behaviour scheme. Later investigation revealed that with the exception of logarithmic factors, mean field exponents are already attained by nearest-neighbour models in four dimensions; this has proved of major importance in the modern theory of critical exponents. We shall discuss these points in §5.4.3.

5.4.1 Initial Susceptibility of $I(\frac{1}{2})$ in Two Dimensions

Shortly after Sykes and I had suggested the value of 7/4 for γ in two dimensions for the $I(\frac{1}{2})$ model, Fisher told us that he could provide a theoretical justification for this value. Fisher's idea was to make use of the relation (4.107) between the susceptibility and the correlations,

$$\frac{\chi_0}{N\beta\mu_0^2} = 1 + \sum_{\mathbf{r}\neq 0} \Gamma(\mathbf{r},T). \tag{5.117}$$

He felt that sufficient information was available from exact calculations of the correlations to construct a plausible form for $\Gamma(\mathbf{r},T)$ from which, replacing the sum by an integral, the critical behaviour of χ_0 could be assessed.

There were three essential points which Fisher incorporated into his construction:

1 At T_c the correlations decay as $A(\theta)r^{-1/4}$ with $A(\theta)$ a slowly varying function of angle θ.
2 Away from T_c the correlations decay exponentially with distance (see e.g. Domb 1960, p. 174).
3 The range of correlation above T_c is inversely proportional to $\ln(\lambda^+_{\max}/\lambda^-_{\max})$ which is of order $(K^* - K)^{-1} \sim (T - T_c)^{-1}$ (§5.1.4.1(ii))

Fisher's initial suggestion (1959) for $\Gamma(\mathbf{r}, T)$ combining these three points was

$$A(\theta)r^{-1/4} \exp[-br(1 - T_c/T)]. \tag{5.118}$$

With this value,

$$\chi_0 \sim \int_0^{2\pi} A(\theta)\,\mathrm{d}\theta \int_0^{\infty} r^{3/4} \exp[-br(1 - T_c/T)]\,\mathrm{d}r \sim (1 - T_c/T)^{-7/4}. \tag{5.119}$$

This was clearly a result of major importance, and I wrote to ask Onsager's opinion about its validity. Onsager, while showing interest in the approach, expressed anxiety about using only the first two eigenvalues of the transfer matrix; the second eigenvalue is the maximum of a continuous spectrum of eigenvalues (Figure 5.7), and he felt that the contribution of this spectrum would be significant.

In fact since the form of the spectrum was known from the calculations of Kaufman and Onsager (1949), Fisher did take it into account very soon after (the calculation is reproduced in Fisher and Burford (1967)) and he modified (5.118) away from T_c to

$$\frac{1}{r^{1/4}}(\kappa r)^{-1/4}\exp(-kr) \quad [\kappa \sim (1 - T_c/T)]. \tag{5.120}$$

Hence he concluded that the correct asymptotic form of $\Gamma(\mathbf{r},T)$ is

$$\frac{1}{r^{1/4}}f(\kappa r) \tag{5.121}$$

where $f(\kappa r)$ takes the value 1 at $\kappa = 0$, and (5.120) for large r and fixed $\kappa \neq 0$. This modification does not affect the conclusion (5.119). It is of interest that when $T \neq T_c$ and r is large, (5.120) is of Ornstein–Zernike form (4.61), (4.62). This point will be discussed further in §5.5.

A number of detailed calculations followed Fisher's work providing a more rigorous basis for the conclusion (5.119). For example, Kadanoff (1966) derived a closed-form expression for $\Gamma(\mathbf{r},T)$ both above and below T_c:

$$\begin{aligned} &\Gamma(\mathbf{r},T) = \epsilon^{1/4}f^{\pm}(\epsilon r), \quad \epsilon = 4[K - K_c] \\ &f^{+}(x) \sim 2^{-3/8}(\pi x)^{-1/2}\exp(-x) \\ &f^{-}(x) \sim 2^{21/8}\pi^{-1}x^{-2}\exp(-2x) \end{aligned} \tag{5.122}$$

which is valid for large ϵr. Using an argument similar to that of Fisher in (5.119), one can conclude that γ' is also equal to 7/4, and is therefore equal to γ.

Wu (1966) and Cheng and Wu (1967) were able to add correction terms to the result (5.122) and these were of Darboux form. The series of papers which they initiated culminated in a remarkable calculation by Wu *et al.* (1976) who succeeded in evaluating the correlations exactly at an arbitrary point of the lattice. The formula they derived is as follows;

$$\langle \sigma_{00}\sigma_{ij}\rangle = r^{-1/4}G_0^{\pm}(\theta r) + r^{-5/4}G_1^{\pm}(\theta r) + O(r^{-5/4}) \tag{5.123}$$

where

$$\theta = (w^2 + 2w - 1)w^{-1/2}(1 - w^2)^{-1/2} \quad (w = \tanh K)$$

and the functions $G_0^{\pm}$, $G_1^{\pm}$ are Painleve functions of the third kind. They then derived the following *exact* formula for the susceptibility:

$$\beta^{-1}\chi_0 = C_0^{\pm}\,|1 - T_c/T|^{-7/4} + C_1^{\pm}\,|1 - T_c/T|^{-3/4} + \cdots$$

$$\begin{aligned} C_0^+ &= 0.962\,581\,732\,2 \\ C_1^+ &= 0.074\,988\,153\,8 \\ C_0^- &= 0.025\,536\,971\,9 \\ C_1^- &= -0.001\,989\,410\,7. \end{aligned} \tag{5.124}$$

This calculation proved an excellent opportunity to test the accuracy of the series expansion method. Estimates of C_0^+ and C_1^+ given by this method previously (Sykes *et al.* 1972) were

$$\begin{aligned} C_0^+ &= 0.962\,59 \pm 0.000\,03 \\ C_1^+ &= 0.0742 \end{aligned} \tag{5.125}$$

so that the correct value of C_0^+ was well within the error bound, the error being $\sim 10^{-5}$. For C_0^- the estimate of Essam and Hunter (1968) was

$$C_0^- = 0.025\,68 \pm 0.000\,12$$

and the correct value was slightly outside the error bound. (We have already pointed out that low-temperature series are less well behaved than their high-temperature counterparts.) Such accuracy could only be achieved because both the critical point and critical exponent were known exactly in this case.

Fisher (1959) also used relation (5.117) to provide a theoretical justification for the form of the antiferromagnetic susceptibility which had been conjectured from series expansions (§5.3.4). For an antiferromagnet the terms in (5.117) alternate is sign, and we write $\omega_{l,m}(T)$ for the correlation $\langle \sigma_{00}\, \sigma_{lm} \rangle$. Fisher (1959) pointed out that the terms in (5.117) could be rearranged in the form

$$\frac{\chi_0^{(a)}}{N\beta\mu_0^2} = 1 - \omega_{01}(T) - \tfrac{1}{4}\sum_{l,m} \nabla^2 \omega_{lm}(T) \tag{5.126}$$

where the sum is taken over values of l and m for which $l + m$ is even, and ∇^2 is the finite difference operator

$$\nabla^2 \omega_{l,m} = \omega_{l,m+1} + \omega_{l,m-1} + \omega_{l+1,m} + \omega_{l-1,m} - 4\omega_{lm}.$$

The nearest-neighbour correlation is effectively the energy, and the other terms are positive and essentially proportional to $\omega_{lm}(T)$ and have the same type of singularity. The first few terms can be summed explicitly, and the remainder to infinity can be evaluated by the divergence theorem. As a result the form of the antiferromagnetic susceptibility is given by

$$\chi_0^{(a)}(T) = \beta N \mu^2 [1 - q(T) U(T)] \tag{5.127}$$

where $U(T)$ is the energy and $q(T)$ a slowly varying function.

In a later paper (1962) Fisher demonstrated that for a variety of models there is a similar close relation between the derivative of antiferromagnetic susceptibility and the specific heat,

$$C_m(T) \sim A(T) \frac{d}{dT} [\chi_0^{(a)}(T)] \tag{5.128}$$

where $A(T)$ is a slowly varying function.

5.4.2 *The Gaussian and Spherical Models**

The interesting history of these two models has been described in detail by Kac (1964). On hearing from Uhlenbeck of Onsager's solution for the Ising model, Kac tried to think of an interesting model for which the partition function could be calculated readily, and the Gaussian model (5.115) was his first suggestion. Unfortunately, the solution turned out to be non-physical for $T < T_c$.

His next proposal, the spherical model (5.116), was much more realistic, but although he reduced the partition function to an integral, he was unable to evaluate it. The method of steepest descents was ingeniously applied by Berlin to this integral and the critical behaviour could then be readily assessed.

Although Berlin and Kac (1952) obtained a formal solution for a d-dimensional hypercubic lattice, they investigated the properties only for $d = 2, 3$. The properties of the model when $d \geq 4$ and its extension to long-range interactions of the form $J(r) \sim 1/r^{d+\sigma}$ $(\sigma > 0)$ in d dimensions were undertaken by Joyce (1966). These solutions provided the first indication of characteristic discontinuities in the behaviour of critical exponents as functions of d and σ.

We shall sketch the mathematics of the two models only briefly, referring for details to the original paper and review articles. But we shall summarize their behaviour in the critical region.

5.4.2.1. *The Gaussian model*

For the Gaussian model we shall follow the original treatment of Berlin and Kac (1952), and that of Watson (1972) who used the model to derive information about surface and boundary effects. The p.f. for the nearest-neighbour model in a magnetic field H can be written as

$$Z^G(T,H,N) = \left(\frac{1}{2\pi}\right)^{N/2} \int_{-\infty}^{\infty} \cdots \int_{-\infty}^{\infty} d\sigma_1 \, d\sigma_2 \ldots d\sigma_N \times \exp\left(-\frac{1}{2}\sum_{i=1}^{N}\sigma_i^2 + K\sum_{\langle i,j\rangle}\sigma_i\sigma_j + L\sum_{i=1}^{N}\sigma_i\right)$$

$$(K = \beta J,\ L = \beta m H) \tag{5.129}$$

($\langle\ \rangle$ denotes the sum over all nearest-neighbour pairs).

* For this section, I have benefited greatly from the advice and help of Dr Geoff Joyce.

If we postulate a cyclic boundary condition, the matrix of interactions which is then cyclic can readily be diagonalized by an orthogonal transformation, and its eigenvalues $\lambda_1, \lambda_2, \ldots, \lambda_N$ can be written down explicitly. For example, for a two-dimensional lattice with n_1 rows and n_2 columns,

$$\lambda_{\mathbf{p}} = 2\cos\frac{2\pi p_1}{n_1} + 2\cos\frac{2\pi p_2}{n_2}, \quad \left.\begin{matrix} p_1 = 1, 2, \ldots, n_1 \\ p_2 = 1, 2, \ldots, n_2 \end{matrix}\right\} N = n_1 n_2 . \tag{5.130}$$

Writing the transformation to the new variables as

$$\sigma_i = \sum_{k=1}^{N} c_{ik}\, y_k \tag{5.131}$$

where the c_{ik} are appropriately related to the eigenvectors of the interaction matrix, we find that

$$-\frac{1}{2}\sum_{i=1}^{N}\sigma_i^2 + K\sum_{\langle i,j\rangle}\sigma_i\sigma_j = -\frac{1}{2}\sum_{\mathbf{p}}(1 - K\lambda_{\mathbf{p}})y_{\mathbf{p}}^2 \quad [\mathbf{p} = (p_1, \ldots, p_d)]. \tag{5.132}$$

Therefore

$$Z^{G}(T,H,N) = \left(\frac{1}{2\pi}\right)^{N/2}\int_{-\infty}^{\infty}\cdots\int_{-\infty}^{\infty} \times \exp\left(-\frac{1}{2}\sum_{\mathbf{p}}(1 - K\lambda_{\mathbf{p}})y_{\mathbf{p}} + L\sum_{\mathbf{p}=1}^{N} c_{\mathbf{p}}\, y_{\mathbf{p}}\right) \mathrm{d}y_1 \ldots \mathrm{d}y_N \tag{5.133}$$

where

$$c_{\mathbf{p}} = \sum_{i=1}^{N} c_{i\mathbf{p}} . \tag{5.134}$$

The multiple integral then factorizes into a product of N one-dimensional integrals, and using the standard form

$$\int_{-\infty}^{\infty}\exp(-\alpha y^2 + \beta y)\,\mathrm{d}y = (\pi/\alpha)^{1/2}\exp(\beta^2/4\alpha) \tag{5.135}$$

we find that

$$Z^{G}(T,H,N) = \prod_{\mathbf{p}=1}^{N}(1 - K\lambda_{\mathbf{p}})^{-1/2}\exp[L^2 c_{\mathbf{p}}^2/2(1 - K\lambda\mathbf{p})]. \tag{5.136}$$

The logarithm of the p.f. is therefore expressed as a sum of terms, which when N becomes large can be replaced by an integral. In zero field for a d-dimensional SC lattice this integral is

$$\frac{1}{(2\pi)^d}\int_0^{2\pi}\cdots\int_0^{2\pi}\ln\left(1 - 2K\sum_{i=1}^{d}\cos\phi_i\right)\mathrm{d}\phi_1\,\mathrm{d}\phi_2\ldots\mathrm{d}\phi_d \tag{5.137}$$

$\left(\mathop{\mathrm{Lt}}_{N\to\infty} (1/N) \ln Z^{G} = -\frac{1}{2}[\text{Integral } (5.137)]\right)$. In the two-dimensional case the form of the integral

$$\frac{1}{(2\pi)^2}\int_0^{2\pi}\cdots\int_0^{2\pi}\ln[1-2K(\cos\phi_1+\cos\phi_2)]\,\mathrm{d}\phi_1\,\mathrm{d}\phi_2 \tag{5.138}$$

bears some resemblance to Onsager's solution (5.28) for $I(\frac{1}{2})$.

The critical point occurs at $K_c = (2d)^{-1}$ because the integrand in (5.137) then becomes infinite when all the ϕ_i are zero. Thus even a one-dimensional model has a critical temperature which is non-zero. However, when $T < T_c$, the argument of the logarithm in (5.137) can have negative values, and the free energy becomes complex and non-physical.

From (5.136) the effect of a non-zero magnetic field is to add a term

$$L^2 \sum_{\mathbf{p}=1}^{N} c_{\mathbf{p}}^2/(1-K\lambda_{\mathbf{p}}) \tag{5.139}$$

to $\ln Z^G$, and this can also be converted into an integral. Hence it can be shown that the susceptibility exponent γ has the mean field value 1 for all d.

The energy is obtained by differentiating (5.137) w.r.t. K, and its singular behaviour* is given by $(K_c - K)^{(d/2)-1}$. The singular behaviour is similar to that of the lattice Green functions to be introduced shortly. The specific heat exponent is $\alpha = 2-(d/2)$ for $d \le 4$; in four dimensions its behaviour is logarithmic.

For $d > 4$ the specific heat is finite at T_c and therefore $\alpha = 0$. (For $d = 4$ there is a logarithmic divergence.) However, we need to extend the definition of the exponent given in §5.3.3 in order to describe the character of the specific heat more precisely. Suppose that a quantity $\psi(\tau)$ is of the form

$$\psi \sim A + B\tau^{\theta_s} \tag{5.140}$$

as τ approaches zero. If θ_s is not an integer, then if ψ is differentiated a sufficient number of times a value will be obtained diverging to infinity. For this reason Fisher (1967) termed θ_s the exponent of the singular part of ψ. For the specific heat of the Gaussian model†

$$\alpha_s = 2-(d/2) \quad (d > 4). \tag{5.141}$$

5.4.2.2 *The spherical model*

Our discussion of the spherical model will be based on the original paper (Berlin and Kac 1952), but also on the excellent review by Joyce (1972), which demonstrates that many features of the spherical model can be expressed in exact mathematical form.

* Strictly this is valid only for d odd. For d even there is a logarithmic factor $\ln(K_c - K)$.
† When α_s is an integer ($d = 6, 8, 10\ldots$) there is an additional logarithmic factor.

The partition function for the spherical model in zero field is given, from (5.116), by

$$Z^s(T,N) = A_N^{-1} \int \cdots \int \mathrm{d}\sigma_1 \ldots \mathrm{d}\sigma_N \exp\Big(K \sum_{\langle i,j\rangle} \sigma_i \sigma_j\Big) \tag{5.142}$$

the integral being taken over all values of σ_i satisfying the spherical constraint (5.116). A_N is a normalizing constant given by

$$A_N = \sum \sigma_i^2 = N \int \cdots \int \mathrm{d}\sigma_1 \ldots \mathrm{d}\sigma_N = 2\pi^{N/2} N^{(N-1)/2}/\Gamma(N/2). \tag{5.143}$$

The condition (5.116) is replaced by a delta function factor $\delta(\sum \sigma_1^2 - N)$ in the integral of (5.142), and using the representation

$$\delta(u) = \frac{1}{2\pi \mathrm{i}} \int_{-i\infty}^{\mathrm{i}\infty} \exp(us)\, \mathrm{d}s \tag{5.144}$$

we find that

$$Z^s(T,N) = \frac{A_N^{-1}}{2\pi \mathrm{i}} \int_{a-\mathrm{i}\infty}^{a+\mathrm{i}\infty} \mathrm{d}s \exp(Ns) \int_{-\infty}^{\infty} \cdots \int_{-\infty}^{\infty} \mathrm{d}\sigma_1 \ldots \mathrm{d}\sigma_N$$
$$\times \exp\Big(-s \sum_{i=1}^{N} \sigma_i^2 + K \sum_{\langle i,j\rangle} \sigma_i \sigma_j\Big) \tag{5.145}$$

where a is an arbitrary constant introduced for technical reasons.

The integral over the σ_i can be evaluated in the same way as for the Gaussian model to give

$$\pi^{N/2} \exp\Big[-\frac{1}{2} \sum_{\mathbf{p}=1}^{N} \ln\Big(s - \frac{K\lambda_{\mathbf{p}}}{2}\Big)\Big] \tag{5.146}$$

and the remaining integral over s in (5.145) which we are interested in for large N can be dealt with by the method of steepest descents.

It is convenient to write $s = K\xi$, and we are then interested in evaluating the integral

$$\int_{a-\mathrm{i}\infty}^{a+\mathrm{i}\infty} \mathrm{d}\xi \exp\Big(NK\xi - \frac{1}{2} \sum_{\mathbf{p}=1}^{N} \ln(K\xi - K\lambda_{\mathbf{p}})\Big). \tag{5.147}$$

As before, the sum in the integrand of (5.147) is replaced by an integral. Let us write

$$F_d(\xi) = \frac{1}{(2\pi)^d} \int_0^{2\pi} \cdots \int_0^{2\pi} \mathrm{d}\phi_1 \ldots \mathrm{d}\phi_d \ln\Big(\xi - \sum_{i=1}^{d} \cos \phi_i\Big) \tag{5.148}$$

$$G_d(\xi) = F_d'(\xi) = \frac{1}{(2\pi)^d} \int_0^{2\pi} \cdots \int_0^{2\pi} \mathrm{d}\phi_1 \ldots \mathrm{d}\phi_d \Big(\xi - \sum_{i=1}^{d} \cos \phi_i\Big)^{-1}. \tag{5.149}$$

$G_d(\xi)$ is usually called the *lattice Green function* and is of importance in the theory of random walks (Montroll and Weiss 1965).

The saddle point is given by

$$2K = F'_d(\xi) = G_d(\xi) \tag{5.150}$$

and if the solution is denoted by ξ^*, the limiting free energy per spin in zero field is given by

$$-\beta\psi_d = -\tfrac{1}{2} - \tfrac{1}{2}\ln(2K) + K\xi^* - \tfrac{1}{2}F_d(\xi^*). \tag{5.151}$$

For $d = 1$ and 2 a solution of (5.150) is possible for all K. This is because $G_d(\xi)$ becomes infinite as ξ approaches the singularity of the function. The free energy is therefore continuous as a function of K. However, for $d = 3$, $G_d(\xi)$ approaches a finite value which we call K_c as ξ approaches its singular value 3. For $T < T_c$ or $K > K_c$ equation (5.150) cannot be satisfied. The saddle point 'sticks' at $\xi = 3$ and the free energy is given by (5.151) with $\xi^* = 3$. This sticking of the saddle point gives rise to discontinuous critical behaviour, and K_c is identified with the Curie temperature. A similar pattern of behaviour occurs when $d > 3$.

In the presence of a magnetic field an additional term

$$L^2/2K(\xi - d)^2 \tag{5.152}$$

must be added to the RHS of the saddle-point equation (5.150), arising from an addition to $\ln Z$ of magnetic energy $L^2/4K(\xi - d)$ (see Joyce (1972) for details).

We summarize the critical behaviour of the spherical model for a d-dimensional lattice with finite range interactions from Joyce (1972).

Susceptibility $\gamma = 2\ (d = 3), \gamma = 1\ (d \geq 4)$.

$$\begin{aligned} &\chi_0 \sim \tau^{-2/(d-2)} && (2 < d < 4) \\ &\chi_0 \sim \tau^{-1}|\ln \tau| && (d = 4) \\ &\chi_0 \sim \tau^{-1} && (d > 4). \end{aligned} \tag{5.153}$$

Specific heat C_H is finite as $T \to T_c^*$, and $C_H = k/2$ for all $T < T_c$.

$$\begin{aligned} &\alpha_s = \frac{d-4}{d-2} && (2 < d < 4) \\ &{}^*\alpha_s = -\tfrac{1}{2}(d-4) && (d \geq 4) \\ &C_H \sim |\ln \tau|^{-1} && (d = 4). \end{aligned} \tag{5.154}$$

Magnetization

$$\delta = 2\gamma + 1 \quad \text{(for all } d). \tag{5.155}$$

Spontaneous magnetization

$$\beta = \tfrac{1}{2} \quad \text{(for all } d > 2). \tag{5.156}$$

* Note that there are logarithmic singularities for $d = 6, 8, 10$ and α_s is *not* defined for $d = 3$.

It is important to note that when d reaches the value 4 the critical exponents attain their mean field values, *and they remain fixed at these values for* $d > 4$. The dependence of critical exponents on d is therefore analytic when $2 < d < 4$ with a singularity occurring at $d = 4$; when $d > 4$ they remain constant. The significance of this 'sticking' of the exponents was properly appreciated only later when Wilson and Fisher (1972) introduced the ϵ expansion for critical exponents ($\epsilon = 4 - d$, see §7.1 and §7.2.4).

Low-temperature behaviour The low-temperature susceptibility is infinite when $d = 3$; this is a general characteristic of the Heisenberg model with classical vector interactions. The infinite entropy for $T < T_c$ associated with the finiteness of C_H as $T \to 0$ is a similar characteristic of these models.

Long-range interactions Joyce (1966) calculated the properties of the spherical model with a long-range interaction of the form

$$J(r) \sim r^{-(d+\sigma)} \tag{5.157}$$

in d dimensions ($\sigma > 0$). This was particularly useful since it gave an indication of the type of behaviour one might expect in the Ising and Heisenberg models with long-range interactions of this type. Joyce found that, for sufficiently small values of σ, critical behaviour is present in one- and two-dimensional models. There are critical values of σ at which the exponents change discontinuously. The details are as follows.

Susceptibility

$$\begin{aligned} \chi_0 &\sim \tau^{-\sigma/(d-\sigma)} && (1 < d/\sigma < 2) \\ \chi_0 &\sim \tau^{-1}|\ln \tau| && (d/\sigma = 2) \\ \chi_0 &\sim \tau^{-1} && (d/\sigma > 2) \end{aligned} \tag{5.158}$$

provided that

$$0 < \sigma < \min\{2, d\}. \tag{5.159}$$

Hence

$$\gamma = \begin{cases} \sigma/(d-\sigma) & (1 < d/\sigma < 2) \\ 1 & (d/\sigma \geq 2) \end{cases} \tag{5.160}$$

provided that (5.159) is satisfied.

For $d > 2$, short-range interaction exponents are obtained when σ reaches 2, and they retain these values when $\sigma > 2$.

Specific heat

$$\begin{aligned} \alpha_s &= -(2\sigma - d)/(d - \sigma) && (1 < d/\sigma < 2) \\ \alpha_s &= -(d - 2\sigma)/\sigma && (d/\sigma \geq 2) \end{aligned} \tag{5.161}$$

provided that (5.159) is satisfied. For the magnetization and spontaneous magnetization relations (5.155) and (5.156) remain valid for all σ.

We note that the pattern of dependence of critical exponents on σ is qualitatively similar to their dependence on d. In one or two dimensions a finite critical temperature arises when $\sigma < d$. The variation is then analytic until σ drops to $d/2$ when mean field values are attained. For $\sigma < d/2$ exponents remain at their mean field values.

For $d = 3$ there is always a finite T_c, and the short-range interaction values apply when $\sigma \geq 2$. When $3/2 < \sigma < 2$ the variation of critical exponents is analytic until mean field values are attained at $\sigma = 3/2$; for $\sigma < 3/2$ exponents retain their mean field values.

For the spherical model it is clear that mean field exponents are attained in two independent ways:

1 for short-range interactions if the dimension is sufficiently high ($d \geq 4$);
2 for long-range interactions in any dimension $d < 4$ if the range of the interaction is sufficiently long ($\sigma \leq d/2$).

We shall see in §7.5.5 that this pattern of behaviour applies to a wide range of models.

5.4.3 Systems with Weak Long-range Interactions

The mathematics of systems with weak long-range forces is sophisticated, and we shall not attempt to reproduce it in this elementary treatise. We shall briefly summarize the conclusions, and refer the reader to the review by Hemmer and Lebowitz (1976) for details of the mathematical arguments and a list of the original papers.

In 1959 Kac introduced a one-dimensional continuum model with long-range forces which seemed to offer the possibility of exact treatment. The molecules interact with a hard-core repulsion of length b, and an exponential attraction $\lambda \exp(-\lambda r)$. For this model he showed that the p.f. can be derived as the maximum eigenvalue of a linear integral equation with a positive definite Hilbert–Schmidt kernel.

The calculations were pursued in two papers by Kac, Uhlenbeck and Hemmer (1963). They first showed that for any finite value of λ the p.f. is continuous, and there is no phase transition. However, in the limit of $\lambda \to 0$ there is a phase transition, and it can be evaluated in closed form. It gives rise, in fact, to the van der Waals equation (2.34), with the Maxwell construction automatically included. At the end of their paper the above authors discussed the three-dimensional model, and advanced arguments to show that in this case also the attractive and repulsive parts of the potential make independent contributions to the free energy. In three dimensions the hard-sphere p.f. is not known in closed form, but Monte Carlo estimates are available. A heuristic derivation of this latter result has already been given in §2.3.

For a one-dimensional Ising model $I(\frac{1}{2})$ Kac and Helfand (1963) used a different approach to show that the Weiss mean field solution (3.31) is derived for the same type of attractive force. For a two-dimensional model they were able to deal with interactions along rows in one dimension, and a selection of interactions in the second dimension; the two- and three-dimensional lattice problems were further explored by Kac and Thompson (1969).

The above calculations suggested that a mean field solution will always be obtained for a model with a weak long-range attractive force of the form

$$J(r) = \operatorname*{Lt}_{\lambda \to 0} \lambda^d \exp(-\lambda r). \tag{5.162}$$

However, the work of Joyce on the spherical model with long-range forces indicated that a mean field solution should result for models with forces of considerably shorter range. The spherical model solution deviates much more from the mean field solution than its Ising counterpart. For example, it shows no critical behaviour in two dimensions, and in three dimensions $\gamma = 2$, whereas for $I(\frac{1}{2})$, $\gamma = 5/4$ (in comparison with $\gamma = 1$ for mean field). But from (5.158)–(5.160) a force of the form $1/r^{d+\sigma}$ is sufficient to give a mean field solution for the spherical model if $\sigma < d/2$. Therefore one should certainly expect to derive a mean field solution with this force for the Ising model. Later research work on the one-dimensional Ising model has provided some support for this conclusion (Thouless 1969, Frohlich and Spencer 1982).

5.5 Correlations

We have already drawn attention in §5.1.5 to the deviation of the exact results of Onsager and Kaufman, relating to spin correlations from the classical Ornstein–Zernike theory. In §5.4.1 we discussed Fisher's proposal (5.121) for the $I(\frac{1}{2})$ model in two dimensions, and its subsequent confirmation.

What should be the modified form for a general model in dimension d? It was Fisher again who proposed in 1964 the appropriate replacement of the O–Z formulae which could take account of non-classical exponents. A detailed numerical investigation was undertaken by Fisher and Burford (1967) for the $I(\frac{1}{2})$ model in three dimensions to check the validity of the proposal and estimate the relevant critical exponents.

It was reasonable to assume that the three points made at the beginning of §5.4.1 would have their analogues for a general model as follows:

1 At T_c the $1/r^{d-2}$ decay given by O–Z (equation (4.59)) should be modified to $1/r^{d-2+\eta}$. The exponent η has the value 0 in classical theory, and $\frac{1}{4}$ for the $I(\frac{1}{2})$ model in two dimensions.
2 Away from T_c the correlations decay exponentially with distance.

3 The range of correlation above T_c, which is $\tau^{-1/2}$ in O–Z theory (equation (4.50)) and τ^{-1} for $I(\frac{1}{2})$ in two dimensions, should in general be written $\tau^{-\nu}$, where ν is a correlation critical exponent.

Hence, by analogy with (5.121), the general form of the spin–spin correlation function should be written

$$\frac{1}{r^{d-2+\eta}} f(\kappa r). \tag{5.163}$$

The new exponents ν,η are not independent. Using the d-dimensional analogue of the integral (5.119) for the initial susceptibility, one finds that

$$\gamma = \nu(2-\eta). \tag{5.164}$$

In the next section we shall see that Fisher advanced theoretical arguments to support the view that the O–Z formula should be valid generally when $T \neq T_c$, and that asymptotically

$$f(\kappa r) \sim (\kappa r)^{(1/2)(d-3)+\eta}[\exp(-\kappa r)]. \tag{5.165}$$

This would lead to (4.61) for large κr, and to (4.62) for fixed κ and large r. We have already seen from (5.120) that this is correct for $I(\frac{1}{2})$ in two dimensions. This result later received rigorous justification in a series of papers by Fisher and Camp (Fisher and Camp 1971, Camp 1972, 1973).

On the low-temperature side an exponent ν' was introduced to parallel ν, with a low-temperature function $f^-(\kappa r)$ to parallel $f(\kappa r)$. A relation analogous to (5.164)

$$\gamma' = \nu'(2-\eta) \tag{5.164'}$$

can readily be derived.

5.5.1 Modification of Light Scattering Formulae

In his 1964 paper Fisher considered the modifications to the O–Z theory of §4.6 and §4.6.1 which result from the proposals of §5.5. He assumed that analogous formulae applied to fluids, $\Gamma(\mathbf{r})$ of (4.107) being replaced by $\rho h(\mathbf{r})$ of (4.23). At the critical temperature T_c,

$$\rho h(\mathbf{r}) \sim D/r^{d-2+\eta} \tag{5.166}$$

and hence from (4.30)

$$\rho\hat{h}(q) \simeq \hat{D}/q^{2-\eta} \tag{5.167}$$

instead of (4.45). We note incidentally that the total scattering intensity in all directions for a three-dimensional model is now given by the integral

$$\int_0^{2\pi} \frac{2\pi \sin\theta \, d\theta}{(2\pi \sin\theta/2)^{2-\eta}} \tag{5.168}$$

instead of (4.46), and this converges. Hence, the total scattering at the critical point is now finite, and the theory is consistent.

Using (4.36) we find that

$$\hat{c}(q) = \hat{c}(0)(1 - c_0 q^{2-\eta} + \cdots) \qquad (5.169)$$

and if η is small and non-zero, $\hat{c}(q)$ is non-analytic at T_c, so that a Taylor expansion of the form (4.40) is no longer possible. This corresponds to the non-analyticity of the free energy at T_c discovered by Onsager.

The form of the direct correlation function $c(\mathbf{r})$ at T_c can be deduced from (5.169) by asymptotic Fourier inversion,

$$c(\mathbf{r}) \sim F/r^{d+2-\eta}. \qquad (5.170)$$

This is of shorter range than $h(\mathbf{r})$ in (5.166), as anticipated by O–Z, but not of sufficiently short range for its second moment, L^2 in (4.41), to exist when $\eta \neq 0$. However, when $T \neq T_c$ it is reasonable to assume that $c(q)$ is analytic, and an expansion of the form (4.40) should be valid if $c(\mathbf{r})$ drops off sufficiently rapidly. This provides a justification for the view that O–Z is then correct.

How should the general Lorentzian formula (4.42) be modified in the revised theory? Near T_c Fisher wrote

$$f(\kappa r) \sim \exp(-\kappa r)[1 + Q(\kappa r)] \qquad (5.171)$$

where $Q(x) \to 0$ as $x \to 1$, and $Q(x)$ does not grow exponentially as $x \to \infty$. If $Q(x)$ is neglected, Fourier transformation of (5.171) gives

$$\chi(q) \sim \hat{D}/(\kappa^2 + q^2)^{1-\eta/2} \qquad (5.172)$$

for small q^2. However, this has only limited validity, and in the next section we shall summarize the comprehensive investigation of Fisher and Burford (1967) into the behaviour of the $I(\frac{1}{2})$ model.

5.5.2 *Estimation of Critical Correlation Exponents and Critical Scattering Function*

To estimate the exponent ν for $I(\frac{1}{2})$ one needs to find an appropriate quantity which can be expanded as a high-temperature series, the divergence of which is related to ν. Fisher and Burford introduced the moments defined by

$$\mu_t(w) = \sum_{\mathbf{r}\neq 0} (r/a)^t \Gamma(\mathbf{r},w) \qquad (5.173)$$

where a is the lattice spacing. Using (5.163), one readily finds on replacing the sum by an integral that

$$\mu_t \sim \kappa^{-(2+t-\eta)} \sim \tau^{-\gamma-t\nu} \qquad (5.174)$$

with the help of relation (5.164).

The susceptibility corresponds to μ_0, and we have already discussed in §5.2.1.1(iii) its diagrammatic expansion in terms of graphs with two free ends embedded in the lattice. For the second moment μ_2 a corresponding expansion can be found which uses the end-to-end distances as well as the numbers of the various graphs.

Alternative expansions introduced by Fisher and Burford which can be directly related to O–Z theory (4.37) and (4.40) are the following:

$$1/\hat{\chi}(\mathbf{q},w) = \chi_0^{-1}[1 + \Lambda_2(w)q^2a^2 - \Lambda_4(w)q^4a^4 + \Lambda_6(w)q^6a^6 \ldots] \tag{5.175}$$

or equivalently

$$1/\hat{\chi}(\mathbf{q},w) = \chi_0(w)^{-1} + L_2(w)q^2a^2 - L_4(w)q^4a^4 + L_6(w)q^6a^6 \ldots \tag{5.176}$$

The coefficients $\Lambda_2(w)$, $L_2(w)$ can readily be related to the moment μ_2 in (5.173). From the definition

$$\hat{\chi}(\mathbf{q},w) = 1 + \sum_{\mathbf{r}\neq 0} \exp(\mathbf{iq}\cdot\mathbf{r})\Gamma(\mathbf{r},w)$$

by expanding the exponential we find that

$$\Lambda_2(w) = \chi_0(w)L_2(w) = \mu_2(w)/2d\chi_0(w) \tag{5.177}$$

for a simple cubic lattice in d dimensions. Higher coefficients Λ_4, L_4, ... depend also on the direction of the wave vector q, and cannot be simply related to the moments.

O–Z theory assumed that an expansion of the type (5.176) is possible with $L_2(w)$ finite at T_c. In fact since the modified theory precludes such an expansion in view of (5.169), we must expect L_2 to diverge at T_c. From (5.177) it is easy to see that near T_c

$$L_2 \sim \tau^{-\gamma-2\nu+2\gamma} = \tau^{\gamma-2\nu} = \tau^{-\eta\nu} \tag{5.178}$$

from (5.174) and (5.164).

For the SC lattice in three dimensions, Fisher and Burford derived high-temperature series for the decay factor $\omega_x = \exp(-\kappa_x a)$, from which it was possible to make an estimate of ν. A second estimate was obtained from the series expansion for the second moment $\mu_2(w)$. Eventually they suggested

$$\nu = 0.6430 \pm 0.0025 \simeq \tfrac{9}{14}. \tag{5.179}$$

The exponent η was then calculated from relation (5.164),

$$\eta = 0.056 \pm 0.008 \simeq \tfrac{1}{18} \tag{5.180}$$

so that it is very small. Shorter series were derived for the BCC and FCC lattices which gave results consistent with (5.179).

For the scattering function near T_c Fisher and Burford proposed the following empirical formula which they found to fit the numerical data very accurately:

$$\hat{\chi}(\mathbf{q},T) = \left(\frac{a}{r_1}\right)^{2-\eta} \frac{[(\kappa_1 a)^2 + \phi^2 a^2 K^2(\mathbf{q})]^{\eta/2}}{[(\kappa_1 a)^2 + \psi a^2 K^2(\mathbf{q})]}. \tag{5.181}$$

Here κ_1 is related to Λ_2 in (5.175) by

$$(\kappa_1 a)^2 = \Lambda_2^{-1} \tag{5.182}$$

K^2 is defined by

$$a^2 K^2 = 2d\left(1 - Q^{-1} \sum \exp(\mathrm{i}\mathbf{q} \cdot \mathbf{r})\right) \tag{5.183}$$

the sum being taken over nearest neighbours of a point of the lattice, and Q being the coordination number of the lattice; r_1, ϕ and ψ are slowly varying functions of temperature.

The modification to the O–Z inverse scattering function plot (Figure 4.4) is shown schematically in Figure 5.22.

The linked cluster method (§5.2.2) is particularly powerful for the calculations involving the pair correlation function since no basic new diagrammatic procedure is required. Ferer, Moore and Wortis (1969) used this method to obtain 12 terms in the series for the SCC, BCC and SC lattices. They suggested the values

$$\nu \simeq 0.638^{+0.002}_{-0.001} \tag{5.179'}$$

$$\eta \simeq 0.041^{+0.006}_{-0.003} \tag{5.180'}$$

as an improvement of (5.179) and (5.180).

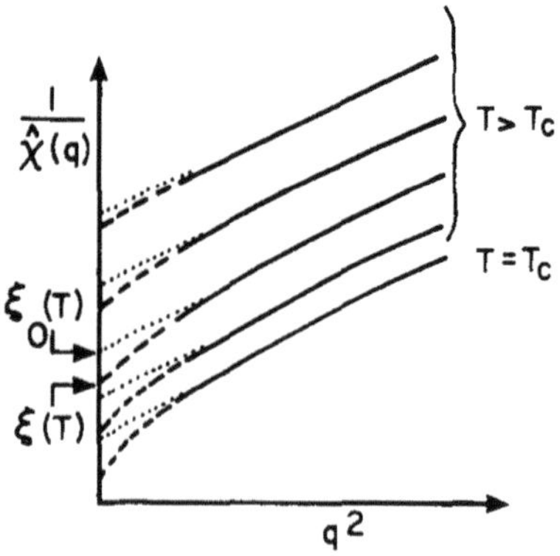

Figure 5.22 Plot of inverse scattering to be expected with a positive value of η (illustrative and larger than that estimated for the Ising or Heisenberg models). Note the difference between the apparent linear intercept $\xi_0(T)$ and the true intercept $\xi(t) = 1/\hat{\chi}(0)$ (after Fisher 1965).

5.6 Fluids

5.6.1 *Historical Notes*

We have already noted in §1.4.1 and §2.3 the reservations of Boltzmann and Maxwell in relation to van der Waals' treatment of the hard-core repulsion. If we expand the hard-sphere equation of state in the form

$$\frac{Pv}{RT} = 1 + \frac{B}{v} + \frac{C}{v^2} + \frac{D}{v^3} + \cdots \tag{5.184}$$

the van der Waals equation requires $B = b$, $C = b^2$, $D = b^3$ .... Boltzmann had calculated C and found it to be $\frac{5}{8}b^2$, and as a result of the evaluation of a key integral by van Laar in 1899, D was also known to be $0.2869b^3$ (Boltzmann 1899). Thus it was clear to Boltzmann and presumably to van der Waals that the equation could be regarded as rigorously correct only for dilute gases.

There began a period of collection of accurate experimental data initiated by Kamerlingh Onnes (1901) who expressed the equation of state in the form

$$\frac{P}{RT} = \frac{1}{v} + \frac{B(T)}{v^2} + \frac{C(T)}{v^3} + \frac{D(T)}{v^4} + \cdots \tag{5.185}$$

where $B(T)$, $C(T)$, $D(T)$ . . . were called the second, third, fourth ... virial coefficients. This then became the standard procedure for recording data on imperfect gases. A challenge was posed to theoreticians to try to find a theoretical description of virial coefficients, but the problem seemed extremely formidable.

Then in 1937 one of the great triumphs of statistical mechanics occurred. J. E. Mayer developed an elegant formalism which enabled him to express all the virial coefficients as integrals over the intermolecular potentials. The formula he derived was

$$\frac{P}{kT} = \rho - \sum_{k=1}^{\infty} \frac{k}{k+1} \beta_k \rho^k \tag{5.186}$$

where the β_k, which he termed *irreducible cluster integrals*, could be represented simply and diagrammatically as *multiply connected* graphs. For example, the first three of the β_k are represented as follows:

$\beta_1 =$ 

$2!\beta_2 =$ 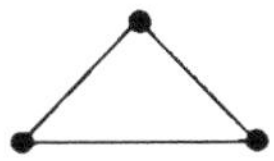 (5.187)

$3!\beta_3 = 3$  $+ 6$ 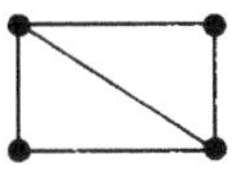 $+$ 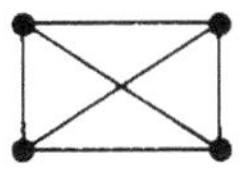 .

If $\phi = \phi(|\mathbf{r}_{ij}|)$ is the intermolecular potential between molecules i and j, and we write

$$f_{ij} = \exp[-\beta\phi(r_{ij})] - 1 \tag{5.188}$$

a diagram connecting k points is a $3(k-1)$-dimensional integral, with an appropriate f_{ij} for each pair of points connected by a line in the diagram. For example,

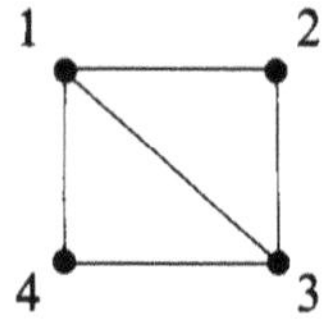

$$= \frac{1}{V}\iiint f_{12}\, f_{23}\, f_{34}\, f_{41}\, f_{13}\, \mathrm{d}\mathbf{r}_1\, \mathrm{d}\mathbf{r}_2\, \mathrm{d}\mathbf{r}_3\, \mathrm{d}\mathbf{r}_4. \tag{5.189}$$

A corresponding series for the g.p.f. in powers of the activity made use of *reducible cluster integrals*, b_l, in the coefficients which were represented by *connected graphs*. But it was generally assumed that the virial series (5.186) which eliminated large numbers of graphs and led directly to comparison with experiment was of greater significance. (For a detailed account of the theory see e.g. Mayer and Mayer (1940), Domb (1974), Ch. 1).) The same development based on correlations between the particles in a fluid was suggested independently by Yvon (1937), but his work was less well known.

My first mentor at Oxford, M. H. L. Pryee, has told me that he was present at the colloquium in Cambridge when Mayer's results were first presented. Fowler, the outstanding world authority on statistical mechanics, was deeply impressed by Mayer's work, and said that he would not have believed that such progress could have been achieved in his lifetime. The way seemed open for a detailed explanation in atomic terms of the liquid and gaseous phases, the critical point, and the striking discontinuities which occur in the equation of state.

In fact, Mayer tried to assess the relevant analytical properties of the series (5.186), combining knowledge of the properties of fluids with conjectures regarding the asymptotic behaviour of the β_k. He was strongly supported by Born, who, in a joint paper with Fuchs published in 1938, said 'we believe that we have succeeded in showing rigorously, and in a somewhat simpler way than Mayer himself, that his statements are completely correct'.

Mayer identified two different temperatures T_m and T_c associated with critical behaviour. T_m defined the flat top of a coexistence curve below which the standard separation into phases took place. T_c was defined by a homogeneous 'Derby hat' region above T_m. The behaviour is illustrated in Figure 5.23 taken from the book by Mayer and Mayer (1940).

From the experimental point of view there were three major difficulties in performing accurate experiments near the critical point: (i) the effect of gravity causing a density inhomogeneity in a region of very high compressibility; (ii)

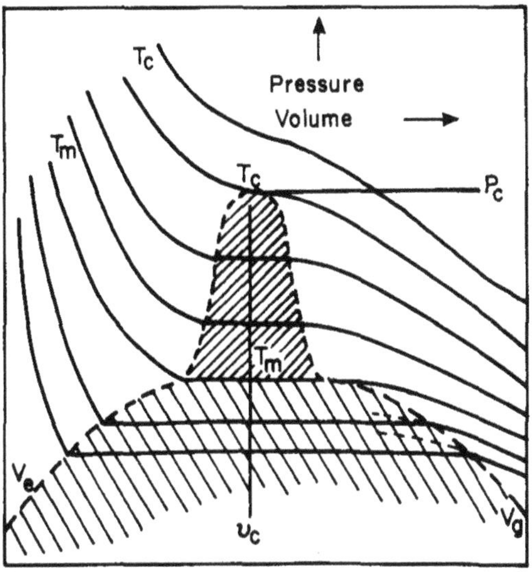

Figure 5.23 Plot of pressure against volume at different temperatures near the critical temperature (after Mayer and Mayer).

the long times needed to reach equilibrium and the problem of eliminating temperature gradients; (iii) the effect of impurities.

The fact that gravity might be important in the critical region was pointed out by Gouy as long ago as 1892. However, in the 1930s it was generally assumed that in the region currently available to experiment the effect was negligible. As we have mentioned the prestige of van der Waals was low, and experimentalists were impressed by the prediction of the cluster integral theory that the coexistence curve had a flat-top structure. In a review in 1938 Maass quoted experimental results to support this prediction, and expressed gratification at the agreement achieved between experiment and theory.

It needed a good deal of courage to counter this formidable combination of experiment and theory. The true nature of the coexistence curve and the importance of gravity were first clearly demonstrated by Schneider and his coworkers (Schneider 1952, Weinberger and Schneider 1952). The results were first presented at the Paris Conference on Phase Transitions in 1952. Schneider argued that if the compressibility were indeed large enough for gravity to be important, then in a long tube one would obtain a flat-topped coexistence curve experimentally even if the true coexistence curve was not flat topped. As the density changed the meniscus would move up and down the tube, so that at the meniscus itself critical conditions would apply. Hence the meniscus would be present over a range of densities. To test this hypothesis Schneider used a long tube and performed the experiment with the tube vertical (height 19.5 cm) and horizontal (height 12 mm). The striking result is shown in Figure 5.24.

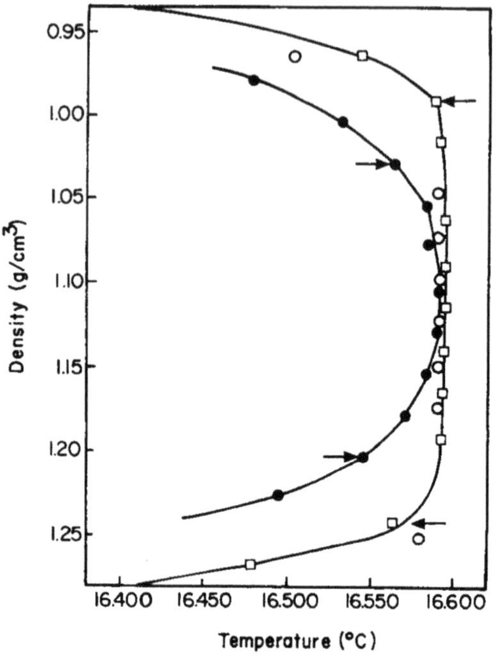

Figure 5.24 Liquid–vapour coexistence curves in the critical region: ○, □, long tube; ●, short tube (after Schneider).

Uhlenbeck and his coworkers (see Domb (1974, Ch. 1)) for detailed references) introduced graph theory into statistical mechanics, and revealed the true nature of the problem to be tackled in evaluating the β_k. They showed that although the number of integrals contributing to β_k is small at first, it becomes large extremely rapidly, as will be seen from Table 5.5. Each integral of order k is in space of $3k$ dimensions. It is not surprising that even for the simplest hard-sphere potential only six terms have been calculated using the most powerful modern computers, and these are inadequate to assess asymptotic behaviour. To account for condensation it is essential to take account of the attractive part of intermolecular forces and even fewer terms are then available. One is reluctantly driven to the conclusion that the problem is intractable.

Table 5.5 Number of integrals contributing to β_k

k	1	2	3	4	5	6	7	8
Number of integrals	1	1	3	10	56	468	7123	194066

5.6.2 *The Lattice Gas Model*

When a theoretical physicist is faced with an intractable problem he does not give up; he tries to find an alternative problem which is tractable and which may cast some light on the original problem. The lattice gas model was introduced by Cernuschi and Eyring in 1939. The cells of an SQ lattice in two dimensions, or an SC lattice in three dimensions, can be either occupied by a *single* molecule or unoccupied (a hole). Any two neighbouring occupied cells have an attractive interaction; the fact that an occupied cell cannot accept another molecule provides a crude representation of a hard-core repulsion, and the possibility of varying the number of holes allows for changes in free volume, one of the essential features of a liquid.

We shall see shortly that the model is isomorphic with $I(\frac{1}{2})$, and we shall trace the relationship in detail. The model restored the magnet–fluid analogy which was first suggested by Pierre Curie in 1895, used by Pierre Weiss in his classic molecular field paper of 1907, and then forgotten for more than 30 years (§1.4.2).

Of course, the model is crude, and I can well remember the scepticism with which my own preliminary efforts in the late 1940s to use it as a model of liquid–gas equilibrium were greeted. In 1950 Kirkwood made the model more respectable by showing the precise nature of the approximation involved. Any continuum model can be converted into a cell model of arbitrary cell size by using appropriate statistical mechanics; if the intermolecular force involves a hard core, the cell size can be chosen so that no cell can be occupied by more than one molecule. But the interaction between neighbouring cells must take into account all the different positions of the molecules within the cells, and is therefore extremely complicated. The lattice gas model approximates this complicated interaction in the simplest possible way.

In fact, we have already traced in §4.8 the relationship between the Ising model and solid solutions. If an atom of type A is identified with a molecule, and an atom of type B with a hole, ϵ_{AB} and ϵ_{BB} are zero, and $\epsilon = -\frac{1}{2}\epsilon_{AA}$ (equation (4.68)). From (4.79) we derive the p.f. for a lattice gas, which we relate to the configurational integral (2.26). We then form the g.p.f. (1.10) which we associate with the Ising model p.f. (4.80).

However, there are a number of details which need to be clarified for the analogy to be developed properly. The Ising model p.f. is related to the number of *spins*, which is equal to the number of *sites* N. The lattice gas p.f. is related to the number of *molecules* N_A. If the volume of a unit cell is denoted by v_0, the volume V of the lattice gas is Nv_0; N is thus related to *volume* in the lattice gas model.

We assume that each molecule can move freely around its cell of volume v_0. Therefore the configurational integral (2.26) becomes, in d dimensions,

$$Z_{N_A} = \frac{v_0^{N_A}}{\Lambda^{dN_A}} z_A^{qN/2} \sum_{N_{AB}} g(N; N_A, N_{AB}) z^{N_{AB}}. \tag{5.190}$$

There is no need to divide by $N!$ since the molecules are not identical, each being characterized by its own cell. If we now form the g.p.f. following the prescription of (1.10), we obtain

$$\zeta(T,V,\mu) = \Lambda_N(y,z), \tag{5.191}$$

where

$$y = \frac{z_{\mathrm{A}}^{q/2} v_0 \exp(\beta\mu)}{\Lambda^d}. \tag{5.192}$$

We then use the fundamental thermodynamic relation (1.11) to give

$$PV = NPv_0 = kT \ln \zeta \tag{5.193}$$

and (1.12) to derive thermodynamic functions. Incidentally, we can derive the magnetic analogue of the pressure, P. From (4.78) and (4.90) we find for the free energy of the Ising model,

$$F^{\mathrm{I}} = -N\mu_0 H - \frac{qN}{2} J - kT \ln \Lambda_N. \tag{5.194}$$

Hence we deduce that

$$-Pv_0 = \mu_0 H + \frac{F^{\mathrm{I}}}{N} + \frac{q}{2} J. \tag{5.195}$$

The density ρ of the fluid is defined by

$$\rho = \frac{N_{\mathrm{A}}}{V} = \frac{N_{\mathrm{A}}}{Nv_0} = \frac{1}{Nv_0}\, y \frac{\partial}{\partial y} [\ln \Lambda_N(y,z)]. \tag{5.196}$$

This is then analogous to the magnetization,

$$\rho v_0 \equiv \tfrac{1}{2}(1 - m/\mu_0) \tag{5.197}$$

from (5.52). Similarly from (5.192) we find for the chemical potential μ that

$$\mu - \mu^* \equiv -2\mu_0 H \quad \left(\mu^* = -q\epsilon_{\mathrm{AA}} + kT \ln \frac{\Lambda^d}{v_0}\right). \tag{5.198}$$

The total configurational entropy S is the same for both models from (5.191). But if we consider the entropy s_e per system, we have

$$\begin{aligned} S &= N_{\mathrm{A}} s_e^{\text{fluid}} = N s_e^{\text{magnet}} \\ \rho v_0 s_e^{\text{fluid}} &= s_e^{\text{magnet}}. \end{aligned} \tag{5.199}$$

It must be remembered that C_V for the fluid corresponds to constant N_{A}, and is therefore analogous to C_M for the magnet. Using (5.199)

$$\rho v_0 C_V = C_M \tag{5.200}$$

C_P and C_H can then be derived by standard thermodynamic formulae, but they make use of χ_0 and K_T, the relationship between which we must now consider.

From (5.193) we find that

$$\frac{\mathrm{d}P}{\mathrm{d}\rho} = \frac{kT}{Nv_0}\frac{\partial}{\partial\rho}(\ln\zeta) = \frac{kT(\partial/\partial y)(\ln\zeta)}{Nv_0\,\partial\rho/\partial y} = \frac{kTNv_0P}{(\partial/\partial y)[y(\partial/\partial y)\ln\Lambda_N]} \tag{5.201}$$

using (5.196). From this we deduce the analogy

$$\rho^2 v_0 K_T \equiv \chi_0/4\mu_0^2. \tag{5.202}$$

Finally, by comparing (4.23) with (4.107) and replacing the integral by a sum over cells we relate the correlation functions

$$\rho v_0 h(\mathbf{r}) \equiv \Gamma(\mathbf{r}). \tag{5.203}$$

In 1952 Yang and Lee published two important papers on the lattice gas model, which demonstrated rigorously that it was capable of accounting for the salient features of liquid–gas equilibrium. At the same time they were able to pinpoint the limitations of the cluster integral development.

The following conclusions resulted from their work:

1 The structure of singularities of the Ising model p.f. is represented in Figure 5.13. The function is analytic everywhere except in a region on the $H = 0$ axis for $T < T_c$. The discontinuity in magnetization on this axis gives rise to the characteristic horizontal sections of the liquid–vapour isotherms, which terminate at the critical point T_c. For temperatures greater than T_c there are no discontinuities.

2 In studying the cluster integral theory they found it more useful to concentrate on the reducible cluster integral series involving the b_l. They pointed out that the b_l are functions of V and T, each of which converges to a definite limit as $V \to \infty$. In the Mayer theory the cluster integrals b_l are replaced from the very beginning by their limiting values $b_l(\infty)$. Yang and Lee demonstrated that this precluded the existence of a liquid phase of finite density. The Mayer theory would therefore give correct results for the gas phase up to the condensation density, but it was incapable of going beyond this density to give horizontal isotherms. It could not therefore describe liquid–gas equilibrium.

From the computational point of view as well the lattice gas model has distinct advantages. The exact particle–hole symmetry introduces simplifying features. The cluster integrals are converted into cluster sums in the calculation of which computers can be of great help (§5.2). Instead of only three or four terms of the series being available, the number of terms can be extended to 10 or 15. It is interesting that attention has been focused largely on the analogue (5.49) of the reducible cluster integral series. For example, the $g_l(z)$ which correspond to the b_l have a simple characteristic pattern of behaviour

(Gaunt 1978). It is difficult to find any corresponding pattern for the β_k series, although the restriction to multiply connected graphs produces greater benefits on a lattice than in a continuum, and the series has been exploited for this purpose (Domb 1974, Ch. 1).

Another advantage of the lattice gas model is the possibility of studying different lattices in a given dimension, and hence distinguishing between properties specific to a particular lattice and properties which are universal. The latter would then be expected to apply equally to a continuum model.

5.6.3 Surface Tension and the Interface

In §2.7.1 we discussed van der Waals' attempt to account for the critical behaviour of the interface between a liquid and a gas. His theory led to the conclusion that the surface tension σ vanishes at T_c as $\tau'^{3/2}$ (equation (2.146)), the width, l, of the interfacial boundary becomes infinite as $\tau'^{-1/2}$ (equation (2.149)) and the interfacial density profile, r, has the form given by (2.148).

For the two-dimensional Ising model Onsager showed that the boundary tension vanishes as τ' (5.30′); and whilst exact calculations were not available for the behaviour of l, it could reasonably be associated with the correlation length which behaves as τ'^{-1}.

In 1965 Widom pointed out that the main defect of the van der Waals theory of the interface was its adoption of classical values for the critical exponents from the van der Waals equation of state. Its basic ideas were essentially correct, and could be taken over and used in conjunction with the newly emerging values of critical exponents to provide a revised theory of the interface. The arguments were given a more refined expression in a subsequent review article (Widom 1972) on which the following discussion is based.

We first introduce an exponent μ to characterize the critical behaviour of surface tension,

$$\sigma \sim \tau'^{\mu} \tag{5.204}$$

and a second exponent ν' to characterize the critical behaviour of the interfacial thickness

$$l \sim \tau'^{-\nu'}. \tag{5.205}$$

However, guided by the suggestion that there is only one length of significance in the critical region we conjecture that (Hart 1961, Widom 1965) interface thickness $\sim$ correlation length, so that ν' is in fact the low-temperature correlation exponent.

The value of the exponent μ can be derived from the following consideration. The surface tension σ represents the interfacial free energy per unit of surface; but every unit of surface is associated with a volume l of fluid. Hence the free-energy density in the interface is σl^{-1}. But if an additional conjecture is made that there is only one free energy of significance for critical behaviour, we can identify the singularity of the interfacial free-energy density with the singularity of the bulk free-energy density, giving

$$\sigma l^{-1} \sim \tau'^{\mu+\nu} \sim \tau'^{2-\alpha'} \tag{5.206}$$

so that

$$\mu + \nu' \sim 2 - \alpha'. \tag{5.207}$$

Relation (5.207) is correct for the two-dimensional Ising model $I(\frac{1}{2})$ ($\mu = \nu' = 1$, $\alpha = 0$). For the three-dimensional model, assuming that the exponents are symmetric as discussed in §5.3.3., and taking $\alpha \simeq \frac{1}{8}$, $\nu \simeq 0.643$, a value of ~ 1.23 was obtained for μ, and this was in fair agreement with experimental observation.

Fisk and Widom (1969) went on to calculate the density profile using the ideas of van der Waals. Basically they adopted the square gradient hypothesis of van der Waals, (2.129), (2.131), and wrote for the free energy of the inhomogeneous layer

$$\Psi(z) = \psi[\rho(z)] + \tfrac{1}{2}B[\rho'(z)]^2. \tag{5.208}$$

But an important new feature which they introduced was the freedom of coefficient B to be singular at T_c.

In §4.9.1 the square gradient hypothesis was justified by expanding the free energy of a fluid with fluctuations (equation (4.140)); but just as the coefficient c is singular (zero) at T_c, so one may envisage f being singular (infinite) at T_c. In fact Fisk and Widom found that

$$B \sim \tau'^{\gamma-2\nu} = \tau'^{-\nu\eta} \tag{5.209}$$

using (5.164). For the two- and three-dimensional Ising models $I(\frac{1}{2})$ with $\eta \neq 0$, B is indeed infinite at T_c.

In order to develop the calculation Fisk and Widom needed to devise a procedure for estimating $\psi[\rho(z)]$, the free energy for densities between ρ_l and ρ_g, to replace van der Waals' use of the metastable and unstable branches of his equation. We defer further discussion until the non-classical equation of state has been considered in the next chapter (§6.7).

5.7 Experimental Results

The experimental investigation of the behaviour of a substance near a critical point is a matter of considerable difficulty. Very accurate temperature measurements are needed since small changes of temperature are accompanied by large differences in equilibrium values. Substances must be very pure; a small concentration of impurity can exercise a significant influence. As T_c is approached the time needed to reach equilibrium becomes large, so that a long time-scale is needed for the performance of experiments. We have already noted in §5.6.1 that for fluids gravity must be taken into account. The theoretical values of several of the thermodynamic properties investigated become infinite at T_c. But such an infinity applies only to an unbounded sample; in practice the finite size of the sample becomes important since it provides a bound on the values which can be obtained. In solids this involves a consideration of grain boundaries.

A comprehensive survey of early experimental investigations on fluids during the lifetime of van der Waals has been given by Levelt-Sengers (1976). In 1893 capillarity experiments were undertaken independently by de Vries in Leiden and Ramsay and Shields in Britain, from which the surface tension exponent, μ (§5.6.3), was estimated. De Vries derived six data points on the capillary rise and surface tension of ether in the temperature range -102 to $+183\,°C$ ($T_c = 193.6\,°C$). The data were analyzed by van der Waals whose own theory gave a value of $\frac{3}{2}$ for μ in contrast to 1 given by the older theories of Gauss or Laplace. Van der Waals remarked as follows:

> The result of experiment although not quite in agreement with $k(1 - T/T_c)^{3/2}$ is in conflict with the older theories. If one tries to calculate the value of the exponent, one finds, at some distance from the critical point, the value 1.23. However, the values increase on approaching T_c and seem to confirm the proposition that the limiting value may be equal to 3/2.

The more extensive data of the British workers on nine organic fluids were subjected to a similar analysis by van der Waals and yielded comparable results. He again expressed the hope that in the immediate vicinity of T_c the exponent might revert to 3/2. It is amazing that this early experimental estimate of μ is virtually identical with that derived theoretically by Widom more than 70 years later (§5.6.3).

In 1894 Verschaeffelt, a young Belgian scientist, joined the staff of Kamerlingh Onnes at the Leiden Laboratory, and at first continued the work on surface tension for CO_2 and N_2O. He concluded that, without the use of accurate measured data of the density near T_c, it would not be possible to decide whether the exponents revert to their classical values or not. A few years previously Amagat (1892) had published new data on CO_2 with measurements of coexisting densities up to 0.35 °C below T_c. Verschaeffelt (1896) analyzed these data *from first principles* for a power-law dependence without any bias towards a parabolic law. He obtained a value for the exponent β of about 0.367.

When accurate data for isopentane became available from the laboratory of S. Young (1894), Verschaeffelt (1900) undertook a similar analysis, and concluded that the entire body of data was well fitted by a power law with exponent $\beta = 0.3434$; between 1.8 and 0.8 °C from T_c, the exponent value was 0.344, and between 0.8 and 0.4 °C it was 0.337. Thus the exponent had a steady non-classical value with no tendency to grow on approaching T_c.

Having found a non-classical value for the coexistence curve, Verschaeffelt proceeded to examine the critical isotherm to see if the cubic curve of van der

Waals needed modification. An investigation of the data of Young on the critical isotherm of isopentane led to the suggestion that $\delta = 4.259$.

Unfortunately, Verschaeffelt's ideas and conclusions made virtually no impact. Kamerlingh Onnes does not seem to have been impressed by his results, and gave him little encouragement to continue his work. Possible reasons for this negative reaction are discussed by Levelt Sengers; 45 years elapsed before the non-classical values of the critical exponents were discussed again. As we have noted in §1.5, Guggenheim (1945) suggested an estimate of about 1/3 for β, and Habgood and Schneider (1954) about 4.2 for δ.

The investigations discussed so far had nothing to do with Onsager, and arose from a desire to test the validity of the van der Waals equation near T_c. By contrast the experiments to be discussed in the rest of this section were all stimulated by the new theoretical developments. We have already referred in §1.5 to some of these experiments, notably to those of Fairbank, Buckingham and Kellers (1957) on the λ-point of liquid helium, and of Heller and Benedek (1965) on the spontaneous magnetization of europium sulfide.

The literature of experimental work during this period is extensive, and the reader interested in pursuing details is referred to review articles (Green and Sengers 1965, Heller 1967, Kadanoff *et al.* 1967; for magnets alone see Domb and Miedema (1964)). We shall give details of a few experiments which played a key role in promoting the emerging ideas on critical behaviour.

Voronel in the USSR was deeply impressed with the experiments on the λ-point of liquid helium. But helium is a quantum fluid – would a corresponding result be obtained for a classical fluid to which van der Waals' theory had usually been applied? After spending a good deal of time and effort designing a suitable apparatus Voronel and his collaborators measured the specific heat of argon along its critical isochore (Bagataskii, Voronel and Gusak 1962). The purity of the argon was 99.99%, and 5–6 hours were allowed for the system to attain equilibrium, continuous stirring being maintained during this period. The resulting specific heat seemed clearly to be moving to infinity. A corresponding experiment was undertaken for oxygen (Voronel *et al.* 1963) with a comparable result (reproduced in Figure 5.25) Voronel plotted his results against $\ln|T - T_c|$ and obtained a relationship which seemed fairly linear. Theory predicted a small positive power, but the experimental results were not sufficiently accurate to distinguish between this and a logarithm.

It is difficult to find a magnet which conforms satisfactorily to the Ising interaction. However, we have seen in §4.7 and §4.8 that an order–disorder transition in a binary alloy can also provide a realization of an Ising model. β-brass (CuZn) was chosen by Als-Nielsen and Dietrich (1966, 1967) for neutron scattering experiments to test the theoretical predictions for the three-dimensional Ising model $I(\frac{1}{2})$ (§4.8.4). Since the scattering powers of the Cu and Zn atoms are different, superlattice peaks appear in neutron scattering below T_c, and diffuse critical scattering is observed in the vicinity of these peaks.

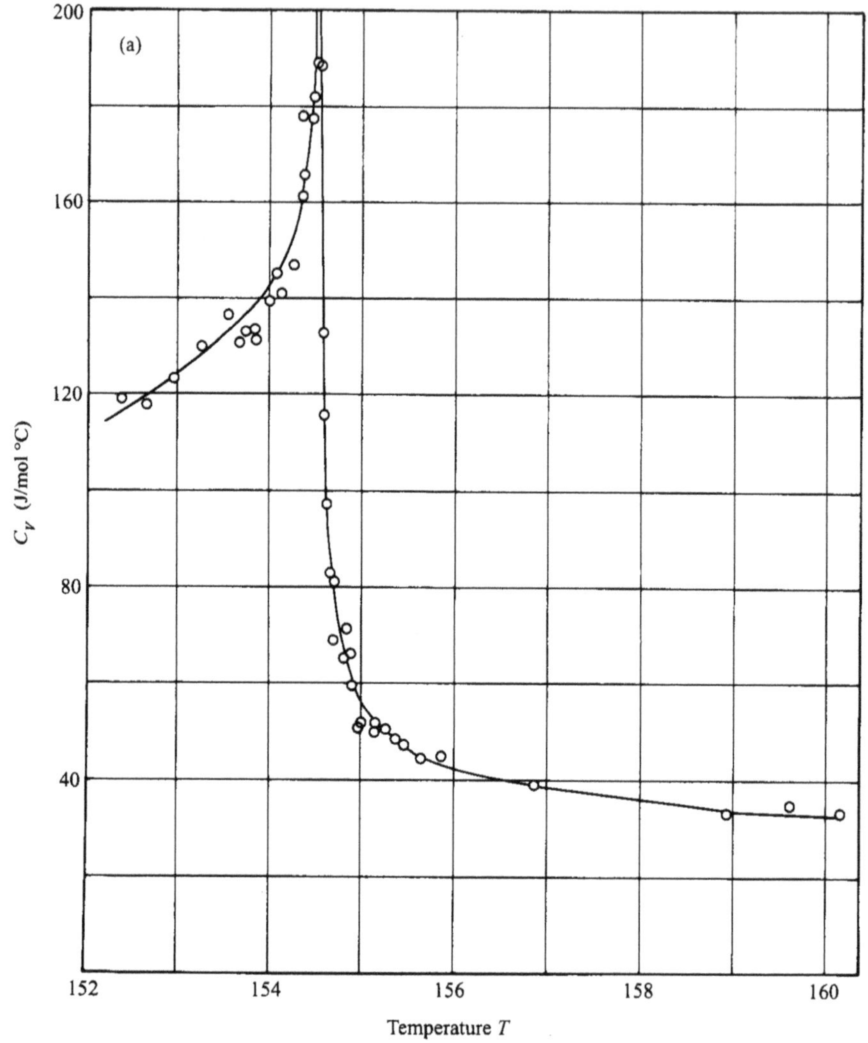

Figure 5.25 (a) Temperature dependence of C_V of oxygen at $\rho \sim \rho_c = 0.408$ g/cm^3; (b) (opposite) dependence of C_V of oxygen on $\ln|T - T_c|$.

From the square root of the superlattice peak intensities below T_c, the temperature dependence of the long-range order parameter, $t(\infty)$ (equation (4.100)), was found to be

$$t(\infty) \sim \tau'^{\beta}, \quad \beta = 0.305 \pm 0.005. \tag{5.210}$$

The behaviour of the diffuse scattering at the position of the superlattice peak from $\tau = 0.003$ to $\tau = 0.03$ was also investigated. This provides informa-

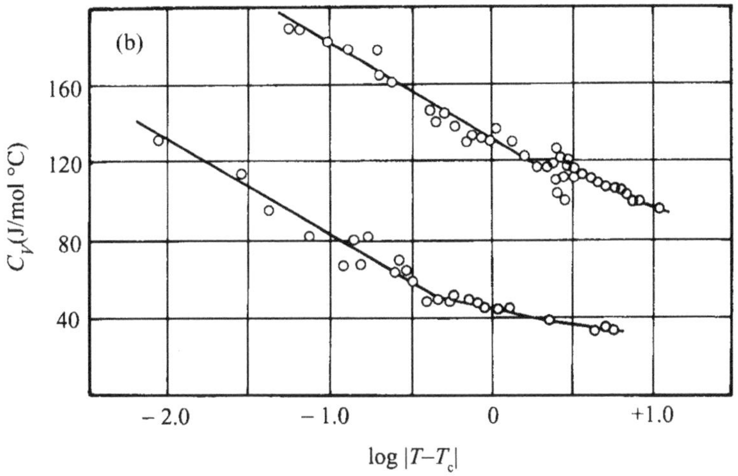

FIG 5.25 (b) *(Continued)*

tion on the temperature dependence of the analogue of the staggered susceptibility (the susceptibility of an antiferromagnet in a field which alternates positively and negatively from lattice site to lattice site). The results gave the estimate

$$\gamma = 1.25 \pm 0.02. \tag{5.211}$$

Finally, the angular dependence of the scattering above T_c was investigated in detail, and from this the inverse correlation range κ, and its associated exponent ν, could be estimated. The result obtained was

$$\nu = 0.65 \pm 0.02. \tag{5.212}$$

All three results, (5.210), (5.211) and (5.212), were consistent with the theoretical predictions for $I(\frac{1}{2})$ in three dimensions, and gave added confidence to the validity of these estimates.

Finally, we shall refer to an experiment which attempted to establish a non-zero value for η in three dimensions. We have seen from (5.180) and (5.180′) that the theoretical prediction of η for $I(\frac{1}{2})$ in three dimensions is small, and for normal experimental purposes it can be ignored. However, Fisher emphasized its theoretical significance, and tried to persuade experimentalists to look for a small downward curvature in the O–Z plots as the temperature approaches T_c. The experimental difficulties in performing a high-precision light scattering experiment are formidable (see Heller 1967). Nevertheless, Chu (1966) chose the mixture n-dodecane-β,β'-dichloroethyl ether as one in which undesirable effects like multiple scattering are minimal. The results of his experiment are reproduced in Figure 5.26 and do show the downward curvature indicative of a non-zero value of η. Despite this positive result the literature of the period

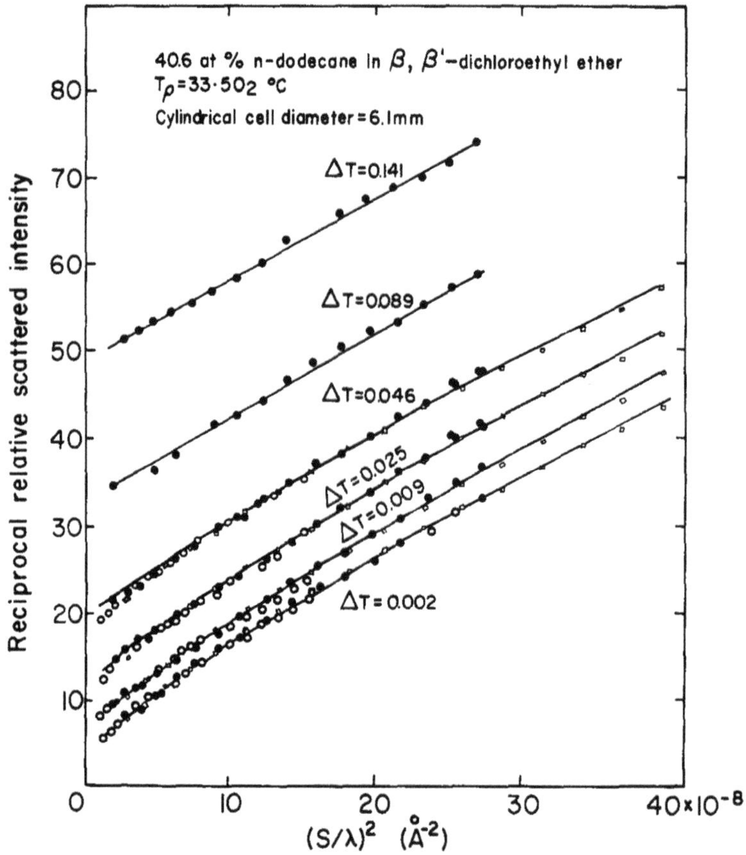

Figure 5.26 Plot of reciprocal relative scattered intensity vs. $(s/\lambda)^2$.

continued to harbour suggestions that η is exactly zero in three dimensions. Fisher's ideas were vindicated with the advent of the renormalization group (Chapter 7) which showed that η attains its classical value (zero) in four dimensions, like other critical exponents. Its small value arises from the fact that the deviation from its classical value is of second order in the parameter ϵ $(= 4 - d)$, whereas the deviations of other critical exponents are first order.

5.8 Conclusions

In this chapter we have shown how the exact calculations by Onsager (1944) and Kaufman and Onsager (1949) on the two-dimensional Ising model $I(\frac{1}{2})$ stimulated a new approach to the investigation of critical behaviour. Methods of calculation were devised which enabled reliable predictions to be made of the critical properties of a variety of magnetic models. Experiments were undertaken on systems which could reasonably be represented by these models, and satisfactory agreement was obtained between theory and experiment.

However, each critical property had to be treated individually, and there was nothing comparable with the coherence of the classical theories in which the whole pattern of critical behaviour stemmed from one basic assumption.

We referred in §1.6 to the Conference on Critical Phenomena at the National Bureau of Standards, Washington, in April 1965, at which many of the new theoretical calculations and experimental measurements were discussed; we have quoted liberally in this chapter from the proceedings of the conference. We also mentioned in §1.6 the remarks of Uhlenbeck in his keynote address on the need to reconcile Onsager with van der Waals.

In the next chapter we shall discuss how this reconciliation began shortly after the conference when an empirical equation of state was postulated from which the critical behaviour of a large variety of models could be derived. In addition to unifying the diverse calculations of this chapter, the new equation of state led naturally to the concept of scaling, and prepared the background for the hypothesis of universality. Scaling and universality play a central role in modern theories of the critical point.

Notation for Chapter 5

See Chapter 4 for preliminary notation of Ising model.
Ising nets: N number of spins, L number of links.
$\nu(r)$ number of low-temperature configurations with r broken links.
$p(r)$ number of closed polygons which can be constructed with r links of the lattice.
$J, J', J'' \ldots$ interaction energies in different lattice directions.
$K = \beta J$, $w = \tanh K$, $\tau = \tanh(\beta\mu_0 H)$, N^*, L^*, K^* relate to dual net.
w_c, K_c critical values.

Quantities related to Onsager's exact solution

$\lambda_1, \lambda_2, \ldots, \lambda_r, \ldots$ eigenvalues of transfer matrix $\mathbf{V}$.
$\gamma_1, \gamma_2, \ldots, \gamma_{2n}$ defined in (5.26) in terms of which the λ_i can be simply expressed.
$\lambda_b \lambda'_b$, p.f.s for boundaries; σ, σ' boundary tensions.

Series expansions

$F_r(N,z)$ defined by (5.42), $f_r(z)$ by (5.45), $g_r(z)$ by (5.49).
$\Psi_r(N,\tau)$ by (5.58).
$\theta(\infty)$ long-range order, $\eta(z)$ mean fraction of overturned spins, related by (5.65).

Linked cluster expansions

$h(i) = \beta\mu_0 H_i$, H_i field at site i.
$v(ij) = \beta J_{ij}$, J_{ij} interaction between spins at sites i, j.
$\mu(i) = s_{zi}/s$.
$K_n(1, 2, \ldots, n)$ = n-point cumulant, $K_1(1)$ magnetization, $K_2(12)$ pair correlation.
Padé approximant $[L,M] = P_L(z)/Q_M(z)$.
$\chi^{(a)}$ susceptibility of antiferromagnet.
$\omega_{lm}(T) = \langle \sigma_{\infty} \sigma_{lm} \rangle$ for SQ lattice.
$L = \beta\mu_0 H$ (spherical and Gaussian models).
Long-range interaction $J(r) \sim r^{-(d+\sigma)}$.
$\mu_t(w)$ moments of pair correlation function defined by (5.173).
$\Lambda_2(w), \Lambda_4(w), \ldots$ defined by (5.175); $L_2(w), L_4(w), \ldots$ by (5.176).
κ_1 defined in (5.182), $K(Q)$ in (5.183).
(Q is used here for coordination number of lattice.)

Fluids

$B(T), C(T), D(T), \ldots$ virial coefficients (see (5.185)).
$\phi = \phi(|\mathbf{r}_{ij}|)$ intermolecular potential, $f_{ij} = \exp[-\beta\phi(r_{ij})] - 1$.
β_k irreducible cluster integrals (see (5.187) and (5.189).
b_l reducible cluster integrals.
Surface tension critical exponents μ, ν' defined in (5.204) and (5.205).

References

Note As stated earlier, DG is used as an abbreviation for the Domb–Green series *Phase Transitions and Critical Phenomena* published by Academic Press, London and New York. DL is used for its continuation by Domb–Lebowitz.

Als-Nielsen, J. and Dietrich, O. W. (1966) *Critical Phenomena*, ed. M. S. Green and J. V. Sengers, NBS Miscellaneous Publication 273, Washington, DC: NBS, p. 144.

Als-Nielsen, J. and Dietrich, O. W. (1967) *Phys. Rev.* **153**, 706, 711.

Amagat, E. H. (1892) *J. Phys.* **3**, (1), 288.

Bagataskii, M. I., Voronel, A. V. and Gusak, V. G. (1962) *Zh. Eksp, Teor. Fiz.* **43**, 728 (English translation (1963) *Sov. Phys. JETP* **18**, 517).

Baker, G. A. (1961) *Phys. Rev.* **124**, 768.

Baker, G. A. (1963) *Phys. Rev.* **130**, 1406.

Baker, G. A. (1965) *Adv. Theor. Phys.* **1**, 1.

Baker, G. A., Gammel, J. L. and Wills, J. G. (1961) *J. Math. Anal. Appl.* **2**, 405.

Baxter, R. J. (1982) *Exactly Solved Models in Statistical Mechanics*, London and New York: Academic Press.

Berlin, T. H. and Kac, M. (1952) *Phys. Rev.* **86**, 821.

Bitter, F. (1937) *Phys. Rev.* **54**, 79.

BLOCH, F. (1930) *Z. Phys.* **61**, 206.
BOLTZMANN, L. (1899) *Verlag. Gewone Vergadering Natuurk. Ned. Akad. Wet.* **7**, 484.
BORN, M. and FUCHS, K. (1938) *Proc. R. Soc. A* **166**, 391.
BRAUER, R. and WEYL, H. (1935) *Am. J. Math.* **57**, 425. (See also MURNAGHAN, F. D. (1938) *The Theory of Group Representations*, Baltimore, MD: Johns Hopkins University Press, Ch. 10.)
BROOKS, J. E. and DOMB, C. (1951) *Proc. R. Soc. A* **207**, 343.
BROUT, R. (1959) *Phys. Rev.* **115**, 824.
BRUSH, S. G. (1967) *Rev. Mod. Phys.* **39**, 883.
CAMP, W. J. (1972) *Phys. Rev. B* **6**, 960.
CAMP, W. J. (1973) *Phys. Rev. B* **7**, 3187.
CERMUSCHI, F. and EYRING, H. (1939) *J. Chem. Phys.* **7**, 547.
CHENG, H. and WU, T. T. (1967) *Phys. Rev.* **164**, 719.
CHU, B. (1966) *Critical Phenomena*, ed. M. S. GREEN and J. V. SENGERS, NBS Miscellaneous Publication, 273, Washington, DC: NBS, p. 123.
DALTON, N. W. and WOOD, D. W. (1967) *Proc. Phys. Soc.* **90**, 459.
DE VRIES, E. C. (1893) *Versl. Kon. Akad. Werensch. Amsterdam* **2** (Feb. 25), 155.
DIENES, P. (1931) *The Taylor Series*, Oxford, reprinted Dover, New York, 1987).
DOMB, C. (1949a) *Proc. R. Soc. A* **196**, 36.
DOMB, C. (1949b) *Proc. R. Soc. A* **199**, 199.
DOMB, C. (1960) *Adv. Phys.* **9**, 149, 171, 173–4, 211–20, 227, 245.
DOMB, C. (1985) *The Wonderful World of Stochastics*, ed. M. F. SHLESINGER and G. H. WEISS, Ch. 2, Amsterdam: North-Holland.
DOMB, C. and GREEN M. S. (1974) DG **3**.
DOMB, C. and MIEDEMA, A. R. (1964) *Progress in Low Temperature Physics*, ed. C. J. GORTER, Amsterdam: North-Holland.
DOMB, C. and SYKES, M. F. (1957a) *Proc. R. Soc. A* **240**, 214.
DOMB, C. and SYKES, M. F. (1957b) *Phys. Rev.* **108**, 1415.
DOMB, C. and SYKES, M. F. (1961) *J. Math. Phys.* **2**, 61.
DOMB, C. and SYKES, M. F. (1962) *Phys. Rev.* **128**, 168.
DYSON, F. J. (1956) *Phys. Rev.* **102**, 1217, 1230.
ENGLERT, F. (1963) *Phys. Rev.* **129**, 567.
ESSAM, J. W. and FISHER, M. E. (1963) *J. Chem. Phys.* **38**, 802.
ESSAM, J. W. and HUNTER, D. L. (1968) *J. Phys. C: Solid State Phys.* **1**, 392.
ESSAM, J. W. and SYKES, M. F. (1963) *Physica*, **29**, 378.
FAIRBANK, W. M., BUCKINGHAM, M. J. and KELLERS, C. F. (1957) *Proc. 5th Int. Conf. on Low Temperature Physics*, University of Wisconsin Press.
FERER, M., MOORE, M. A. and WORTIS, M. (1969) *Phys. Rev. Lett.* **22**, 1382.
FISHER, M. E. (1959) *Physica*, **25**, 521.
FISHER, M. E. (1962) *Philos. Mag.* **7**, 1732.
FISHER, M. E. (1964) *J. Math. Phys.* **5**, 944.
FISHER, M. E. (1965) *Critical Phenomena*, ed. M. S. GREEN and J. V. SENGERS, NBS Miscellaneous Publication 273, Washington, DC: NBS, pp. 22–3.
FISHER, M. E. (1966) *Phys. Rev. Lett.* **16**, 11.
FISHER, M. E. (1967) *Rep. Prog. Phys.* **30**, 615.
FISHER, M. E. and BURFORD, R. J. (1967) *Phys. Rev.* **156**, 583.
FISHER, M. E. and CAMP, W. J. (1971) *Phys. Rev. Lett.* **26**, 73.
FISHER, M. E. and SYKES, M. F. (1962) *Physica* **28**, 939.

FISK, S. and WIDOM, B. (1969) *J. Chem. Phys.* **50**, 3219.
FOX, P. F. and GAUNT, D. S. (1972) *J. Phys. C: Solid State Phys.* **5**, 3085.
FOX, P. F. and GUTTMAN, A. J. (1970) *Phys. Lett.* **31A**, 234.
FOX, P. F. and GUTTMAN, A. J. (1973) *J. Phys. C: Solid State Phys.* **6**, 913.
FROHLICH, J. and SPENCER, T. (1982) *Commun. Math. Phys.* **84**, 87.
GAUNT, D. S. (1978) *J. Phys. A: Math. Gen.* **11**, 1991.
GAUNT, D. S. and GUTTMAN, A. J. (1974) DG **6**, Ch. 4.
GAUNT, D. S. and SYKES, M. F. (1972) *J. Phys. C: Solid State Phys.* **5**, 1429.
GAUNT, D. S. and SYKES, M. F. (1973) *J. Phys. A: Math. Gen.* **6**, 1517.
GAUNT, D. S., FISHER, M. E., SYKES, M. F. and ESSAM, J. W. (1964) *Phys. Rev. Lett.* **13**, 713.
GOODWIN, E. T. (1949) *Proc. Cambridge Philos. Soc.* **45**, 241.
GOUY, A. (1892) *Comptes Rendu* **115**, 720.
GREEN, M. S. and SENGERS, J. V. (1965) *Critical Phenomena*, ed. M. S. GREEN and J. V. SENGERS, NBS Miscellaneous Publication 273, Washington, DC: NBS.
GUGGENHEIM, E. A. (1945) *J. Chem. Phys.* **13**, 253.
GUTTMAN, A. J. (1989) DL **13**, Ch. 1.
GUTTMAN, A. J., DOMB, C. and FOX, P. F. (1971) *Proc. 1970 Grenoble Conf. on Magnetism, J. Phys.* **32**, 354.
HABGOOD, H. W. and SCHNEIDER, W. G. (1954) *Can. J. Chem.* **32**, 98.
HART, E. W. (1961) *J. Chem. Phys.* **34**, 1471.
HEISENBERG, W. (1928) *Z. Phys.* **49**, 619.
HELLER, P. (1967) *Rep. Prog. Phys.* **30**, 731.
HELLER, P. and BENEDEK, G. B. (1965) *Phys. Rev. Lett.* **14**, 71.
HEMMER, P. C. and LEBOWITZ, J. L. (1976) DG **5b**, Ch. 2.
HORWITZ, G. and CALLEN, H. B. (1961) *Phys. Rev.* **124**, 1757.
HOUTAPPEL, R. M. F. (1950) *Physica* **16**, 425.
HUSIMI, K. and SYOZI, I. (1950) *Prog. Theor. Phys.* **5**, 177, 341.
JASNOW, D. and WORTIS, M. (1968) *Phys. Rev.* **176**, 739.
JOYCE, G. S. (1966) *Phys. Rev.* **146**, 349.
JOYCE, G. S. (1972) DG **2**, Ch. 10.
KAC, M. (1959) *Phys. Fluids* **2**, 8.
KAC, M. (1964) *Phys. Today* **17**, 40.
KAC, M. and HELFAND, E. (1963) *J. Math. Phys.* **4**, 1078.
KAC, M. and THOMPSON, C. J. (1969) *J. Math. Phys.* **10**, 1373.
KAC, M., UHLENBECK, G. E. and HEMMER, P. C. (1963) *J. Math. Phys.* **4**, 216.
KADANOFF, L. P. (1966) *Nuovo Cimento B* **44**, 276.
KADANOFF, L. P., GOTZE, W., HAMBLEN, D., HECHT, R., LEWIS, E. A. S., PALCIAUSKAS, V. V., RAYL, M. and SWIFT, J. (1967) *Rev. Mod. Phys.* **39**, 395.
KAMERLINGH ONNES, H. (1901) *Versl. Kon. Akad. Werensch. Amsterdam* **10** (June 29), 136 (*Commun. Phys. Lab. Leiden* No. 71).
KAUFMAN, B. (1949) *Phys. Rev.* **76**, 1232.
KAUFMAN, B. and ONSAGER, L. (1949) *Phys. Rev.* **76**, 1244.
KIRKWOOD, J. G. (1950) *J. Chem. Phys.* **18**, 3890.
KOSTERLITZ, J. M. and THOULESS, D. J. (1973) *J. Phys. C: Solid State Phys.* **6**, 1181.

KRAMERS, H. A. and WANNIER, G. H. (1941) *Phys. Rev.* **60**, 252, 263.
LEVELT-SENGERS, J. M. H. (1976) *Physica* **82A**, 319.
LIEB, E. (1967) *Phys. Rev. Lett.* **18**, 692, 1046; **19**, 108; *Phys. Rev.* **162**, 162.
MAASS, O. (1938) *Chem. Rev.* **23**, 17.
MARSHALL, W., GAMMEL, J. and MORGAN, L. (1963) *Proc. R. Soc. A* **275**, 257.
MAYER, J. E. (1937) *J. Chem. Phys.* **5**, 67.
MAYER, J. E. and MAYER, M. G. (1940) *Statistical Mechanics*, New York: John Wiley.
MERMIN, N. D. and WAGNER, H. (1966) *Phys. Rev. Lett.* **17**, 1133.
MILLS, R. E., ASCHER, E. and JAFFEE, R. F. (1971) *Critical Phenomena in Alloys, Magnets and Superconductors*, New York: McGraw-Hill.
MONTROLL, E. W. and WEISS, G. H. (1965) *J. Math. Phys.* **6**, 167.
MOORE, M. A. (1969) *Phys. Rev. Lett.* **23**, 861.
MOORE, M. A., JASNOW, D. and WORTIS, M. (1969) *Phys. Rev. Lett.* **22**, 940.
NELSON, D. R. (1983) DL **7**, Ch. 1.
NEWELL, G. F. (1950) *Phys. Rev.* **79**, 876.
NINHAM, B. W. (1963) *J. Math. Phys.* **4**, 679.
ONSAGER, L. (1944) *Phys. Rev.* **65**, 117.
ONSAGER, L. (1949) *Nuovo Cimento (Suppl.)* **6**, 261.
OPECHOWSKI, W. (1937) *Physica* **4**, 181.
PADÉ, H. (1892) *Thesis, Ann. Sci. de l'Ecole, Norm. Sup. Suppl. (3)* **9**, 1.
RAMSAY, W. and SHIELDS, J. (1893) *Z. Phys. Chem.* **12**, 433.
ROWLINSON, J. S. and WIDOM, B. (1982) *Molecular Theory of Capillarity*, Oxford.
RUSHBROOKE, G. S. (1963) *J. Chem. Phys.* **39**, 8942.
RUSHBROOKE, G. S. and WOOD, P. J. (1958) *Mol. Phys.* **1**, 257.
RUSHBROOKE, G. S., BAKER, G. A. and WOOD, P. J. (1974) DG **3**, Ch. 5.
SAUL, D. M., WORTIS, M. and JASNOW, D. (1975) *Phys. Rev. B* **11**, 7, 2541.
SCHNEIDER, W. G. (1952) *Changements de Phases, CR de la 2me Reun. Annu., Soc. Chim. Phys.*, p. 69.
SCHULTZ, T. D., MATTIS, D. C. and LIEB, E. (1964) *Rev. Mod. Phys.* **36**, 856.
SHEDLOVSKY, T. and MONTROLL, E. W. (1963) *J. Math. Phys.* **4**, (unnumbered pages after p. 146).
STANLEY, H. E. (1968a) *Phys. Rev. Lett.* **20**, 150.
STANLEY, H. E. (1968b) *Phys. Rev. Lett.* **20**, 589; *Phys. Rev.* **176**, 718.
STANLEY, H. E. (1974) DG **3**, Ch. 7.
STANLEY, H. E. and KAPLAN, T. A. (1966) *Phys. Rev. Lett.* **17**, 913.
SYKES, M. F. (1961) *J. Math. Phys.* **2**, 52.
SYKES, M. F. and FISHER, M. E. (1958) *Phys. Rev. Lett.* **1**, 321.
SYKES, M. F. and FISHER, M. E. (1962) *Physica* **28**, 919.
SYKES, M. F., GAUNT, D. S., ROBERTS, P. D. and WYLES, J. A. (1972) *J. Phys. A: Math. Gen.* **5**, 624, 640.
SYOZI, I. (1972) DG **1**, Ch. 7.
TEMPERLEY, H. N. V. (1950) *Proc. R. Soc A* **202**, 202.
TEMPERLEY, H. N. V. (1972) DG **1**, Ch. 6.
THOULESS, D. J. (1969) *Phys. Rev.* **187**, 732.
VAN LAAR, J. J. (1899) *Proc. Akad. Sci. Amsterdam* **1**, 273.
VERSCHAEFFELT, J. E. (1896) *Versl. Kon. Akad. Werensch. Amsterdam* **5** (June 27), 94; *Commun. Phys. Lab. Leiden* No. 28.

Verschaeffelt, J. E. (1900) *Versl. Kon. Akad. Werensch. Amsterdam* **8** (March 31), 651; *Proc. Soc. Sci. Kon. Akad. Amsterdam* **2**, 588; *Commun. Phys. Lab. Leiden* No. 55.

Voronel, A. V., Chashkin, Yu. R., Popov, V. A. and Simkin, V. G. (1963) *Zh. Eksp. Teor. Phys.* **45**, 828 (English translation (1964) *Sov. Phys. JETP* **18**, 568).

Wannier, G. H. (1945) *Rev. Mod. Phys.* **17**, 50.

Wannier, G. H. (1950) *Phys. Rev.* **79**, 357.

Watson, P. G. (1972) DG **2**, Ch. 4.

Weinberger, M. A. and Schneider, W. G. (1952) *Can. J. Chem.* **30**, 422.

Widom, B. (1965) *J. Chem. Phys.* **43**, 3898.

Widom, B. (1972) DG **2**, Ch. 3.

Wilson, K. G. (1971) *Phys. Rev. B* **4**, 3174, 3184.

Wilson, K. G. and Fisher, M. E. (1972) *Phys. Rev. Lett.* **28**, 248.

Wu, T. T. (1966) *Phys. Rev.* **149**, 380.

Wu, T. T., McCoy, B. M., Tracy, C. A., Craig, H. and Barouch, E. (1976) *Phys. Rev.* **13**, 316.

Yang, C. N. (1952) *Phys. Rev.* **85**, 808.

Yang, C. N. and Lee, T. D. (1952) *Phys. Rev.* **87**, 404, 410.

Young, S. (1894) *Philos. Mag.* **38**, 569.

Yvon, J. (1937) *Actualités Scientifiques et Industrielles*, No. 542, Paris: Herman.

6

Reconciliation

In this chapter we will describe in more detail the developments summarized in §1.6.

6.1 Relations between Critical Exponents

6.1.1 The Droplet Model

We have already referred in §5.3.3.5 to a paper by Essam and Fisher (1963) on Padé approximant studies of critical exponents for temperatures below T_c. In the final section of this paper they raised the question of possible relations which might exist between critical exponents, and introduced 'with due caution' a heuristic model which they hoped might have 'some suggestive value' in relation to critical behaviour. The model was elaborated by Fisher in a lecture given at the Centennial Conference on Phase Transitions held at the University of Kentucky in 1965, and published in 1967.

The basic idea behind the model was put forward some 25 years previously in independent papers by Bijl (1938), Frenkel (1939) and Band (1939). These authors endeavoured to provide a physical interpretation of the formalism of the Mayer theory of condensation (§5.6.1) and they suggested that the cluster integrals b_l in the g.p.f. expansion correspond to droplets of l molecules. The rigorous theory of Mayer takes account of the volume exclusion of the different droplets, but at low densities it is reasonable to ignore this and the problem can then be treated as one of chemical equilibrium between all the different types of droplet. It is possible to construct an approximate p.f. and to extend the equilibrium theory to the kinetics of growth of droplets (Becker and Döring 1935).

The analogue of the Mayer g.p.f. expansion for the lattice gas or $I(\frac{1}{2})$ model is the density expansion described in §5.2.1.1(i). Starting from the ground state of fully aligned spins, excited states of 1, 2, 3, ..., r, ... overturned spins are considered, and taking the logarithm of the p.f. ensures that only connected clusters survive. The result is (equation (5.49))

$$\ln \Lambda(y,z) = \sum_r y^r g_r(z). \tag{6.1}$$

The detailed calculation of $g_r(z)$ is complex since volume exclusion is taken into account, but if this is ignored, we are led to the following approximation for $g_r(z)$:

$$g_r(z) = \sum_s w(r,s) z^s. \tag{6.2}$$

Here the sum is taken over all different connected clusters of r spins, s is the surface area of the cluster, and $w(r,s)$ is a weight or entropy term representing the number of clusters with a given r,s. For any cluster we must have

$$s = Ar^{\sigma} \tag{6.3}$$

where the minimum value of σ (1/2 in two dimensions, 2/3 in three dimensions) corresponds to the most *compact* cluster, a circle or sphere, whereas the maximum value of σ, 1, relates to *ramified** clusters extending over long distances and perforated with holes. Following standard procedure in statistical mechanics Essam and Fisher assumed that the sum (6.2) can be replaced by its maximum term corresponding to $s = \bar{s}$, and equal to $\bar{A}r^{\bar{\sigma}}$, where it was conjectured that $\bar{\sigma} < 1$. Exact enumerations for loops in two dimensions (Hiley and Sykes 1961) provided evidence for the assumption

$$w(r,s) \sim \lambda^{\bar{s}}/(\bar{s})^{\theta} = \lambda^{\bar{A}r^{\bar{\sigma}}}/r^{\bar{\sigma}\theta} \tag{6.4}$$

where λ is a constant characteristic of the lattice, and θ an exponent characteristic of dimension. This led to the mimic p.f.

$$\ln \Lambda(y,z) = \sum_r y^r z^{r^{\bar{\sigma}}} \lambda^{Ar^{\bar{\sigma}}} r^{-\bar{\sigma}\theta}. \tag{6.5}$$

Essam and Fisher noted the following properties of the mimic function (6.5):

1 It converges for all z when $|y| < 1$ so that there are no singularities except in zero field ($y = 1$). This is a property established by Yang and Lee (1952) for the true p.f. (see §5.6.2).
2 When $y = 1$ there is a singularity at $z_c = 1/\lambda$ which they assumed to correspond to the critical point.
3 Taking successive derivatives of ln Λ w.r.t. magnetic field H, and writing

* This term was first used by Domb (1976) following the suggestion of Sir Charles Frank.

$$\Psi_k(y,z) = (y\partial/\partial y)^k \ln \Lambda(y,z) = \sum_r r^k y^r z^{r\bar{\sigma}} \lambda^{\bar{A}r^{\bar{\sigma}}} r^{-\bar{\sigma}\theta} \tag{6.6}$$

it is easy to see that near the critical point, z_c,

$$\Psi_k(1,z) \sim C_k(1 - z/z_c)^{\phi(k)} \tag{6.7}$$

where C_k is a constant, and

$$\phi(k) = \theta - (k + 1)/\bar{\sigma}. \tag{6.8}$$

The exponent α' of the specific heat (the second derivative with respect to temperature of ln Λ) is related to $\phi(0)$

$$\alpha' = 2 - \phi(0), \tag{6.9}$$

the magnetization exponent β is equal to $\phi(1)$, and the susceptibility exponent, γ', is equal to $-\phi(2)$. Hence

$$\alpha' + 2\beta + \gamma' = 2. \tag{6.10}$$

Also from (6.7) and (6.8) there is a constant 'gap' exponent, Δ', between successive derivatives w.r.t. H of ln Λ, equal to $1/\bar{\sigma}$.

In his subsequent elaboration of the model Fisher (1967) pointed out that the radius of convergence of the y series in (6.5) is equal to 1, so that the function lnΛ (y,z) has a singularity when $y = 1$ ($H = 0$). However, all the derivatives w.r.t. y at $y = 1$ converge when $z < z_c$ (see (6.6)). Hence ln $\Lambda(y,z)$ has an *essential singularity* at $y = 1$, a very weak singularity which could certainly not be detected experimentally, but which prevents the p.f. from having an analytic continuation into the region $y > 1$. (As an example of an essential singularity consider the function $\exp(-1/x)$ at $x = 0$ for which all derivatives on the real axis vanish.)

This led to an important difference from the classical model (Chapter 2) and Gibbs' interpretation (§2.2) in relation to metastable states. For the van der Waals model we have seen that there is a unique analytic continuation of the equation of state into the metastable region (§2.3.1). According to Fisher this arises only because of the long-range character of the intermolecular forces which are associated with this model (§5.4.3). For a short-range force model the essential singularity prevents any analytic continuation, and there can be no unique metastable state. Metastability arises when droplets greater than a specified size are prevented from forming, and the shape of the metastable extension of the equation of the state will depend on where the cut-off occurs.

There are a number of defects in the above treatment. Whilst it is a reasonable approximation model at low temperatures it cannot be regarded as adequate near the critical point where excluded volume effects are of major importance. The p.f. (6.5) has a line of singularities *above* T_c which is certainly incorrect (Gaunt and Baker 1970) (Figure 5.13). When numerical estimates of the critical exponents in three dimensions are substituted, $\bar{\sigma}$ has the value 0.640 which is geometrically impossible.

Nevertheless the important suggestion regarding the essential singularity at $y = 1$ is supported by a rigorous theorem of Lanford and Ruelle (1969). These authors showed that for intermolecular forces of finite ranges if the *stable* thermodynamic state is not analytic at a given point (as is the case in a first-order transition) then the multiple correlation functions cannot all be analytic at this point. We should then expect a singularity in one of the multiple correlation functions to be reflected by a singularity in the p.f.

Despite its limitations the droplet model gave a clear indication that there might be exact relations between critical exponents, and suggested that these exponents might depend on only two parameters $\bar{\sigma},\theta$. It was thus the precursor of the more exact scaling theories to be considered shortly.

6.1.2 Thermodynamic Restrictions on Critical Exponents

The paper by Essam and Fisher (1963) which we have just discussed led to an important application of thermodynamics to critical exponents by Rushbrooke (1963). The standard thermodynamic formula for the difference between C_P and C_V (Pippard 1959, p. 61) when the work term $P\,dV$ is replaced by the work term $-H\,dM$ becomes

$$C_H - C_M = [T(\partial M/\partial T)_H^2]/(\partial M/\partial H)_T . \tag{6.11}$$

Rushbrooke felt that this formula might be usefully applied in the critical region, and the investigation by Essam and Fisher suggested that an appropriate application could be for $H \to 0^+$ and $T \to T_c^-$. One would then have

$$M(T) \sim (T_c - T)^\beta \quad (\partial M/\partial H)_T \sim (T_c - T)^{-\gamma'}. \tag{6.12}$$

Hence the right-hand side of (6.11) behaves like $(T_c - T)^{-\epsilon}$, where

$$\epsilon = 2 - 2\beta - \gamma'. \tag{6.13}$$

If $\epsilon \leq 0$, equation (6.11) tells nothing about C_H except that C_H and C_M have equal singularities. On the other hand if $\epsilon > 0$, then since C_M is necessarily positive being essentially a mean-square temperature fluctuation (cf. (4.8)), we can conclude that C_H has a singularity at T_c at least as strong as $(T_c - T)^{-\epsilon}$. Since C_H behaves like $(T_c - T)^{-\alpha'}$, we can conclude

$$\alpha' \geq 2 - 2\beta - \gamma' \quad \alpha' + 2\beta + \gamma' \geq 2. \tag{6.14}$$

As we have already mentioned in §5.3.3.5, Rushbrooke used this rigorous thermodynamic inequality to criticize the currently suggested estimates for critical exponents, and these were eventually modified.

Rushbrooke's work stimulated a search for other inequalities of a similar kind, and quite a number were quickly discovered (for more details and references see Stanley (1971, Ch. 4)). The best known of these, due to Griffiths (1965),

$$\alpha' + \beta(1 + \delta) \geq 2 \tag{6.15}$$

was established by using the convexity properties of the free energy $A(T,M)$. For the scaling form of equation of state to be introduced shortly all of these inequalities become equalities.

6.2 Equation of State in the Critical Region

Of the three independent attempts to generalize the classical equation of state described in §1.6, two were concerned with magnets (Domb and Hunter 1965, Patashinskii and Pokrovskii 1966) and one with fluids (Widom 1965). In our discussion of the classical theories in Chapters 2 and 3 more attention was devoted to fluids than to magnets for historical reasons. Now, however, we shall find it more convenient to develop the new ideas in magnetic language, and then adapt it to fluids. First the symmetry of the free energy in relation to reversal of magnetic field or magnetization is exact for magnetic systems, whilst the corresponding symmetry is only approximate for fluids. Second the interacting magnetic spin systems offer a variety of different possible internal symmetries (Ising, Heisenberg, x–y, etc.), and the study of the critical behaviour of the many different possible models led to important progress in relation to the concept of universality.

6.2.1 *Magnets*

The basic idea of Domb and Hunter was to use a high-temperature expansion for the $I(\frac{1}{2})$ model as a function of magnetic field to determine the equation of state. Equation (5.58) can be written in the form

$$\ln Z(H,T) = \ln(2 \cosh \beta\mu_0 H) + \tfrac{1}{2}q \ln(\cosh \beta J) + E_0(T) + H^2E_2(T) + H^4E_4(T) + \cdots. \tag{6.16}$$

Here $Z(H,T)$ represents the p.f. per spin, $E_0(T)$ the free energy in zero field, and $E_2(T)$ the magnetic susceptibility whose critical behaviour had been well established,

$$E_2(T) \sim e_2\tau^{-\gamma} \quad (\tau = T/T_c - 1). \tag{6.17}$$

Domb and Hunter assessed the behaviour near T_c of $E_4(T)$, $E_6(T)$, as follows:

$$E_4(T) \sim e_4\tau^{-\gamma-2\Delta} \qquad E_6(T) \sim e_6\tau^{-\gamma-4\Delta}. \tag{6.18}$$

They assumed that this general pattern of behaviour would be followed by later terms in the series, i.e. that

$$E_{2r}(T) \sim e_2 \tau^{-\gamma - 2(r-1)\Delta}. \tag{6.19}$$

The behaviour was confirmed by a more detailed investigation by Essam and Hunter (1968) who estimated several of the coefficients e_r for two- and three-dimensional models. Δ was termed the *gap* exponent (Domb 1966).

Patashinskii and Pokrovskii also started from an expansion of the form (6.16) but they expressed the functions $E_{2r}(T)$ in terms of multiple spin correlations. Using an empirical picture of critical behaviour they were able to obtain a justification of formula (6.19), and an estimate of Δ in terms of the correlation exponent ν as follows.

Using (4.105) the spin pair correlation function between spins at $\mathbf{r}_1$ and $\mathbf{r}_2$ can be defined as

$$\Gamma(\mathbf{r}_1,\mathbf{r}_2) = \langle \sigma_{\mathbf{r}_1}\sigma_{\mathbf{r}_2}\rangle - \langle \sigma_{\mathbf{r}_1}\rangle\langle \sigma_{\mathbf{r}_2}\rangle \tag{6.20}$$

and from (4.106) and (4.107) we see that the second derivative of $\ln Z(H,T)$ w.r.t. H is equivalent to the sum of $\Gamma(\mathbf{r}_1, \mathbf{r}_2)$ over all pairs of lattice sites $\mathbf{r}_1,\mathbf{r}_2$. Similarly we can define a four-point correlation function $\Gamma(\mathbf{r}_1,\mathbf{r}_2,\mathbf{r}_3,\mathbf{r}_4)$ and show that the fourth derivative of $\ln Z(H,T)$ w.r.t. H is equivalent to the sum of $\Gamma(\mathbf{r}_1,\mathbf{r}_2,\mathbf{r}_3,\mathbf{r}_4)$ over all four lattice sites $\mathbf{r}_1,\mathbf{r}_2,\mathbf{r}_3,\mathbf{r}_4$. The $2n$th derivative of $\ln(H,T)$ will be analogously related to the sum of the $2n$-point correlation functions $\Gamma(\mathbf{r}_1,\mathbf{r}_2,\ldots,\mathbf{r}_{2n})$.

Near the critical point Patashinskii and Pokrovskii envisaged that the system would consist of a number of regions of linear dimension ξ, each of which would have a non-zero magnetic moment (ξ is the coherence length $\sim\tau^{-\nu}$). However, these regions would have on average equal positive and negative orientations so that the average total magnetic moment of the system would be zero. When estimating the multiple correlation function $\Gamma(\mathbf{r}_1,\mathbf{r}_2,\ldots,\mathbf{r}_{2r})$, the points $\mathbf{r}_1,\mathbf{r}_2,\ldots,\mathbf{r}_{2r}$ within each coherent region would contribute a quantity of the order m^{2r} whilst those in different coherent regions would average to zero. The number of coherent regions contained in a volume V would be V/ξ^d in a system of dimension d. Thus they obtained the estimate

$$E_{2r}(T) \sim m^{2r}V/\xi^d. \tag{6.21}$$

But the second derivative corresponds to the susceptibility, so that

$$E_2(T) \sim m^2 V/\xi^d \sim \tau^{-\gamma}. \tag{6.22}$$

Hence the asymptotic behaviour of m^2 is

$$m^2 \sim \tau^{-\gamma-d\nu}. \tag{6.23}$$

Inserting this value in (6.21), a justification is derived for formula (6.19) with

$$2\Delta = \gamma + d\nu. \tag{6.24}$$

Proceeding to the equation of state, (6.16) must be differentiated w.r.t. H to obtain the magnetization, m, and the approximations (6.18) and (6.19) used near T_c. In the critical region we can therefore write

$$m \simeq 2He_2\tau^{-\gamma} + 4H^3e_4\tau^{-\gamma-2\Delta} + 6H^5e_6\tau^{-\gamma-4\Delta} + \cdots$$
$$+ 2rH^{2r+1}e_{2r}\tau^{-\gamma-2r\Delta} + \cdots = H\tau^{-\gamma}X(H^2\tau^{-2\Delta}). \tag{6.25}$$

However, Domb and Hunter noted that for the *mean field* $I(\frac{1}{2})$ model (§3.4.2) the $m(H,T)$ equation of state is implicit (equation (3.44))

$$m/\mu_0 = \tanh(\beta\mu_0 H + t^{-1}m/\mu_0) \quad (t = T/T_c) \tag{6.26}$$

and a series expansion is difficult to derive, whereas the $H(m,T)$ equation of state is explicit (equation (3.46))

$$\beta\mu_0 H = \tanh^{-1}(m/\mu_0) - t^{-1}m/\mu_0 \tag{6.27}$$

and a series expansion is immediate. Moreover, this series expansion can be used for $T < T_c$ to derive a value for the spontaneous magnetization. They therefore decided to invert series (6.25) to obtain

$$H \simeq f_2 m\tau^{\gamma} + f_4 m^3\tau^{3\gamma-2\Delta} + f_6 m^5\tau^{5\gamma-4\Delta} + \cdots$$
$$+ f_{2r} m^{2r-1}\tau^{(2r-1)\gamma-(2r-2)\Delta} + \cdots = m\tau^{\gamma}Y(m^2\tau^{2\gamma-2\Delta}). \tag{6.28}$$

By analogy with (6.27) one would expect to be able to obtain a spontaneous magnetization $m_0 \neq 0$ when $H = 0$, which must correspond to a zero of the function $Y(u)$. Also from the Yang–Lee theorem (§5.6.2) the equations of state represented by (6.28) should be analytic when $m,H \neq 0$. The variable $\tau^{2\gamma-2\Delta}$ is convenient from this point of view since it does not allow an easy passage from $\tau > 0$ to $\tau < 0$ when $2\gamma - 2\Delta$ is non-integral.

Equation (6.28) was reformulated by Griffiths (1967) to remove this defect. Griffiths chose as operating variable

$$x = \tau m^{-1/\beta} \tag{6.29}$$

where for the time being β is defined as $\Delta - \gamma$; there is now no difficulty with the variable x in switching from τ positive to τ negative. In terms of this variable equation of state (6.28) becomes

$$H \simeq \tau^{\Delta}x^{-\beta}Y(x^{-2\beta}) = \tau^{\Delta}W(x). \tag{6.30}$$

We now consider the H–M behaviour near the critical isotherm $x = 0$. From the Yang–Lee theorem (§5.6.2) the equation of state is analytic when $m,H \neq 0$, and we can therefore generate a Taylor expansion

$$H = D_0(m) + D_1(m)\tau + D_2(m)\tau^2 + \cdots \tag{6.31}$$

where $D_0(m)$, $D_1(m)$, ... are analytic except at $m = 0$. We have also, by definition of the exponent δ,

$$D_0(m) \sim d_0 m^{\delta}. \tag{6.32}$$

Comparing (6.31) and (6.32) with (6.30) we see that near $x = 0$ we must have

$$W(x) \sim d_0 x^{-\Delta} \tag{6.33}$$

with the auxiliary condition

$$\delta = \Delta/\beta. \tag{6.34}$$

Griffiths thus rewrote equation (6.30) in the form

$$H \simeq m^{\delta} h(x) \tag{6.35}$$

where

$$h(x) = x^{\Delta} W(x). \tag{6.36}$$

$h(x)$ is clearly finite at $x = 0$, and from (6.31) it has an expansion of the form

$$h(x) = h_0 + h_1 x + h_2 x^2 + \cdots. \tag{6.37}$$

Equation (6.37) then implies that the leading terms in $D_1(m)$, $D_2(m)$, ... should be of order $m^{\delta - 1/\beta}$, $m^{\delta - 2/\beta}$, ... etc. Corresponding to a spontaneous magnetization m_0, $h(x)$ must have a zero at $x = -x_0$, and

$$m_0 \sim (\tau'/x_0)^{\beta} \tag{6.38}$$

so that β defined as $\Delta - \gamma$ is identical with the critical exponent for spontaneous magnetization. Expansion (6.28) can be rewritten

$$Hm^{-\delta} = f_2 x^{\gamma} + f_4 x^{\gamma - 2\beta} + f_6 x^{\gamma - 4\beta} + \cdots \tag{6.39}$$

and gives the asymptotic behaviour of $h(x)$ for large x.

Our discussion so far has latently assumed in defining quantities like $m^{-1/\beta}$ that $H \geq 0$, $m \geq 0$. But symmetry requires that a change in sign of both m and H does not affect the behaviour of the system. Hence to be more precise in dealing with all values of m,H we should write instead of (6.35)

$$H = m|m|^{\delta - 1} h(\tau |m|^{-1/\beta}). \tag{6.35'}$$

For conciseness we will use (6.35), but if m,H become less than 0 (6.35′) will be implied. Also, since by the Yang–Lee theorem there are no singularities in the p.f. when $H \neq 0$, we should expect $h(x)$ to be analytic when $-x_0 < x < \infty$.

The above equation of state was derived for the $I(\frac{1}{2})$ model. But since the only assumption entering was the existence of a gap exponent, (6.19), there were good reasons for expecting that it could apply to a large variety of other magnetic models. It could also be tested on the exact solutions for the mean field and spherical models, as we shall see shortly.

6.2.2 *Scaling Properties*

The data represented by an (H,m,T) equation of state are usually two-dimensional in character and should form a continuum in the plane. But equations like (6.30) or (6.35) imply that by a suitable choice of variables all of these data collapse on to a single curve. Such a relation is therefore termed a *scaling law* and variables like $Hm^{-\delta}$, $\tau m^{-1/\beta}$ in (6.35), or $H\tau^{-\Delta}$ in (6.30), are called *scaling variables.*

We are interested in the region where τ, m and H are all small. But in this region the scaling variables can take all values depending on the relative magnitudes of these quantities. The question of how close to T_c one must be for an equation of type (6.35) to be valid depends on neglected terms which give rise to *corrections to scaling.* These will be discussed in §6.2.5.

It was noted by Hankey and Stanley (1972) that scaling laws are closely related to homogeneous functions. A homogeneous function of degree p in two variables u,v is defined by the relation

$$g(\lambda u,\lambda v) = \lambda^p g(u,v) \tag{6.40}$$

for all λ. But if we use (6.40) with the value $\lambda = (1/u)$, we deduce that

$$g(u,v) = u^p g(1,v/u) = u^p G(v/u). \tag{6.41}$$

Hence a homogeneous function of two variables scales to a function of a single variable.

To take account of the types of exponent arising in relations like (6.30) or (6.35), Hankey and Stanley introduced the concept of a *generalized homogeneous function* which satisfies the relation

$$g(\lambda^a u,\lambda^b v) = \lambda g(u,v). \tag{6.42}$$

A standard homogeneous function (6.40) corresponds to $a = b$. Taking λ to be $u^{-1/a}$ we find as a generalization of (6.41)

$$g(u,v) = u^{1/a} g(1,v/u^{b/a}) = u^{1/a} G(v/u^{b/a}) \tag{6.43}$$

and this is of the same form as (6.30) or (6.35). From (6.35), for example, we have

$$a = 1/\delta, \quad b = -1/\beta\delta. \tag{6.44}$$

The scaling equation (6.35) involves two parameters β,δ and one function $h(x)$, and these completely characterize the critical behaviour. All critical exponents are determined by β and δ. We have already dealt with the two exponents Δ,γ

$$\Delta = \beta\delta \tag{6.34'}$$

$$\gamma = \Delta - \beta = \beta(\delta - 1). \tag{6.45}$$

We can readily derive the low-temperature counterparts which will be determined by the derivatives of m w.r.t. H at $m = m_0$ (the spontaneous magnetization) or $x = -x_0$. Differentiating (6.35) we find for the low-temperature susceptibility χ_0'

$$
\begin{aligned}
1/\chi_0' = \frac{\partial H}{\partial m} &= \delta m^{\delta-1}h(x) - \frac{1}{\beta}\, m^{\delta}h'(x)\tau m^{-1/\beta-1} \\
&= \delta m^{\delta-1}\left(h(x) - \frac{x}{\beta}\, h'(x)\right)
\end{aligned}
\tag{6.46}
$$

evaluated at $x = -x_0$, i.e. $m = (\tau'/x_0)^{\beta}$, from (6.38). The first term vanishes, and the second term reduces to

$$\tau'^{\gamma}[h'(-x_0)]/\beta x_0^{\gamma-1} \tag{6.47}$$

using (6.45). We thus find that $\gamma' = \gamma$, and that the amplitude of χ_0' is determined by $h'(-x_0)$. If we now differentiate (6.46) to obtain $\partial^2 H/\partial m^2$ we find that the terms independent of x are of order $m^{\delta-2}$. Using the standard relation

$$\partial^2 m/\partial H^2 = -(\partial^2 H/\partial m^2)/(\partial H/\partial m)^3 \tag{6.48}$$

we obtain

$$(\partial\chi'/\partial H)_0 \sim m_0^{-2\delta-1} \sim \tau'^{-\gamma-\Delta} \tag{6.49}$$

using (6.34). Hence $\Delta' = \Delta$. Proceeding similarly for higher derivatives a constant gap index is obtained equal to Δ. The critical amplitudes of the higher derivatives are related to the higher derivatives of $h(x)$ at $x = -x_0$.

Relation (6.24) derived by Patashinskii and Pokrovskii is independent of (6.35), and enables the correlation exponent ν to be expressed in terms of β and δ,

$$d\nu = \beta(\delta + 1). \tag{6.50}$$

From the Fisher relation $\gamma = \nu(2 - \eta)$ (equation (5.164)) the second correlation exponent can also be derived,

$$\eta = 2 - [d(\delta - 1)]/(\delta + 1). \tag{6.51}$$

Also from the parallel relation below T_c, $\gamma' = \nu'(2 - \eta)$ (equation (5.164′)) we deduce that $\nu' = \nu$, since $\gamma' = \gamma$. Scaling relations (6.50) and (6.51) involve the dimension d directly, and are associated with a characteristic correlation length. They were usually termed *length scaling* relations as opposed to *thermodynamic scaling* relations like (6.34′) and (6.45).

To determine the specific heat exponents α and α' the equation of state must be integrated to derive an expression for the free energy. For this purpose it is convenient to use (6.25) rewritten in the form

$$m \simeq \tau^{\beta}U(H\tau^{-\Delta}) \tag{6.52}$$

where we have used (6.45), and

$$U(y) = yX(y^2). \tag{6.53}$$

By means of an expansion analogous to (6.31) for m as a function of H, we can show that near $y = 0, U(y)$ is of order $y^{1/\delta}$.

The free energy $F(H,T)$ is determined by integrating the equation $m = -\partial F/\partial H$, to give

$$F(H, \tau) \simeq A_0(T) - \tau^{\beta+\Delta} V(H\tau^{-\Delta}) \tag{6.54}$$

where $V'(y) = U(y)$, and $A_0(T)$ is an arbitrary function assumed to be analytic. Since the specific heat is derived by differentiating $F(H,\tau)$ twice with respect to τ, and $V(y)$ is finite at $y = 0$, we can identify $\beta + \Delta$ with $2 - \alpha$ $(\alpha \geq 0)$ and using (6.45) and (6.34′) deduce that

$$\alpha = 2 - 2\beta - \gamma = 2 - \beta(\delta + 1). \tag{6.55}$$

By rearranging the variables as in the previous section it can be demonstrated that a relation similar to (6.54) applies for $\tau < 0$, and we then find that $\alpha' = \alpha$.

A more fundamental approach, following Griffiths (1967), is to evaluate the free energy $a(m,\tau)$ by integrating the equation $H = \partial a/\partial m$, using (6.35). The form of $\partial H/\partial m$ in (6.46) and $\partial^2 H/\partial m^2$ in (6.49) leads us to suggest

$$a(m,\tau) = A_0(T) + m^{\delta+1} k(x). \tag{6.56}$$

Substituting in (6.35) we find that $k(x)$ satisfies the differential equation

$$-xk'(x) + (2 - \alpha)k(x) = \beta h(x). \tag{6.57}$$

The solution of this equation, particularly when $\alpha = 0$, involves some tricky points, and the reader is referred to the original paper. When $\alpha = 0$ the interesting and important conclusion is drawn that the most general form of specific heat singularity is a *symmetric* logarithmic infinity with a discontinuity. This result was first derived by Widom (1965). Special cases discussed previously are the two-dimensional Ising model for which the amplitude of the discontinuity is zero (equation (5.31)), and the mean field solution for which the amplitude of the logarithmic term is zero (§3.4.5). The experimental results for the specific heat of liquid helium (Figure 1.4) involve both terms.

For convenience of reference the scaling relations for critical exponents in terms of β,δ are presented in Table 6.1.

6.2.3 *Calculations for Specific Models*

The scaling relations between critical exponents derived in the previous section were satisfied by the known values for the two-dimensional Ising model. Many of the exponents (e.g. γ, ν, α) were known exactly, and the others (e.g. δ, Δ) were estimated numerically (see Chapter 5). They are satisfied exactly by the standard three-dimensional spherical model, and also by the general

Table 6.1 Scaling relations between critical exponents

Thermodynamic quantity	Exponent	$I(\frac{1}{2})$ in two dimensions	Mean field	Spherical model (three dimensions)	Spherical model (long-range interactions)*
Spontaneous magnetization	β	1/8	1/2	1/2	1/2
Magnetization vs. field at T_c	δ	15	3	5	$(d+\sigma)/(d-\sigma)$
High-temperature susceptibility	$\gamma = \beta(\delta - 1)$	7/4	1	2	$\sigma/(d-\sigma)$
High-temperature gap	$\Delta = \beta\delta$	15/8	3/2	5/2	$(d+\sigma)/2(d-\sigma)$
High-temperature specific heat	$2 - \beta(\delta + 1)$	0	0	-1	$(d-2\sigma)/(d-\sigma)$
Low-temperature susceptibility	$\gamma' = \gamma$	7/4	1	–	–
Low-temperature gap	$\Delta' = \Delta$	15/8	3/2	–	–
Low-temperature specific heat	$\alpha' = \alpha$	0	0	–	–
Length scaling					
High-temperature coherence length	$\nu = \beta(\delta + 1)/d$	1	1/2	1	$1/(d-\sigma)$
Low-temperature coherence length	$\nu' = \nu$	1	1/2	–	–
Spin pair correlation function at T_c	$\eta = 2 - [d(\delta - 1)/(\delta + 1)]$	1/4	0	0	$2 - \sigma$

* Interaction $J(r) \sim r^{-(d+\sigma)}$ in d dimensions ($1 < d/\sigma < 2$, $0 < \sigma < \min\{2, d\}$).

spherical model with long-range forces (§5.4.2). Numerical values of the exponents which demonstrate this are given in Table 6.1 (values not quoted in §5.4.2 have been taken from the review by Joyce (1972)).

Incidentally a study of the final column of table provides useful qualitative insight regarding the dependence of different critical exponents on the strength of the cooperative force. Mean field exponents correspond to the strongest cooperative force and are achieved at $\sigma = \frac{1}{2}d$. As σ increases, and the values of γ and δ increase steadily, however, the value of β remains unchanged When σ reaches d the transition disappears. The first column for the Ising model indicates a similar pattern of behaviour for γ and δ but β now decreases with decreasing cooperative force.

The major focus of interest was the three-dimensional Ising model. The widely accepted estimates were 5/4 for γ and 5/16 for β (§5.3.3). The scaling relations then led to the suggestion of 5 for δ and $\frac{1}{8}$ for α. The former value fitted reasonably well with numerical estimates (Gaunt 1967). A special effort was mounted by Sykes and his collaborators to add terms to the high-temperature series for the specific heat, and their result, which seemed to provide dramatic support for $\alpha = \frac{1}{8}$, is reproduced in Figure 6.1.

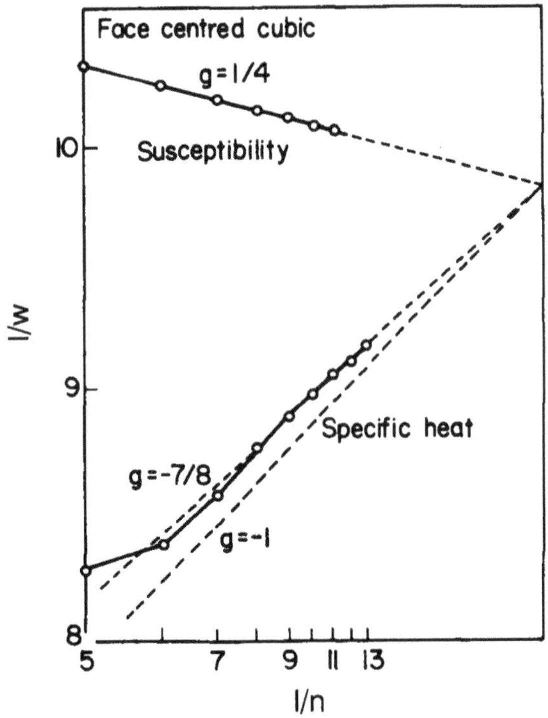

Figure 6.1 Ratio plot of a_n/a_{n-1} versus $1/n$ for the susceptibility and specific heat of the FCC lattice; g represents the exponent (after Sykes).

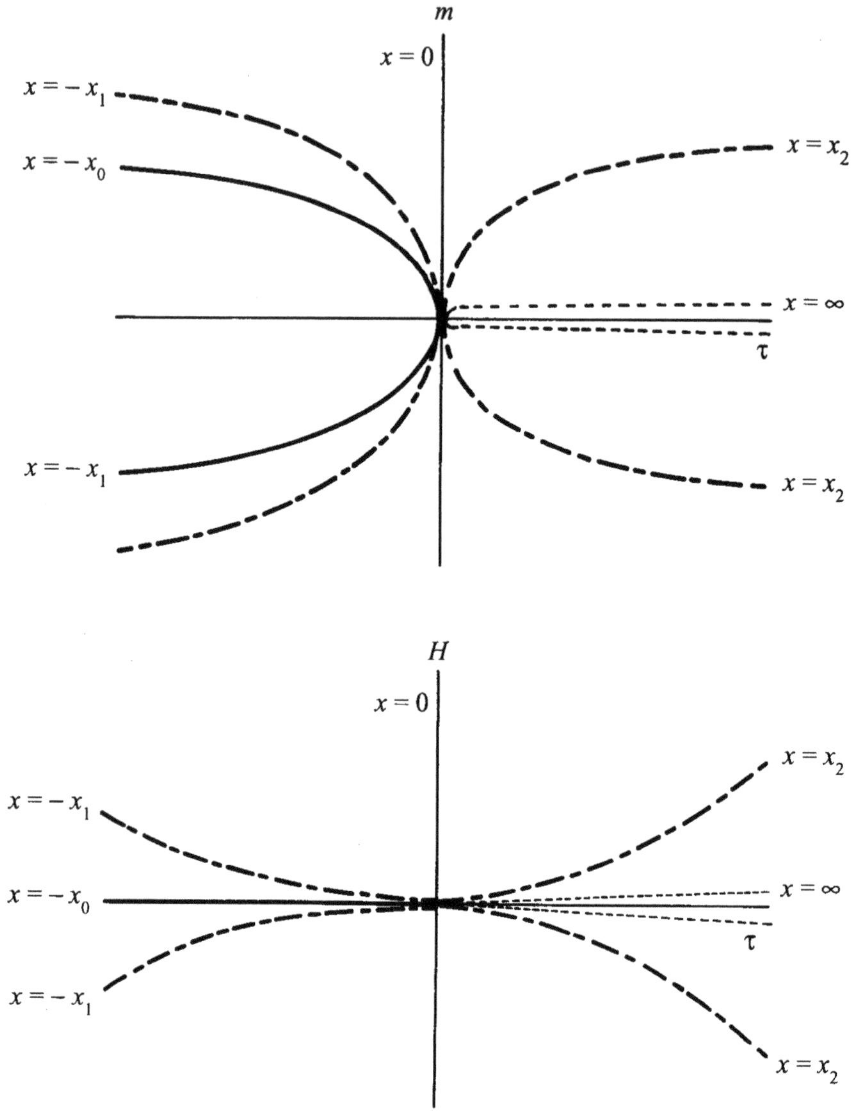

Figure 6.2 Curves of the constant x in the (m,τ) and (H,τ) planes.

On the low-temperature side there was some difficulty in achieving symmetry in the values of critical exponents (Baker and Gaunt 1967). It was generally assumed that this arose from slow convergence, and the hypothesis of thermodynamic scaling was widely accepted.

However, length scaling seemed to pose significant difficulties. The numerical estimates of critical exponents quoted above led to the suggestion that

$\nu = 0.625$ and $\eta = 0$; but the careful numerical analysis discussed in §5.5.2 gave strong indication of a non-zero value for η and supported a value of $\nu \simeq 0.64$. The dilemma took several years to resolve.

Turning now to the function $h(x)$, we first show schematically in Figure 6.2 the regions of the (m,τ) and (H,τ) planes which correspond to a constant value of x; we then indicate in Figure 6.3 how the different values of x relate to different parts of the critical region.

For the mean field solution corresponding to the strongest cooperative interaction, $h(x)$ is linear, of the form $a(x + x_0)$ (see equation (3.77)).

For the nearest-neighbour spherical model in three dimensions, we have from (5.150) and (5.152)

$$\begin{aligned} 4K &= G_d(\xi) + h^2/4K(\xi - 1)^2 \\ m/\mu_0 &= L/2K(\xi - 1). \end{aligned} \tag{6.58}$$

But it is well known from the theory of lattice Green functions (Joyce 1972) that $G_d(\xi)$ can be expanded near $\xi = 1$ in the form

$$G_d(\xi) = a_0 + a_1(\xi - 1)^{1/2} + a_2(\xi - 1) + a_3(\xi - 1)^{3/2} + \cdots. \tag{6.59}$$

From (6.58) and (6.59) we deduce the equation of state

$$H/m = b_0(4\tau + m^2)^2 + b_1(4\tau + m^2)^3 + b_2(4\tau + m^2)^4 + \cdots. \tag{6.60}$$

Comparing with (6.35) and remembering that $\delta = 5$ and $\beta = 1/2$, we find that $h(x)$ is of the form

$$h(x) = b_0(1 + 4x)^2. \tag{6.61}$$

The correction terms in (6.60) will be discussed in §6.2.5.

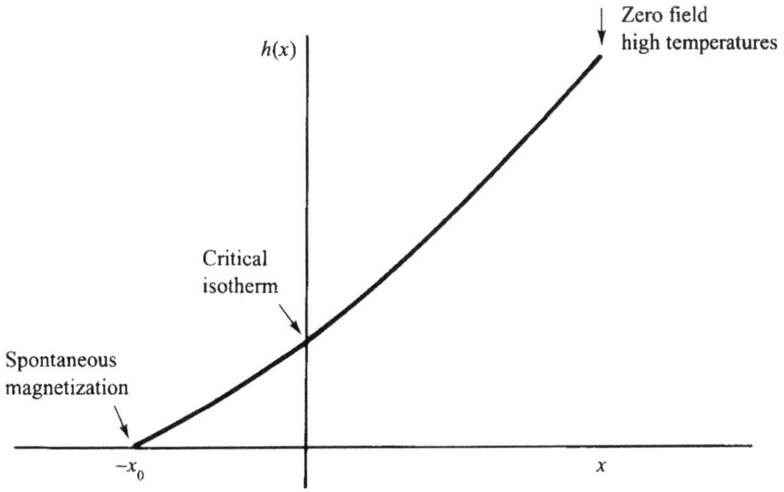

Figure 6.3 Form of $h(x)$ in the scaling hypothesis for the critical equation of state.

For the nearest-neighbour spherical model with $d = 4$ Joyce found that

$$h(x) = (x + x_0)[\ln(x + x_0)]^{-1} \tag{6.62}$$

whilst for $d > 4$ the mean field form is obtained.

For the spherical model with long-range interactions, a similar analysis leads to the general conclusion (Joyce 1972) that

$$h(x) = c_0(x + x_0)^{\gamma} \tag{6.63}$$

where γ, the susceptibility exponent, is equal to $\sigma/(d - \sigma)$ (Table 6.1). From (6.62) we see that the curvature of $h(x)$ increases as the strength of the cooperative interaction decreases.

Numerical investigation of the two- and three-dimensional Ising models by Gaunt and Domb (1970) led to a similar qualitative pattern of behaviour (Figure 6.4). It will be seen that the two-dimensional function has considerably more curvature than the three-dimensional function. It is important to note that $h(x)$ for the Ising model has a finite slope at $x = -x_0$ corresponding to a finite susceptibility as $T \to T_{c-}$. For the spherical model with $\gamma > 1$ the zero slope given by (6.63) corresponds to an infinite susceptibility. Fisher's conjecture that the p.f. for the Ising model has an essential singularity at the phase boundary (§6.1.1) would lead to the conclusions that $h(x)$ has an essential singularity at $x = -x_0$.

For the Heisenberg model numerical data were much less accurate, particularly at low temperatures. Attention should nevertheless be drawn to an attempt by Milosevic and Stanley (1972) to calculate $h(x)$ for spin $\frac{1}{2}$ and spin ∞. The results are shown in Figure 6.5.

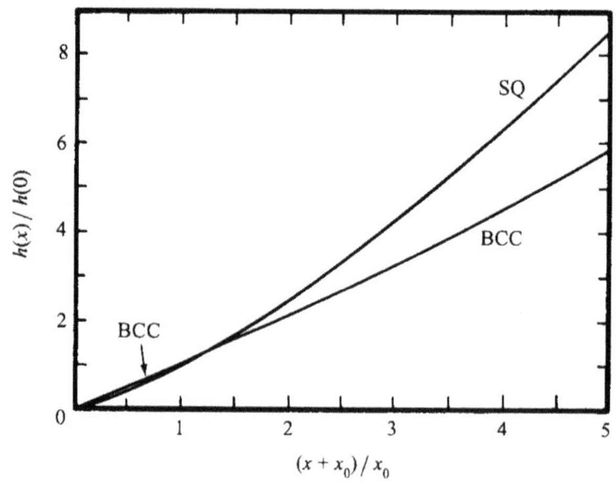

Figure 6.4 Plots of $h(x)/h(0)$ against $(x + x_0)/x_0$ for the SQ and BCC lattices for $(x + x_0)/x_0 \leq 5$.

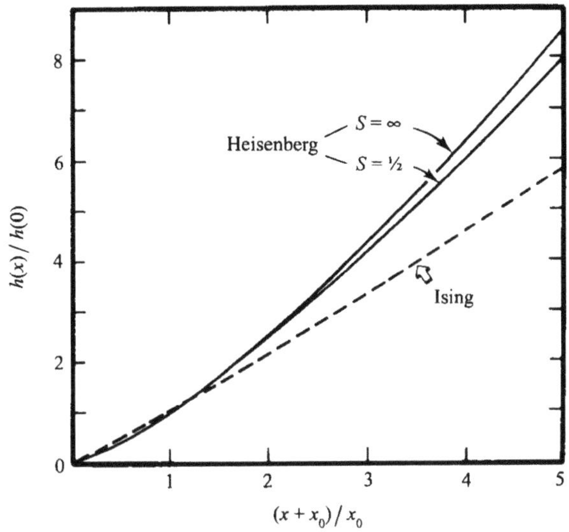

Figure 6.5 Comparison of calculations for the FCC lattice for the Ising model and the $S = \frac{1}{2}$ and $S = \infty$ Heisenberg models.

6.2.4 *Fluids*

There are two alternative methods of developing the magnet–fluid analogy, and formulating a critical equation of state for fluids analogous to (6.35). The magnetic free energy $a(T,m)$ is expressed in terms of an intensive variable T and an extensive variable m, satisfying $\mathrm{d}a = -s\,\mathrm{d}T + H\,\mathrm{d}m$. If we take the extensive variable v as the analogue of m, we are led to the Helmholtz free energy $f(T,v)$ which satisfies

$$\mathrm{d}f = -s\,\mathrm{d}T - P\,\mathrm{d}v.$$

We then find that $-P$ is the analogue of H, and we are led to a standard $P(T, v)$ equation of state.

If, however, we take ρ as the analogue of m, we are led to the free energy $a(T,\rho)$ of §2.1 which satisfies (equation (2.17))

$$\mathrm{d}a = -\rho s\,\mathrm{d}T + \mu\,\mathrm{d}\rho.$$

We then find that μ is the analogue of H, and we are led to a $\mu(T,\rho)$ equation of state, which was used by Widom (1965) in his original formulation of the scaling hypothesis. We considered both alternatives in Chapter 2; for the van der Waals equation, to the first order we found that both are equivalent, but when higher-order terms are taken into account the $\mu(\rho,T)$ equation provides a closer fit.

In addition to exact symmetry between $\pm m$, a magnet has the simplifying feature that when m has its critical value at zero, H is also zero. For a fluid, however, we must introduce functions of temperature

$$P_0(\tau) = P(\rho_c, \tau) \qquad \mu_0(\tau) = \mu(\rho_c, \tau) \tag{6.64}$$

which we assume to be analytic. By analogy with the notation of Chapter 2, we write for the first alternative

$$P = P_c(1 + \pi), \qquad P_0 = P_c(1 + \pi_0) \tag{6.65}$$

and express the critical equation of state in the form

$$\pi - \pi_0 = \omega^\delta h(\tau\omega^{-1/\beta}). \tag{6.66}$$

Similarly for the second alternative, by analogy with (2.53′) we write

$$\frac{\mu - \mu_0(T)}{P_c v_c} \simeq r^\delta h(\tau r^{-1/\beta}). \tag{6.67}$$

The passage from a $\mu(\rho,T)$ to a $P(\rho,T)$ equation of state can be undertaken by the standard thermodynamic formulae given at the end of §2.1. We must first determine the free energy by integration:

$$a(\rho,T) = \int \mu(\rho,T)\,\mathrm{d}\rho. \tag{6.68}$$

For the scaling equation of state (6.67) this gives a result similar to (6.56) and 6.57). The pressure is then given by

$$P(\rho,T) = \rho\mu(\rho,T) - a(\rho,T) \tag{6.69}$$

and in terms of reduced variables, we find that

$$\pi - \pi_0 = r^\delta h(x) + r^{\delta+1}[h(x) - k(x)] \quad (x = \tau r^{-1/\beta}). \tag{6.69′}$$

Equations (6.66) and (6.69′) are equivalent to the first order, but differences arise if corrections to the simple scaling form are taken into account.

For a van der Waals fluid we were able to proceed beyond the approximation of exact symmetry, and we found in §2.5.1 that the diameter of the coexistence boundary curve varies approximately linearly with temperature near T_c. In trying to formulate a non-classical equation of state which could take asymmetry into account we would wish to provide for the possibility of a singularity in the diameter near T_c. The simplest generalization to asymmetry was made by Rehr and Mermin (1973). Using an alternative formulation of scaling based on an equation of type (6.54)

$$\pi(\mu,\tau) - \pi_0(\tau) \simeq \tau^{2-\alpha} Z_{\pm}[(\mu - \mu_0)\tau^{-\Delta}] \tag{6.70}$$

they retained analyticity in $\pi(\mu,\tau)$ except at the critical point, but allowed the analytic functions Z_+ and Z_- to be different. They were led to the conclusion that, for the diameter,

$$\rho_d = \tfrac{1}{2}(\rho_{\text{liquid}} + \rho_{\text{gas}}) \simeq \rho_c + c_0(\tau')^{1-\alpha}. \tag{6.71}$$

A similar result had been obtained in calculations for specific asymmetric models.

6.2.5 *Correction Terms*

In the analysis of series expansions for the behaviour of $I(\frac{1}{2})$ and other models near T_c it was usual to assume that the function has a Darboux form (equation (5.97)),

$$\chi_0 = A(w)(1 - w/w_c)^{-\gamma} + B(w) \tag{6.7.2}$$

where $A(w)$ and $B(w)$ are analytic. This was based on the known exact solutions for $I(\frac{1}{2})$ in two dimensions, and on the behaviour of the spherical model. One of the important conclusions of the renormalization group (RG) treatment (Chapter 7) was that this assumption is not generally valid and must be replaced by

$$\chi_0 = A_0(w)(1 - w/w_c)^{-\gamma} + A_1(w)(1 - w/w_c)^{-\gamma+\theta} + A_2(w)/(1 - w/w_c)^{-\gamma+\theta'} + \cdots. \tag{6.73}$$

Even if the form (6.73) had been postulated, it would have been difficult to fit an additional parameter θ from the limited data available; fortunately, as we shall see in Chapter 7, the RG method provides a reasonable estimate of θ. All of the estimates of critical exponents before the RG were based on the assumption that $\theta = 1$.

Similarly, attempts to include correction terms to the scaling equation of state (Domb 1971) based themselves on the Darboux form (6.72), and led to an extension of (6.30) to

$$H = \tau^{\Delta}W(x) + \tau^{\Delta+1}W_1(x) + \tau^{\Delta+2}W_2(x) + \cdots + \tau^{\Delta+\gamma}W^{(1)}(x) + \cdots. \tag{6.74}$$

When this is transformed to the Griffiths form (6.35) the correction terms are as follows:

$$H = m^{\delta}h(x) + m^{\delta+1/\beta}h_1(x) + m^{\delta+2/\beta}h_2(x) + \cdots + m^{\delta+\gamma/\beta}h^{(1)}(x) + \cdots. \tag{6.75}$$

No exact equation of state was available for the $I(\frac{1}{2})$ model in two dimensions, but the exact calculations for the spherical model (6.60) conform to the pattern (6.75), as do calculations for the mean field model (equations (3.46)–(3.48)).

6.2.6 *Experimental Results*

To undertake an experimental test of the scaling hypothesis (6.35), data in the critical region should be plotted in the form of $Hm^{-\delta}$ as a function of $m^{-1/\beta}$ to see if they lie on a one-dimensional curve. As a preliminary, two independent critical exponents must be estimated from which β and δ can be determined. We have seen an example of such a test in Figure 1.10 which made use of the (P,V,T) data of Roach (1968) on ^{4}He. The estimates of the parameters were $\beta = 0.359$, $\delta = 4.45$, and the values of $(\mu - \mu_c)$ were derived by graphical integration along a P–V isotherm.

A similar plot for the insulating ferromagnet $CrBr_3$ using the data of Ho and Litster (1969) is shown in Figure 6.6. In this case estimates of the parameters were $\beta = 0.368$, $\gamma = 1.22$, $\delta = 4.32$. It will be seen that the data again fall on a single curve; the solid curve drawn to fit the data is an empirical formula proposed by Levelt-Sengers, Green and Vicentini-Missoni (1969) which fits reasonably well except at the upper end.

Quite remarkably the old data on nickel measured by Weiss and Forrer (1926) were reanalyzed by Arrott and Noakes (1967) and shown to provide good support for the scaling hypothesis. Kouvel and Fisher (1964) had already initiated such an analysis for the critical exponents, and had been led to the non-classical values. Arrott and Noakes adopted the values $\gamma = 1.31$, $\delta = 4.39$, $\beta = 0.3854$ and obtained a good fit to the (M,H,T) data in the critical region with a function $h(x)$ of the form $(1 + x)^{\tau}$.

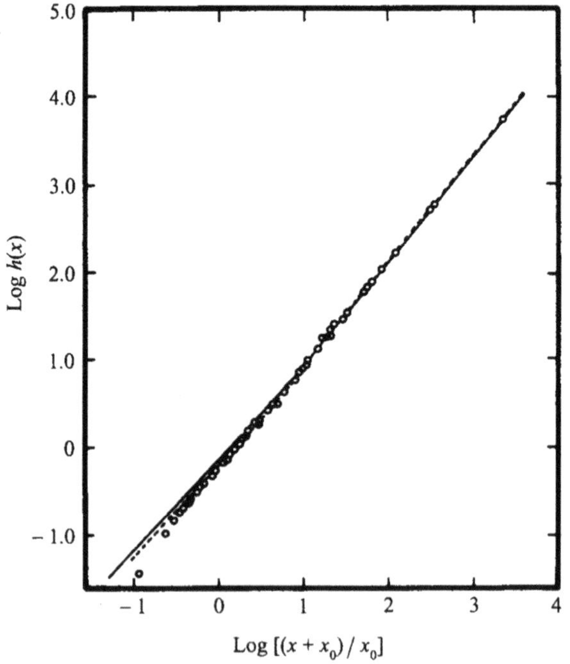

Figure 6.6 Logarithmic plot of the scaled function $h(x)$ versus $(x + x_0)/x_0$ for the data on $CrBr_3$. The symbols represent experimental points, the dashed line and the solid line attempted equations of state.

Data for other fluids and magnets were discussed in a review article by Vicentini-Missoni (1972); in all cases in agreement with the scaling hypothesis was reasonably satisfactory.

When we look back at the above analyses from the vantage point of later, more comprehensive knowledge, an interesting point arises. RG theory to be discussed in Chapter 7 requires the critical exponents for a fluid to be identical with those for an $I(\frac{1}{2})$ magnet, which are currently estimated as $\beta = 0.325 \pm 0.0015$, $\delta = 4.82 \pm 0.02$. The experimental estimates for ^{4}He differ significantly from these *true* values and we must assume that they are *effective* values obtained because the experimental range did not approach closely enough to T_c. Similar remarks apply to the magnetic data. This seems to indicate empirically that the scaling hypothesis is well satisfied even with effective exponents. A theoretical justification of this behaviour was later provided by Aharony and Ahlers (1980) and Chang and Houghton (1980);* see also the general experimental review by Ahlers (1980). For later experimental work on β for fluids see Hocken and Moldover (1976).

6.3 Kadanoff's Approach

Shortly after the advent of scaling, Kadanoff (1966) tried to provide a theoretical justification by introducing the concept of a new *block spin* which could replace a cell of L^d spins of the original lattice. The value of L should be large compared with unity, but small compared with $\xi(\sim\kappa^{-1})$, the range of correlation; since ξ tends to ∞ as T tends to T_c these conditions can be satisfied for T sufficiently close to T_c. He then suggested that within a cell the spins would tend to line up so that they mostly point either up or down, and could be approximately represented by a new Ising spin taking the values ± 1.

The free energy of the original model, $f(\tau,h)$, is expressed in terms of two variables τ $(=(T/T_c) - 1)$ and h $(=\beta\mu_0 H)$. The interaction of the cell of L^d spins with the magnetic field contributes a term to $-\beta\boldsymbol{H}$ of the form

$$h \sum_{L^d} \sigma_i . \tag{6.76}$$

Kadanoff assumed that the sum could be replaced by

$$\sum_{L^d} \sigma_i = \langle\sigma\rangle_L L^d \tilde{\sigma}_I \tag{6.77}$$

where the average $\langle\sigma\rangle_L$ depends strongly on L but not on τ or h. The interaction between neighbouring cells is quite complex, but Kadanoff assumed that it could be replaced by $\tilde{K}\sigma_I\sigma_J$ where $\tilde{K}$ is a new parameter which again depends strongly on L but not on τ or h. Hence the original Ising model can

* I am grateful to Professor Michael E. Fisher for reminding me of these references.

be replaced by a new Ising model of block spins, and since the total free energy is unchanged, we can write for the free energy per spin

$$f(h,\tau) = L^{-d}f(\tilde{h},\tilde{\tau}) \tag{6.78}$$

where

$$\tilde{h} = L^{d}\langle\sigma\rangle_L h \tag{6.79}$$

and $\tilde{\tau}$, which is derived from $\tilde{K}$, bears a similar relationship as τ to K.

The correlations in the original model can be related to the block spin correlations in a similar manner,

$$\langle\sigma_i\sigma_j\rangle = \langle\sigma\rangle_L^2\langle\tilde{\sigma}_I\tilde{\sigma}_J\rangle \tag{6.80}$$

leading to the relation (in the notation of equation (4.105))

$$\Gamma(r,h,\tau) = \langle\sigma\rangle_L^2\Gamma(L^{-1}r,\tilde{h},\tilde{\tau}). \tag{6.81}$$

From equations (6.78) and (6.81) the formulae of equilibrium and length scaling can readily be deduced. In zero field the singular term in the free energy is of the form $\tau^{2-\alpha}$. Therefore, from the RHS of (6.78)

$$L^{-d}\tilde{\tau}^{2-\alpha} = \tau^{2-\alpha} \tag{6.82}$$

independently of L. Hence we find that

$$\tilde{\tau} = L^y\tau, \quad (2-\alpha)y = d. \tag{6.83}$$

In a non-zero field for $T < T_c$ the first magnetic singular term is $(-\tau)^\beta h$, and hence from the RHS of (6.78)

$$L^{-d}(-\tilde{\tau})^\beta\tilde{h} = (-\tau)^\beta h \tag{6.84}$$

independently of L. Hence we must have

$$\tilde{h} = L^x h, \quad x = d - \beta y. \tag{6.85}$$

Equation (6.78) can now be put in the form

$$f(h,\tau) = L^{-d}f(L^x h, L^y\tau) \tag{6.86}$$

which, being satisfied for arbitrary L, shows that the free energy is a generalized homogeneous function (§6.2.2). Taking $L^y\tau = 1$, we are led to

$$f(h,\tau) = \tau^{d/y}f(h\tau^{-x/y},1) \tag{6.87}$$

which should be compared with (6.87′)

$$f(H,\tau) \sim a_0(T) - \tau^{2-\alpha}V(H\tau^{-\Delta}) \tag{6.87'}$$

(derived from (6.54)). There are two free parameters x,y in terms of which all other exponents can be expressed, and in terms of our previous basic exponents β and δ

$$x = d\delta/(1 + \delta), \quad y = d/\beta(1 + \delta). \tag{6.88}$$

Also from (6.79) we can evaluate $\langle\sigma\rangle_L$ as L^{x-d} so that

$$\langle\sigma\rangle_L = L^{-d/(1+\delta)}. \tag{6.89}$$

The correlation scaling relation (6.81) can now be written in the form

$$\Gamma(r,h,\tau) = L^{2(x-d)}\Gamma(L^{-1}r,L^{x}h,L^{y}\tau) \tag{6.90}$$

which is again a generalized homogeneous function. Taking $L^{-1}r = 1$, we find, using Table 6.1, that

$$\Gamma(r,h,\tau) = (1/r^{d-2+\eta})\Gamma(1,r^{x}h,r^{y}\tau). \tag{6.91}$$

This can be rewritten in the form

$$\Gamma(r, h, \tau) = (1/r^{d-2+\eta})G(rh^{\bar{\nu}}, r\tau^{\nu}) \tag{6.92}$$

where

$$\nu = 1/y = \beta(\delta + 1)/d, \quad \bar{\nu} = 1/x = (\delta + 1)/\delta d = \nu/\beta\delta. \tag{6.93}$$

Hence the length scaling relations are also derived, and the new exponent $\bar{\nu}$ introduced to describe the behaviour of the correlations in a non-zero field is also readily expressed in terms of the two basic exponents β and δ.

By successive differentiation of (6.87′) we obtain scaling laws for the magnetization, susceptibility, and higher derivatives in the form

$$M = \tau^{\beta}V'(H\tau^{-\Delta}) \tag{6.94}$$

$$\chi = \tau^{-\gamma}V''(H\tau^{-\Delta}) \tag{6.95}$$

and so on. If we sum over the correlations in (6.92), i.e. if we form the integral

$$\int r^{d-1}\Gamma(r,h,\tau)\,\mathrm{d}r, \tag{6.96}$$

we should re-derive the susceptibility. We readily find on substituting $z = r\tau^{\nu}$ that this gives

$$\tau^{-\gamma}\int_0^{\infty} z^{1-\eta}G(zh^{\bar{\nu}}\tau^{-\nu},z)\,\mathrm{d}z \tag{6.97}$$

where we have used relation (5.164), $\gamma = \nu(2 - \eta)$. Since $\nu/\bar{\nu}$ is equal to Δ, we verify the scaling form (6.95). Using the same substitution for the Fourier transform of the correlations

$$\hat{\Gamma}(q,h,\tau) = \int \Gamma(r,h,\tau) \exp(-i\mathbf{q}\cdot\mathbf{r})\, d\mathbf{r}$$

we find the scaling relation

$$\hat{\Gamma}(q,h,\tau) = \tau^{-\gamma}\hat{G}(q\tau^{-\nu}, h\tau^{-\Delta}). \tag{6.97'}$$

Kadanoff's idea that a block of L^d spins can be effectively replaced by a single new Ising spin does not stand up to critical examination. But the concept of a block spin which he introduced was taken up by Wilson (1971) and proved to be of major importance in the development of the renormalization group.

The relations which he derived, (6.78) and (6.81), lead simply and directly to the scaling formulae for the free energy and for the correlations. These relations can be seen in a new perspective following the pioneering work of Mandelbrot (1977) on systems which display self-similarity, for which a change of scale does not change the basic character of the system. The relations express the fact that magnets or fluids near their critical points are self-similar.

6.4 Cross-over Behaviour*

6.4.1 *Introduction*

We have seen in §5.3.3.1. that critical exponents for a given model depend on dimension and not on lattice structure. If we consider a Hamiltonian of the form

$$\mathscr{H} = -J\sum_{\mathbf{i},\mathbf{j}} \sigma_{\mathbf{i}}\sigma_{\mathbf{j}} - gJ\sum_{\boldsymbol{\lambda},\boldsymbol{\mu}} \sigma_{\boldsymbol{\lambda}}\sigma_{\boldsymbol{\mu}} - = \mathscr{H}_0 + g\mathscr{H}_1 \tag{6.98}$$

for a simple cubic lattice, where **i**,**j** refer to nearest neighbours in horizontal planes, and $\boldsymbol{\lambda},\boldsymbol{\mu}$ to nearest neighbours in the vertical direction, then for $g = 0$ the behaviour will be two-dimensional in character, but for $g \neq 0$ it will be three-dimensional. How does the transition or *cross-over* depend on the parameter g when g is small?

A similar question can be asked in relation to the anisotropic Heisenberg model (5.109) in relation to the nature of the transition in the presence of a small anisotropy parameter g ($J' = J(1 + g)$). The conclusion drawn in §5.3.3.1 was that critical exponents depend on the symmetry of the order parameter in the ground state. Hence for $g = 0$ the behaviour is Heisenberg-like, for a small negative g it is Ising-like, and for a small positive g it is x–y-like.

* I am grateful to Professor D. Mukamel for a helpful discussion in relation to this section.

Although the latter problem was the first to be tackled historically, by Riedel and Wegner (1969), we shall find it more convenient to deal first with the dimensional cross-over. Spin anisotropy involves some subtle considerations, and dimensional anisotropy has the advantage of yielding important exact results.

6.4.2 Dimensional Anisotropy

For the Hamiltonian (6.98), by analogy with (6.87′), we postulate for the singular part of the free energy

$$f(g,\tau) \simeq \tau^{2-\alpha_2} V(g\tau^{-\phi}) \tag{6.99}$$

where the subscript 2 is to remind us that we are dealing with the two-dimensional critical exponent, and ϕ is a new *cross-over* exponent analogous to the gap exponent Δ in (6.87′). In fact, if we develop the free energy in powers of g in a similar manner to the development (6.16) in powers of H, our basic assumption in postulating (6.99) is analogous to (6.19). In the presence of a magnetic field H, (6.99) would generalize to

$$f(g,H,\tau) \simeq \tau^{2-\alpha_2} V(g\tau^{-\phi}, H\tau^{-\Delta_2}). \tag{6.100}$$

By differentiating (6.100) we conclude that thermodynamic quantities corresponding to the derivatives of the free energy w.r.t. H satisfy scaling relations similar to (6.99); for example, for the initial susceptibility we can write

$$\chi_0(g,\tau) \simeq \tau^{-\gamma_2} Z(g\tau^{-\phi}). \tag{6.101}$$

However, there is an important difference between the scaling relations w.r.t. g and those w.r.t. H of §6.2.2. For the latter the transition disappears when $H \neq 0$, whereas for the former a new transition point $T_c(g)$ appears with new critical exponents. For this to be possible the scaling function $Z(x)$ in (6.101) must have a singularity at some point x_c, and the singular behaviour must be such as to enable the susceptibility when $g \neq 0$ to correspond to the three-dimensional Ising model:

$$\chi_0(g,\dot{\tau}) \simeq \dot{A}(g)\dot{\tau}^{-\gamma_3} \quad (\dot{\tau} = T/T_c(g) - 1). \tag{6.102}$$

From this we first deduce that the new critical point is given by $g\tau^{-\phi} = x_c$, or

$$\tau_g = [T_c(g)]/[T_c(0)] - 1 = (g/x_c)^{1/\phi}. \tag{6.103}$$

To ensure (6.102) the scaling function near $x = x_c$ must behave as

$$Z(x) = Z_c(1 - x/x_c)^{-\gamma_3}. \tag{6.104}$$

We now use (6.102) and (6.103) to give

$$\dot{\tau} = \frac{1+\tau}{1+\tau_g} - 1 \simeq \tau - \tau_g \tag{6.105}$$

and comparing (6.101) with (6.102), with the help of (6.104), we deduce that

$$\dot{A}(g) = Z_c \phi^{-\gamma_3}(g/x_c)^{(\gamma_3-\gamma_2)/\phi}. \tag{6.106}$$

This dependence of the new amplitude on the parameter g was first derived by Abe (1970), who also examined the cross-over from one to two dimensions in Onsager's solution for the specific heat of an asymmetric net (Figure 5.9, §5.1.4.2). In the latter case since $T_c(0) = 0$ the variables τ and τ_g cannot be used, but using K $(=\beta J)$ he found that

$$K_c \simeq -\tfrac{1}{2}[\ln g + \ln(-\tfrac{1}{2}\ln g)] \tag{6.107}$$

and for the specific heat

$$C/Nk \simeq g(\ln g)^3 \ln \dot{\tau}/2\pi. \tag{6.108}$$

To divide the cross-over region more precisely into areas of temperature where the behaviour is two-dimensional in character, and those where it is three-dimensional, for a given g, we require knowledge of the scaling function $Z(x)$. But the general pattern of these areas can readily be outlined (Liu and Stanley 1973). Let us fix arbitrarily a percentage deviation p from two-dimensional behaviour which we regard as significant, and let x_A be the value of x for which $Z(x_A)$ deviates from $Z(0)$ by p%. Then for $0 < x < x_A$ we can reasonably consider the behaviour as two-dimensional, i.e.

$$\tau > (g/x_A)^{1/\phi}. \tag{6.109}$$

Let us now determine x_B at which the deviation of $X(x)$ from its singular behaviour (6.104) is less than p%. For $x_B < x < x_c$ we can reasonably consider the behaviour as three-dimensional, i.e. for

$$(g/x_B)^{1/\phi} > \tau > (g/x_c)^{1/\phi}. \tag{6.110}$$

The different areas defined by (6.109) and (6.110) are illustrated diagrammatically in Figure 6.7 (following Liu and Stanley 1973) To conform with our notation, R should be replaced by g.

For the correlation functions the appropriate scaling assumption by analogy with (6.92) is

$$\Gamma(r,h,\tau,g) = (1/r^{d-2+\eta_2})G(rh^{\bar{\nu}_2}, r\tau^{\nu_2}, rg^{\bar{\phi}}). \tag{6.111}$$

Integrating to determine the susceptibility in the manner of (6.96) and (6.97), and comparing with (6.101), we find that

$$\nu_2/\bar{\phi} = \phi, \quad \bar{\phi} = \nu_2/\phi. \tag{6.112}$$

Finally, Liu and Stanley (1972) calculated some exact results in relation to dimensional cross-over which enable the value of ϕ to be conjectured with

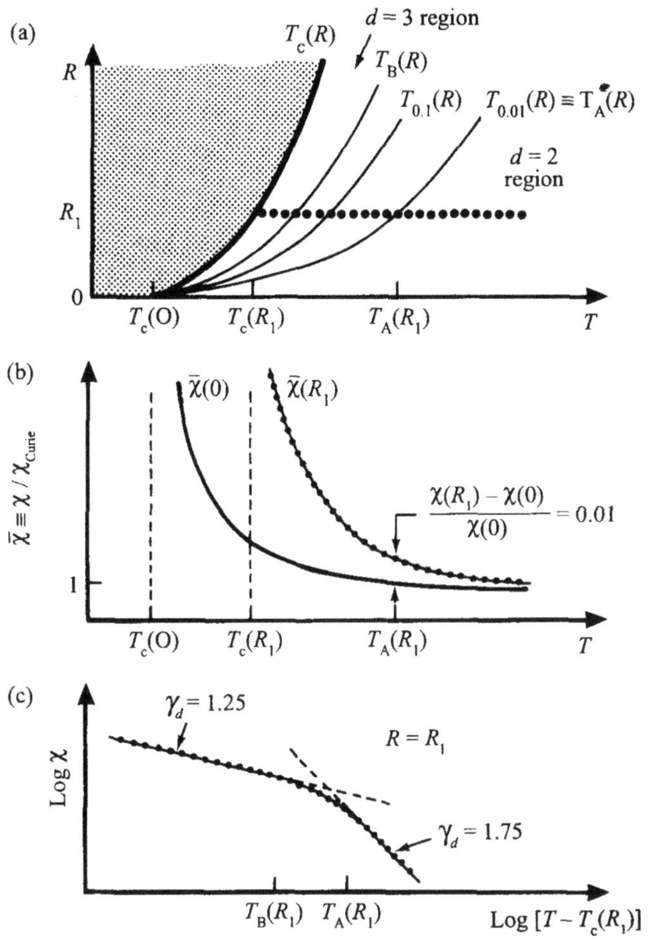

Figure 6.7 Schematic diagram of the cross-over behaviour. (a) The cross-over region (shaded area) is bounded by $T_A(R)$ and $T_B(R)$. ($T_A(R) = T_{0.01}(R)$ is the temperature at which the system differs appreciably (1%) from being two-dimensional.) $T_c(R)$ is the critical temperature. The generalized scaling hypothesis predicts that all curves should approach $T_c(0)$ via the power law $R^{1/\phi}$. (b) Dependence of reduced susceptibility $\bar{\chi}$ upon T for $R = 0$ and for $R = R_1$, indicating the definition of $T_A(R)$. Note that this drawing is not to scale. (c) Sketch of hypothetical experimental data, plotted in the conventional log–log plot, for a system which is described by the Hamiltonian with $R = R_1$.

some confidence. To give a brief account of their argument, we use the relation for the susceptibility (equations (4.105) and (4.107)),

$$\chi/N\beta\mu_0^2 = 1 + \sum_{\mathbf{r}\neq 0} \Gamma(\mathbf{r}) = 1 + \sum_{\mathbf{r}\neq 0} [\langle\sigma_0\,\sigma_{\mathbf{r}}\rangle - \langle\sigma_0\, x\sigma_{\mathbf{r}}\rangle]. \tag{6.113}$$

$\chi/\beta\mu_0^2$ is usually called the reduced susceptibility, and will be denoted by $\bar{\chi}$. If we apply (6.113) in zero magnetic field to determine $\chi(0)$ $T > T_c(g)$, then

$$\langle \sigma_{\mathbf{l}} \rangle = 0. \tag{6.114}$$

$\langle \sigma_{\mathbf{k}} \sigma_{\mathbf{l}} \rangle$ is defined by

$$\langle \sigma_{\mathbf{k}} \sigma_{\mathbf{l}} \rangle = \frac{\sum_{\sigma=\pm 1} \sigma_{\mathbf{k}} \sigma_{\mathbf{l}} \exp(-\beta\mathscr{H})}{\sum_{\sigma=\pm 1} \exp(-\beta\mathscr{H})} \tag{6.115}$$

and is a function of T, $|\mathbf{k}-\mathbf{l}|$, g. When $g=0$ two spins at sites $\mathbf{k}$,$\mathbf{l}$ are uncoupled if they are on different horizontal layers, i.e. if $k_z \neq l_z$. Hence

$$\langle \sigma_{\mathbf{k}} \sigma_{\mathbf{l}} \rangle_{g=0} = \langle \sigma_{\mathbf{k}} \sigma_{\mathbf{l}} \rangle_0 \quad (k_z = l_z)$$

$$= 0 \qquad (k_z \neq l_z),$$

$$\langle \sigma_{\mathbf{k}} \sigma_{\mathbf{l}} \rangle_{g=0} = \delta(k_z, l_z) \langle \sigma_{\mathbf{k}} \sigma_{\mathbf{l}} \rangle_0 \tag{6.116}$$

where the suffix 0 refers to the two-dimensional average. Since $\lambda_z \neq \mu_z$ in (6.98) we see that

$$\langle \mathscr{H}_1 \rangle_0 = 0. \tag{6.117}$$

Expanding (6.115) in powers of g, and using (6.117), we find for $\bar{\chi}_1(g) = [\partial\bar{\chi}(g)/\partial g]$

$$-\chi_1(0)/\beta = \sum \langle \sigma_0 \sigma_{\mathbf{r}} \mathscr{H}_1 \rangle_0. \tag{6.118}$$

To deal with (6.118) we need to consider properties of the four spin averages $\langle \sigma_0 \sigma_{\mathbf{r}} \sigma_{\boldsymbol{\lambda}} \sigma_{\boldsymbol{\mu}} \rangle_0$. Using (6.114) and the fact that spins on different layers are uncoupled when $g=0$, we find that

$$\langle \sigma_0 \sigma_{\mathbf{r}} \sigma_{\boldsymbol{\lambda}} \sigma_{\boldsymbol{\mu}} \rangle_0 = \delta(0,\lambda_z)\delta(z,\mu_z) \langle \sigma_0 \sigma_{\boldsymbol{\lambda}} \rangle_0 \langle \sigma_{\mathbf{r}} \sigma_{\boldsymbol{\mu}} \rangle_0$$

$$+ \delta(0,\mu_z)\delta(z,\lambda_z) \langle \sigma_0 \sigma_{\boldsymbol{\mu}} \rangle_0 \langle \sigma_{\mathbf{r}} \sigma_{\boldsymbol{\lambda}} \rangle_0. \tag{6.119}$$

Equation (6.119) expresses the fact that the only non-zero values of the four spin averages correspond to one pair in the layer containing the origin, and the second pair in the layer either above or below. It is then straightforward to show that

$$\bar{\chi}_1(0) = 2\beta J [\bar{\chi}_0(0)]^2. \tag{6.120}$$

Liu and Stanley checked relation (6.120) numerically from series expansions. They point out that the argument applies equally to any space dimension d and to spins in any dimension n.

They went on to calculate higher derivatives $\bar{\chi}_2(0)$ and $\bar{\chi}_3(0)$ for which analyses of six spin and eight spin averages are required. It is no longer possible to establish exact relations like (6.120), but inequalities can be derived. If we write

$$\bar{\chi}_s(0) = \mathrm{d}^s \chi(0)/\mathrm{d}g^s \sim \tau^{-\gamma_s}, \tag{6.121}$$

then from (6.120)

$$\gamma_1 = 2\gamma \tag{6.122}$$

and from the higher spin correlations they showed that

$$\gamma_2 \geq 3\gamma, \quad \gamma_3 \geq 4\gamma. \tag{6.123}$$

Equations (6.122) and (6.123) provide support for the scaling hypothesis, and if this hypothesis is accepted, then from (6.122)

$$\phi = \gamma. \tag{6.124}$$

6.4.3 Spin Exchange Anisotropy

The definitive paper on spin exchange anisotropy is due to Pfeuty, Jasnow and Fisher (1974), who draw attention to the care needed in choosing appropriately the perturbation terms in the Hamiltonian.

Starting with an isotropic Heisenberg interaction

$$\mathscr{H} = -J \sum_{\mathbf{i},\mathbf{j}} (\sigma_{x\mathbf{i}}\sigma_{x\mathbf{j}} + \sigma_{y\mathbf{i}}\sigma_{y\mathbf{j}} + \sigma_{z\mathbf{i}}\sigma_{z\mathbf{j}}) \tag{6.125}$$

where **i**, **j** are nearest neighbours, one might consider adding a term

$$-gJ\sigma_{z\mathbf{i}}\sigma_{z\mathbf{j}} \tag{6.126}$$

which, for $g \geq 0$, would change the ground state from Heisenberg to Ising symmetry. This would be unsatisfactory for the following reason. The most general perturbation is of the form

$$\epsilon_1\sigma_{x\mathbf{i}}\sigma_{x\mathbf{j}} + \epsilon_2\sigma_{y\mathbf{i}}\sigma_{y\mathbf{j}} + \epsilon_3\sigma_{z\mathbf{i}}\sigma_{z\mathbf{j}}. \tag{6.127}$$

But when $\epsilon_1 = \epsilon_2 = \epsilon_3 = \epsilon$ there is no change in symmetry of the ground state, and hence no change in critical behaviour. However, there will be a change of critical temperature given by

$$kT_c = \theta J(1 + \epsilon). \tag{6.128}$$

Hence the scaling relations (6.101) to (6.103) will remain valid with a crossover exponent ϕ equal to 1.

Now any general perturbation which has a component along the (1,1,1) direction will contain a contribution from the $\phi = 1$ scaling behaviour which may mask the true ϕ value for which we are searching. Therefore we must choose perturbations whose directions are orthogonal to (1,1,1). Two convenient possibilities are $(1,-1,0)$ and $(-\frac{1}{2},-\frac{1}{2},1)$, and these are also orthogonal to one another. The corresponding anisotropic perturbations, termed *rhombic* and *axial*, are

$$-g_1 J(\sigma_{x\mathbf{i}}\sigma_{x\mathbf{j}} - \sigma_{y\mathbf{i}}\sigma_{y\mathbf{j}}) \tag{6.129}$$

and

$$-g_2 J[\sigma_{zi}\sigma_{zj} - \tfrac{1}{2}(\sigma_{xi}\sigma_{xj} + \sigma_{yi}\sigma_{yj})]. \qquad (6.130)$$

The direction (0,0,1) which we suggested initially is unsatisfactory because it has a component of magnitude $\frac{1}{3}$ in the direction (1,1,1) and a component of magnitude $\frac{2}{3}$ in the direction $(-\frac{1}{2},-\frac{1}{2}, 1)$.

Both (6.129) and (6.130) lead from classical Heisenberg to Ising symmetry, and Pfeuty, Jasnow and Fisher (1974) estimated the cross-over exponent ϕ, which is the same for both, as 1.25 ± 0.015. The scaling functions $X(x)$ differ in the two cases, and details are given in the original paper.

Starting from a classical x–y model the only possible perturbation is (6.129) which again leads to Ising symmetry. For this cross-over Pfeuty, Jasnow and Fisher (1974) estimated the exponent ϕ as 1.175 ± 0.015.

6.5 Universality

Whilst the developments of the previous section were taking place two key publications appeared which converted the background assumptions into a coherent formal hypothesis. The individual ideas in these publications were not new, but their synthesis provided a stimulus to further theoretical and experimental investigation to check whether the hypothesis was indeed widely satisfied. More importantly they summarized clearly and precisely results which had been derived empirically, and which any *theory* of critical phenomena must explain.

In a paper entitled 'Dependence of critical indices on a parameter' Griffiths (1970) drew attention to the fact that critical exponents remain constant for wide variations of the parameters in the Hamiltonian, and change only at certain critical values of these parameters. The critical values are associated with (i) a change of lattice dimensionality, (ii) a change of symmetry of the order parameter, (iii) a change in the range of interaction of the intermolecular force.

Result (i) has already been considered in detail. It is convenient to characterize result (ii) in terms of a magnetic classical vector interaction model whose spin is n-dimensional:

$$-J\boldsymbol{\sigma}_i^{(n)}\boldsymbol{\sigma}_j^{(n)} - \mu_0 H\sigma_{iz}^{(n)}. \qquad (6.131)$$

$n = 1$ corresponds to the Ising model, $n = 2$ to the x–y model, $n = 3$ to the classical Heisenberg model. This n-vector model was first introduced by Stanley (1968), who used the symbol D for spin dimension. In his development of the RG, Wilson (1972) changed D to n, and this notation has been widely adopted. By developing high-temperature series expansions for general n, Stanley was led to the important conclusion that as $n \to \infty$ the series are identical with those for the spherical model. Thus the spherical model ceased to be an *odd man out*, and could be fitted into a general framework of magnetic interaction models.

Following Joyce's work on the spherical model for long-range forces (§5.4.2) it is convenient to characterize (iii) by a parameter σ where the interaction $J(r)$ between spins a distance r apart is $\simeq r^{-(d+\sigma)}$ (equation (5.157)).

The critical exponents of the magnetic interaction model (6.131) are thus determined by the parameters (d,n,σ). The parameters d and n change discontinuously, as do the corresponding critical exponents. The parameter σ changes continuously, and so do the critical exponents, but following Joyce's exact results for the spherical model, one might anticipate key values of σ at which a discontinuity in the slope of the exponents arises. For short-range interactions the change from one set of values of (d,n) to another set of values is characterized by the cross-over behaviour described in the previous section.

Similar ideas to the above were put forward independently by Kadanoff in 1970 at the Summer School on Critical Phenomena in Varenna organized by Mel Green (Kadanoff 1971). The descriptive term *hypothesis of universality* was coined by Kadanoff, and each set of values of (d,n,σ) was said to characterize a different *universality class*. One of the first triumphs of the RG was to provide a theoretical basis for the hypothesis of universality and the concept of universality classes.

6.5.1 Lattice–Lattice Scaling

In the previous section we were concerned with the universality of critical *exponents*. But the view had been gaining ground that the shape of scaling functions like $h(x)$ in (6.36) might also follow the pattern of universality. The suggestion seems to have been first stated explicitly by Fisk and Widom (1969). The functions could certainly contain a constant multiplying factor, and the hypothesis was therefore that $h(x)/h(1)$ is a universal function.

This proposal was immediately put to a stringent test by Watson (1969) who pointed out that if it was valid certain combinations of critical amplitudes of thermodynamic quantities should be invariant of lattice structure for a given model in a given dimension. For example, if A_+, A_- are the amplitudes of the specific heat above and below T_c, i.e.

$$C_H \sim \begin{cases} A_+[(\tau^{-\alpha} - 1)/\alpha] & (\tau > 0) \\ A_-[(\tau'^{-\alpha} - 1)/\alpha] & (\tau' > 0) \end{cases} \tag{6.132}$$

($\alpha = 0$ giving rise to a logarithm), B_- the amplitude of the spontaneous magnetization,

$$m_0 \sim B_- \tau'^{\beta}, \tag{6.133}$$

and C_+ the amplitude of the initial susceptibility,

$$\chi_0 \sim C_+ \tau^{-\gamma}, \tag{6.134}$$

then A_+/A_- and $A_+ C_+/B_-^2$ should be invariant. For the $I(\frac{1}{2})$ model in two dimensions, the first ratio was known to be exactly 1 for all lattices, and A_+ and B_- were known exactly for the SQ, triangular, and honeycomb lattices, and C_+ was known accurately from high-temperature series expansions for these lattices. Watson found the same value 0.3186 to four figures for all three lattices, and considered this strong support for the hypothesis. He also calculated the values of a number of other invariants for the $I(\frac{1}{2})$ model in two and three dimensions, and found further evidence in favour of the suggestion.

A more comprehensive hypothesis relating to critical behaviour in individual lattices was put forward by Betts, Guttman and Joyce (1971) in a paper entitled 'Lattice–lattice scaling and the generalized law of corresponding states'. One must remember that the parameter x_0 in the definition of $h(x)$ is also lattice dependent, so that in relating the shape of $h(x)$ to an individual lattice there are two free parameters. Betts and his collaborators suggested that for a given universality class there is a unique function which defines the critical equation of state,

$$m = \psi(\tau,h). \tag{6.135}$$

For any individual lattice there are two parameters τ_0, h_0 relating to temperature and magnetic field respectively which connect its critical behaviour with (6.135); for this lattice the equation of state is

$$m = \psi(\tau_0 \tau, h_0 h). \tag{6.136}$$

If one wishes to deal with the free energy, the unique function being $f(\tau,h)$, the individual lattice free energy will be

$$h_0 f(\tau_0 \tau, h_0 h) \tag{6.137}$$

using the thermodynamic relation $m = -\partial f/\partial h$.

Betts and his collaborators were led to their hypothesis by a consideration of the exact star–triangle transformation (5.17) and its effect on the critical behaviour of the triangular and honeycomb lattices. The hypothesis automatically leads to the invariants introduced by Watson (1969). Numerical estimates of critical amplitudes and exact calculations for the spherical model and for $I(\frac{1}{2})$ in two dimensions fitted in well with the hypothesis.

One would have expected the parameters τ_0 and h_0 to change from one universality class to another. Betts, Guttman and Joyce were surprised to find that while τ_0 changed, h_0 did not, and remained constant for all universality classes.

As an extension of these ideas the cross-over functions introduced in §6.4 (like $V(x)$ in (6.99) or $Z(x)$ in (6.101)) might be expected to show universality; and a parameter g_0 should be related to g for each lattice to provide a generalization of the lattice–lattice scaling assumption (6.136),

$$m = \psi(\tau_0 \tau, h_0 h, g_0 g). \tag{6.138}$$

6.6 Finite Size and Surface Effects

For a comprehensive discussion of the topics in this section see Binder (1983) and Barber (1983).

6.6.1 *Introduction*

We have so far been concerned largely with assemblies which are homogeneous and infinite in extent, and we found that critical points are associated with characteristic singularities. All the assemblies which occur in practice are large on the molecular scale, but finite, and the partition function of a finite assembly is continuous and non-singular. It is important, therefore, to establish for what range of parameters like temperature and magnetic field a large assembly can be treated as infinite, and for what range finite size effects become significant. An additional tool introduced into the exploration of critical behaviour was that of *Monte Carlo computer simulations* of particular models (see e.g. Binder 1976), and the relationship between finite and infinite systems plays a crucial role in the proper interpretation of such simulations.

It is instructive to consider first a d-dimensional Ising system in which one of the dimensions becomes finite of size L. For $d = 2$, we are then dealing with the behaviour of a strip of infinite length but finite width L, and for $d = 3$, we are dealing with a slab infinite in two horizontal directions, but of finite height L. The parameter $1/L$ is analogous to the parameter g in §6.4.2; for $1/L = 0$ the critical behaviour corresponds to d dimensions, for sufficiently large $1/L$ to $(d - 1)$ dimensions, and we are dealing with a cross-over phenomenon from d to $(d - 1)$ dimensions for which we might reasonably expect a parallel treatment to that of §6.4.2 to be appropriate. For any value of L and $d > 2$ there will be a precisely defined $T_c(L)$, and the cross-over from d-dimensional to $(d - 1)$-dimensional critical behaviour will be governed by a cross-over exponent ϕ.

Moving now to a finite system of size L^d with free surfaces, there will no longer be a clearly defined critical temperature, since the finiteness of the assembly will cause any infinities to be rounded off to finite maxima. The behaviour of the specific heat would be expected to follow the pattern illustrated in Figure 6.8; the high-temperature initial susceptibility follows a similar pattern.

We can conveniently accept the temperature of the maximum, $T_{max}(L)$, as the analogue of the critical temperature for a finite assembly, and following

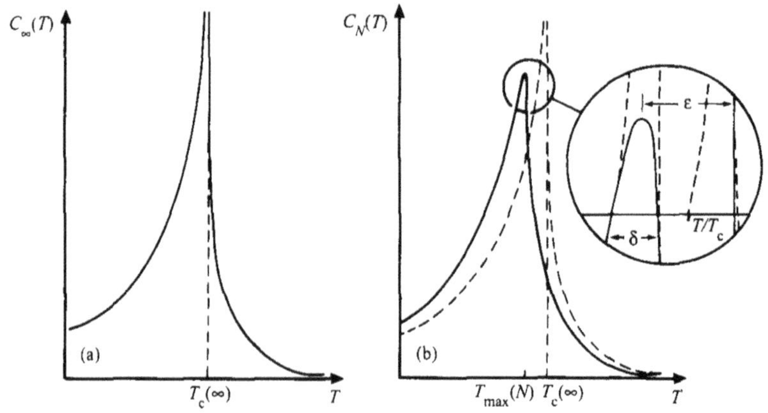

Figure 6.8 (a) Specific heat anomaly in an infinite system; (b) corresponding anomaly in a large finite system illustrating the fractional shift ϵ and rounding g.

(6.103) we define the *fractional shift* as

$$\tau_L = \frac{T_c(L)}{T_c(\infty)} - 1. \tag{6.139}$$

To deal with the rounding, we are guided by Fisher (1971) in defining ΔT, the temperature range of the rounding. First shift the temperature scale for the finite system so that $T_{max}(L) \simeq T_c$; under these circumstances one might expect the *wings* of the plot for $C_L(T)$ or $X_L(T)$ to match fairly closely the wings for the unshifted plot for $C_\infty(T)$ or $X_\infty(T)$. Then locate temperatures above and below T_c at which the deviation reaches some arbitrary value, say 5%, and identify ΔT as the difference between these temperatures. The *fractional rounding* is then defined by

$$\delta_L = \Delta T / T_c(\infty). \tag{6.140}$$

One of the major aims of the discussion will be to characterize the dependence of τ_L and δ_L on L.

For such a finite assembly we can derive useful information from general thermodynamic principles. The total free energy of an assembly containing N bulk molecules and N_s surface molecules is expected to be of the form

$$Nf(T) + N_s f_s(T) \tag{6.141}$$

where $f(T)$ represents the bulk free energy, and $f_s(T)$ the *surface free energy*. $f_s(T)$ is made up of the deviations from homogeneity arising for molecules at and near the surface layer. In our particular case $N \simeq L^d$, $N_s \simeq L^{d-1}$, so that (6.141) can be written in the form

$$N[f(T) + L^{-1} f_s(T)]. \tag{6.142}$$

We shall see that the behaviour of τ_L and δ_L is closely related to the critical behaviour of $f_s(T)$.

The surfaces envisaged in the previous paragraphs arose because layers of spins terminated in a rigid boundary. An alternative type of surface, an *interface*, arises in a ferromagnet at the boundary between two regions of bulk spins oriented in opposite directions. Such a boundary, which is free to change its shape, provides the analogue of the free surface of a fluid, and the $f_s(T)$ corresponding to such an interface gives rise to the surface tension discussed in §2.7 and §5.6.3.

Finally we can consider a finite assembly of L^d molecules with no free surfaces satisfying a cyclic boundary condition. We should not expect (6.142) to apply to such an assembly, and it is of interest to determine by what it is replaced.

6.6.2 *Historical Notes*

Among the results derived by Onsager in his classic solution was the specific heat of a finite crystal of L rows, relation (5.35),

$$C_H/Nk \simeq 0.4945 \ln L \tag{6.143}$$

the amplitude being identical with that of the specific heat as a function of temperature near T_c, relation (5.31),

$$C_H/Nk \simeq -0.4945 \ln \tau.$$

A diagrammatic analysis of the difference between high-temperature series expansions for infinite and finite lattices was undertaken by Domb (1965). The graphs are identical until a value M is reached when the graphs for a finite lattice tail off. Making an apparently plausible assumption that the value of M is of order L, Domb was able to explain the identity of the amplitudes in the above two relations. The same assumption leads to the suggestion that the maximum in the susceptibility in two dimensions is $\sim L^{7/4}$. However, in extending the results to three dimensions, a significant error occurred. The correct assumption on the value of M at which the diagrams start to be united is $L^{\nu} \sim M$ (Domb 1970) and this leads to the general conclusion that any quantity whose critical behaviour as a function of temperature is $\sim \tau^{-\theta}$ will have a maximum $\sim L^{\theta/\nu}$ (the error was not traced in two dimensions since $\nu = 1$).

Fisher and Ferdinand (1967) substantially extended the work of Onsager for the two-dimensional Ising model. They first introduced a neat and powerful device which enabled them to deal simultaneously with several different types of standard boundary condition. Consider a standard ferromagnetic lattice whose ends are connected cyclically, but in which the interaction of an array of bonds has been changed from J to aJ. The array can have a variety of sizes and shapes, and the most important used by Fisher and Ferdinand are

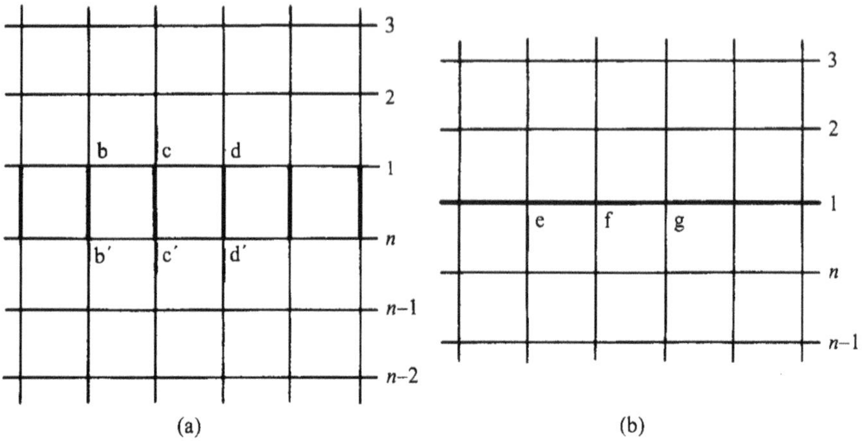

Figure 6.9 Two types of modified square lattice. All bonds have interaction constants J, except for the thicker bonds whose interaction constants are aJ. Both lattices have n rows, each containing m spins, and are wrapped on toruses.

illustrated in Figures 6.9(a) and 6.9(b). In Figure 6.9(a), $a = 0$ leads to an assembly with two rigid free surfaces, $a = -1$ to an assembly with a boundary tension (analogous to surface tension). In Figure 6.9(b), $a = \infty$ requires all the spins in the row to point in the same direction, and this is similar in behaviour to the wall of the containing vessel of a fluid.

A more general statement of the expression for the free energy (6.142) applicable to the modified lattices of Figures 6.9(a) and 6.9(b) is

$$N[f(T) + L^{-1}f(a,T)] \tag{6.144}$$

with an analogous formula for any other thermodynamic property. Fisher and Ferdinand developed an exact treatment for the calculation of $f(a,T)$ for the two-dimensional Ising model. Their results will be quoted in the context of the general theory to be developed.

For general dimension d Watson (1968) was able to establish some important rigorous results relating to the modification in the susceptibility $\chi(a, T)$. Assuming only the general form of the correlation function (5.163) Watson calculated the critical exponent $\gamma(-1)$ exactly

$$\gamma(-1) = \gamma + \nu \tag{6.145}$$

and established the inequality

$$\gamma(0) \geq \gamma + \nu. \tag{6.146}$$

He also undertook extensive calculations for the Gaussian and spherical models, and developed series expansions for the Ising model for a variety of lattices.

Putting together the results of all the calculations available, Watson (1972) suggested four simple empirical rules which govern the critical behaviour of the change $P(a,T)$ in thermodynamic property $P(T)$ caused by the modification:

1 The critical temperature at which the singularity in $P(a)$ occurs is exactly equal to the bulk critical temperature T_c.
2 If $\theta(a)$ represents the high-temperature critical exponent of any property $P(a,T)$, and θ the critical exponent of $P(T)$, then $\theta(a) - \theta$ is independent of P. A similar property holds for the low-temperature critical exponents $\theta'(a)$ and θ'.
3 If the bulk critical exponents θ and θ' obey scaling laws so do the corresponding exponents of the modification $\theta(a)$ and $\theta'(a)$.
4 For short-range force models the exponents of $P(a,T)$ are independent of the value of a.

These rules can be understood if we assume that near T_c there is only one length, ξ, which characterizes critical behaviour, and that the modification arising from any change in the bulk system extends to a range ξ from the region modified.

Watson (1972) therefore suggested the following exponent scheme:

$$\begin{aligned} \gamma(a) &= \gamma'(a) = \gamma + \nu \\ \Delta(a) &= \Delta'(a) = \delta \\ \delta(a) &= \Delta/(\beta - \nu) \\ \beta(a) &= \beta - \nu \\ \alpha_s(a) &= \alpha'_s(a) = \alpha_s + \nu \end{aligned} \tag{6.147}$$

(α_s refers to the singular portion).

Finally Fisher and his collaborators (see Fisher 1971, Barber 1983) formulated a comprehensive theory of finite size scaling which will now be discussed in more detail.

6.6.3 *Finite Size Scaling*

Following the pattern of §6.6.1 we first consider a three-dimensional slab, and adapt the discussion of §6.4.2. The parameter L^{-1} replaces g, and $L^{-1} = 0$ corresponds to a three-dimensional assembly; thus the cross-over is from a three-dimensional to a two-dimensional assembly, and the numbers 2 and 3 must be interchanged.

For the free energy we have, by analogy with (6.100),

$$f(L,H,\tau) = \tau^{2-\gamma_3'}(L^{-1}\tau^{-\phi}, H\tau^{-\Delta_3}) \tag{6.148}$$

and for the susceptibility in the transition to a two-dimensional slab

$$\chi_0(L,\tau) = \tau^{-\gamma_3} Z(L^{-1}\tau^{-\phi}) \tag{6.149}$$

by analogy with (6.101). When L^{-1} is sufficiently large for the behaviour to become two-dimensional, the susceptibility is given by (see (6.102))

$$\chi_0(L,\dot{\tau}) = \dot{A}(L)\dot{\tau}^{-\gamma_2}. \tag{6.150}$$

The new critical temperature deviates from the bulk value by (see (6.103))

$$\tau_L = \frac{T_c(L^{-1})}{T_c(0)} - 1 = (L^{-1}/x_c)^{1/\phi} \sim L^{-1/\phi} \tag{6.151}$$

with a scaling function of the form (6.104),

$$Z(x) = Z_c(1 - x/x_c)^{-\gamma_2}. \tag{6.152}$$

The amplitude $\dot{A}(L)$ is given by (6.106),

$$\dot{A}(L) = Z_c \phi^{-\gamma_2}(L^{-1}x_c)^{(\gamma_2-\gamma_3)/\phi} \sim L^{(\gamma_3-\gamma_2)/\phi}. \tag{6.153}$$

γ_2 is greater than γ_3, so that $\dot{A}(L)$ goes to zero for large L.

The assembly is expected to manifest two-dimensional behaviour when $\tau L^{1/\phi}$ reaches a particular value (see (6.110)). From this a plausible physical argument can be used to identify ϕ. The general philosophy of the scaling hypothesis has been that the only length of relevance in the critical region is the correlation range $\xi(T)$. We should therefore not expect two-dimensional effects to be manifest as long as $\xi(T)$ is small compared with L; but we should expect them to enter when $\xi(T)$ becomes comparable with L, i.e. when

$$\tau^{-\nu_3} \sim L \tag{6.154}$$

or when $\tau L^{1/\nu_3}$ reaches a specific value. From this discussion ϕ can reasonably be identified with ν_3.

The same conclusion follows from Watson's empirical rules for surface exponents given at the end of the previous section. This suggested that

$$\begin{aligned} P(a,T) &= N[P(T) + L^{-1}P(a,T)] \\ &= N[\tau^{-\theta} + L^{-1}\tau^{-(\theta+\nu_3)}] \end{aligned} \tag{6.155}$$

and the second term becomes comparable with the first when $\tau L^{1/\nu_3}$ reaches a significant value.

All of the above arguments could be reasonably applied to a finite system bounded by free surfaces, except that the critical temperature is no longer a precise concept, as discussed in §6.6.1. Because of this it is more convenient to rephrase the scaling hypothesis (e.g. (6.149)) so that the length L rather than the temperature is outside the bracket. Equation (6.149) can readily be rewritten in the form

$$\chi_0(L,\tau) = L^{\gamma_3/\nu_3} Q(\tau L^{1/\nu_3}) = L^{2-\eta_3} Q(\tau L^{1/\nu_3}) \tag{6.156}$$

where

$$Q(u) = u^{-\nu_3}Z(u^{-\nu_3}). \tag{6.157}$$

From (6.156) we can derive the maximum suggested at the beginning of §6.6.2. If scaling of this type is followed both τ_L in (6.139) and δ_L in (6.140) will be of order L^{-1/ν_3}.

For a cyclic boundary condition there will no longer be any surface terms, and the argument based on (6.155) ceases to be relevant. It was noted empirically in early enumerations that convergence to the bulk critical temperature was more rapid with such boundary conditions and for a while the idea was current that the exponent for τ_L might in some cases be larger than that for δ_L (Fisher 1971). More extensive investigation showed this to be incorrect, and confirmed the general validity of scaling relations of the form (6.156). Apparent deviations from this general picture noted for the spherical model and Bose–Einstein fluids (Fisher and Barber 1972a) were traced to the fact that the spherical model and n-vector model as $n \to \infty$ are equivalent only for bulk properties. For a layer system with free surfaces, the $n \to \infty$ limit of an n-component fixed system is *not* the spherical model as generally formulated, but rather a spherical model with distinct constraint fields applied to each layer parallel to the surface (Knops 1973).

Finally it is instructive to express the scaling relation (6.156) in an alternative form which sometimes gives it wider scope (Barber 1983). Our basic philosophy has been that only the correlation length ξ has relevance near T_c. For a thermodynamic quantity $P(\tau)$ whose bulk critical behaviour is given by

$$P(\tau) \sim C_\infty \tau^{-\theta} \tag{6.158}$$

it is plausible to assume a finite size behaviour

$$P_L(\tau) \sim L^\omega R(L/\xi) = L^\omega R(L\tau^{\nu_3}). \tag{6.159}$$

In order to satisfy (6.158) we must have asymptotically

$$R(u) \sim C_\infty u^{-\theta/\nu_3}$$

with $\omega = \theta/\nu_3$. This confirms the empirical suggestion mentioned at the beginning of §6.6.2 that the maximum of $P(\tau)$ for a finite system is $\sim L^{\theta/\nu}$.

In the case of a logarithmic divergence, where

$$P(\tau) \sim C_\infty \ln \tau, \tag{6.160}$$

Fisher (1971) suggested that a simple modification of (6.159) to the form

$$P_L(\tau) - P_L(\tau_0) = R(L\tau^\nu) - R(L\tau_0^\nu), \tag{6.161}$$

where τ_0 is any convenient fixed temperature, fulfils the necessary requirements. To satisfy (6.160) we must now have asymptotically

$$R(u) \sim (C_\infty/\nu) \ln u \tag{6.162}$$

and we find that the maximum of $P_L(\tau)$ is given by

$$(C_\infty/\nu)\ln L. \tag{6.163}$$

This is in agreement with Onsager's result (6.143).

The first relation in (6.159) can be applied to systems for which the dependence of ξ on τ is not governed by a power law. For example, for the Kosterlitz–Thouless transition mentioned in §5.3.3.1, it can be shown that (Kosterlitz and Thouless 1973)

$$\xi(\tau) \sim \xi_0 \exp(b\tau^{-1/2}). \tag{6.164}$$

6.6.4 Exact Calculations

Reference has been made in §6.6.2 to the calculations of Fisher and Ferdinand (1967) for lattices with a single layer of modified bonds, as illustrated in Figures 6.9(a) and 6.9(b). Calculations were undertaken for a number of two-dimensional lattices, the most extensive being for the SQ lattice. We shall draw on the review article by Watson (1972) to summarize the most important conclusions.

The excess energy $e(a,\tau)$ for the array of Figure 6.9(a) is given by

$$e(a,\tau) \simeq [2J(w_c'^2 - w_c^2)]/[\pi(w_c'^2 + w_c^2)]\ln|\tau| \tag{6.165}$$

where w_c' and w_c are the values of tanh aK and tanh K at $T = T_c$. The excess specific heat $C_H(a,\tau)$ due to the modified bonds is obtained by differentiating $e(a,\tau)$ with respect to T, and is given by

$$[C_H(a,\tau)/k] \simeq [2K_c(w_c'^2 - w_c^2)/\pi(w_c'^2 + w_c^2)]\tau^{-1}. \tag{6.166}$$

Relations (6.165) and (6.166) show that the singularities in $e(a,\tau)$ and $C_H(a,\tau)$ occur at exactly the same temperature ($\tau = 0$) as do those of the bulk properties $E(\tau)$ and $C_H(\tau)$. $e(a,\tau)$ has a logarithmic infinity to T_c, which is in marked contrast to the bulk contribution $E(\tau)$ which remains finite at T_c. The critical exponents of $C_H(a,\tau)$ above and below T_c are $\alpha_s(a) = \alpha_s'(a) = 1$, and hence, since the bulk exponents are $\alpha_s = \alpha_s' = 0$,

$$\alpha_s(a) - \alpha_s = \alpha_s'(a) - \alpha_s' = 1. \tag{6.167}$$

This is in accord with Watson's pattern (6.147), since $\nu = 1$.

An important point to notice is that the critical exponents are independent of a. However, the amplitudes of the singularities depend on a, and the amplitude multiplying $\ln|\tau|$ in (6.165) increases slowly and monotonically from $-2J/\pi$ at $a = 0$ to $2^{1/2}J/\pi$ at $a = \infty$.

For the modified array of Figure 6.9(b) the critical behaviour of $e(a,\tau)$ is

$$e(a,\tau) \simeq \frac{4J(1 - w_c w_c')(w_c' - w_c)}{\pi[(1 - w_c w_c')^2 + (w_c' - w_c)^2]}\ln|\tau| \tag{6.168}$$

showing that the nature of the singularity, which is again independent of a, is the same as for the modified array of Figure 6.9(a). The amplitude again increases slowly and monotonically with a, having the values $-2J/\pi$ at $a = -\infty$ and $2J/\pi$ at $a = \infty$.

The behaviour of $e(a,\tau)$, $C_H(a,\tau)$ for other two-dimensional lattices follows the same pattern, so that critical exponents for surface properties depend only on dimensionality and not on lattice structure.

The asymptotic analysis can be extended beyond the leading term. For example, for the SQ lattice with free boundaries

$$C_H(0,\tau) \simeq -A(0)\tau^{-1} + A_1(0)\ln|\tau| + A_2^{\pm}(0) + O(\tau \ln|\tau|) \quad (\tau \to \pm 0). \tag{6.169}$$

Ferdinand and Fisher (1969) also studied the exact form of the specific heat of an $L \times M$ SQ Ising lattice with periodic boundary conditions. An exact explicit expression for the p.f. was first given by Kaufman (1949), but much manipulation of generalized theta functions and other new functions (later known as remnant functions, Fisher and Barber (1972b)) was required in order to derive the asymptotic form of the specific heat. The result, for a fixed shape factor $f = M/L$, is as follows:

$$\begin{aligned} C_{LM}(\tau)/k = {} & 0.4945 \ln L + B(L\tau; f) + B_1(L\tau)(\ln L)/L \\ & + B_2(L\tau; f)/L + O([(\ln L)^3/L^2]. \end{aligned} \tag{6.170}$$

The first two terms confirm the scaling form (6.161) and the maximum (6.163); the shift and rounding exponents are, as expected from the previous discussion, equal to $1/\nu$ with $\nu = 1$.

6.7 Interfacial Density Profile

This section continues the discussion initiated in §5.6.3 and draws further on the review by Widom (1972). Attention was drawn in §5.6.3 to the suggestion by Widom that the basic features of the van der Waals theory of the interface could be taken over unchanged; only the critical exponents and critical equation of state needed modification to accord with contemporary developments.

The basic equations of the van der Waals theory, (2.133) and (2.136), are therefore adopted; but following (5.208) the constant c_2 is changed to B to move away from identification of this constant as an average of intermolecular forces. We thus write instead of (2.133)

$$B[\mathrm{d}^2\rho/\mathrm{d}z^2] = \mu(\rho,T) - \mu_\infty(T) \tag{6.171}$$

and instead of (2.136), for the surface free energy,

$$\sigma = \tfrac{1}{2}B \int_{-\infty}^{\infty} (\mathrm{d}\rho/\mathrm{d}z)^2 \,\mathrm{d}z. \tag{6.172}$$

To make the theory explicit Fisk and Widom (1969) used a scaling type of equation of state in the following form:

$$\mu(\rho,T) - \mu_{\infty}(T) = \Delta\rho\tau'^{\gamma-1}[T - t(\rho)]j(-x_0/x). \tag{6.173}$$

Here x is the variable used in (6.29) ($\tau\Delta\rho^{-1/\beta}$ for fluids), $-x_0$ is the zero of $h(x)$ in (6.36), and $t(\rho)$ is the temperature at which the phase transition occurs for the density ρ. Thus, by analogy with (6.36), $T_c - t(\rho)$ is equal to $x_0|\Delta\rho|^{1/\beta}$. It is easy to identify (6.173) with (6.35) if we write

$$h(x) = x^{\gamma}(1 + x_0/x)j(-x_0/x). \tag{6.174}$$

For a van der Waals fluid ($\gamma = 1$, $h(x) = a(x + x_0)$, see §6.2.3) the function $j(u)$ is a constant. The value $u = 1$ corresponding to $x = -x_0$ represents the phase boundary; the range $1 < u < \infty$ corresponding to $-x_0 < x < 0$ represents the stable phase for $T < T_c$; the range $0 < u < 1$ corresponding to $-\infty < x < -x_0$ represents an extension into the metastable and unstable regions. For the van der Waals theory this extension consists in assuming that $j(u)$ is equal to the same constant for $u < 1$ as for $u > 1$.

Fisk and Widom assumed such an extension to be possible for a general model. Fisher's conjecture (§6.1.1) that the p.f. for the Ising model has an essential singularity at the phase boundary implies that no *true* analytic continuation is possible for $j(u)$ at $u = 1$. We have quoted at the end of §6.1.1 the argument of Lanford and Ruelle (1969) which suggests that this is a general property of short-range force models. Fisk and Widom would presumably be satisfied with pseudo-analytic extension generated by a restrictive condition which prevents clusters above a certain arbitrary size from forming. In fact, their results do not depend in any vital way on the form of $j(u)$ but only on the scaling form of the extension of the equation of state.

Once equation (6.173) has been adopted it is a straightforward matter to integrate equation (6.171) to determine the density profile. For details of the calculation the reader is referred to the original paper of Fisk and Widom, or to Widom's review. The most important conclusions are as follows.

The symmetry in $\pm\Delta\rho$ of equation of state (6.173) ensures that $d\mu/d\rho$ is the same for coexisting liquid and gaseous phases; hence using the thermodynamic relations (2.16) and (2.17),

$$\rho_g^2\chi_g = \rho_l^2\chi_l \tag{6.175}$$

where χ_g, χ_l are the isothermal compressibilities. The density profile is then of the form

$$|\rho(z) - \rho_c| = \tfrac{1}{2}(\rho_l - \rho_g)q(|z|/\xi) \tag{6.176}$$

where

$$\xi = (B\rho_g^2\chi_g)^{1/2} = (B\rho_l^2\chi_l)^{1/2}, \tag{6.177}$$

while the surface tension is given by

$$\sigma = b\xi^{-1}B(\rho_l - \rho_g)^2 \tag{6.178}$$

with b a pure number related to $j(u)$.

The form (6.176) is precisely what would be expected for an assembly with one characteristic length, and it is natural to identify ξ with the range of correlation. From (6.177) we verify the dependence of B on τ' quoted in (5.209), i.e. $B \sim \tau'^{\gamma - 2\nu} = \tau'^{-\nu\eta}$. From (6.178) we then, using (6.55), re-derive the critical exponent for the surface tension

$$\mu = 2 - \alpha - \nu \tag{6.179}$$

in agreement with (5.207).

An important and characteristic property of the profile (6.176) is its exponential approach to the densities of the bulk phases at $z \to \pm\infty$

$$\rho(z) - \rho(\pm\infty) \sim c(\rho_l - \rho_g)\exp(-|z|/\xi), \tag{6.180}$$

with c a pure number related to $j(u)$.

6.8 Conclusions

In this chapter a coherent pattern has been put forward to which both theoretical and experimental results on critical behaviour seem to conform. The pattern was constructed empirically, although there were significant initial gropings towards a theoretical explanation of its basis. In the next chapter we shall see how concepts taken from field theory were used to convert these gropings into a viable mathematical tool, to furnish important further insight into the nature of critical phenomena, and to provide an adequate understanding of the origin of the empirical pattern which has been described.

Notation for Chapter 6

$\bar{s}$, $\bar{\sigma}$, θ, ϕ parameters in the droplet model.
$E_0(T)$, $E_2(T)$, $E_4(T)$ ... coefficients in the high-temperature expansion of ln Z in powers of H (6.16).
e_{2r} coefficients describing the critical behaviour of $E_{2r}(T)$ (6.19).
$\Gamma(\mathbf{r}_1, \mathbf{r}_2, \ldots, \mathbf{r}_{2n})$ $2n$-point cumulant correlation function.
$X(u)$ see (6.25), $Y(u)$ see (6.28), $W(u)$ see (6.30), $h(u)$ see (6.35), $U(u)$ see (6.53), $V(u)$ see (6.54), $k(u)$ see (6.56), functions defining the critical equation of state.
Generalized homogeneous functions $g(u,v)$ defined in (6.42) and (6.43).
L characteristic length in Kadanoff's approach.

$\langle\sigma\rangle_L$ average defined in (6.77).
$\tilde{K}$, $\tilde{h}$, $\tilde{\tau}$ parameters relating to block spin.
$\Gamma(r, h,\tau)$ pair correlation function in the presence of a field.
y, x exponents in Kadanoff's theory defined in (6.83) and (6.85).
g cross-over parameter (6.98), ϕ cross-over exponent.
$Z(u)$ cross-over function defined in (6.101).
$\dot{t} = T/T_c(g) - 1$ (see (6.102)).
$T_c(L)$ analogue of critical temperature for a finite assembly (e.g. temperature of specific heat maximum).
$\tau_L = T_c(L)/T_c(\infty) - 1$, δ_L is defined in (6.140).
$Q(u)$ see (6.157), function defining thermodynamic behaviour of finite systems.
$e(a,\tau)$ excess energy for an array with modified bonds (6.165).
$C_H(a,\tau)$ excess specific heat for an array with modified bonds (6.166).
$j(u)$ function defining scaling equation of state for density profile (6.173) and related to $h(u)$ by (6.174).

References

ABE, R. (1970) *Prog. Theor. Phys.* **44**, 339.
AHARONY, A. and AHLERS, G. (1980) *Phys. Rev. Lett.* **44**, 782.
AHLERS, G. (1980) *Rev. Mod. Phys.* **52**, 489.
ARROT, A. and NOAKES, J. E. (1967) *Phys. Rev. Lett.* **19**, 786.
BAKER, G. A. and GAUNT, D. S. (1967) *Phys. Rev.* **155**, 545.
BAND, W. (1939) *J. Chem. Phys.* **7**, 324.
BARBER, M. N. (1983) DL **8**, Ch. 2.
BECKER, R. and DÖRING , W. (1935) *Ann. Phys., Leipzig* **24**, 719.
BETTS, D. D., GUTTMAN, A. J. and JOYCE, G. S. (1971) *J. Phys. C: Solid State Phys.* **4**, 1994.
BIJL, A. (1938) Doctoral Dissertation, Leiden.
BINDER, K. (1976) DG **5b**, Ch. 1.
BINDER, K. (1983) DL **8**, Ch. 1.
CHANG, M. C. and HOUGHTON, A. (1980) *Phys. Rev. Lett.* **44**, 785.
DOMB, C. (1965) *Proc. Phys. Soc.* **86**, 933.
DOMB, C. (1966) *Proc. Low Temperature Calorimetry Conf., Ann. Acad. Sci. Fennicae A* **VI**, 167.
DOMB, C. (1970) *J. Phys. C: Solid State Phys.* **3**, 256.
DOMB, C. (1971) *Proc. Int. School of Physics Enrico Fermi, Course 51*, ed. M. S. GREEN, p. 207.
DOMB, C. (1976) *J. Phys. A: Math. Gen.* **9**, 283.
DOMB, C. and HUNTER, D. L. (1965) *Proc. Phys. Soc.* **86**, 1147.
ESSAM, J. and FISHER, M. E. (1963) *J. Chem. Phys.* **38**, 147.
ESSAM, J. and HUNTER, D. L. (1968) *J. Phys. C: Solid State Phys.* **1**, 392.
FERDINAND, A. and FISHER, M. E. (1969) *Phys. Rev.* **185**, 832.
FISHER, M. E. (1967) *Physics* **3**, 255.
FISHER, M. E. (1971) *Proc. Enrico Fermi School on Critical Phenomena*, ed. M. S. GREEN, London and New York: Academic Press, p. 1.

FISHER, M. E. and BARBER, M. N. (1972a) *Phys. Rev. Lett.* **28**, 1516.
FISHER, M. E. and BARBER, M. N. (1972b) *Arch. Rat. Mech. Anal.* **47**, 205.
FISHER, M. E. and FERDINAND, A. (1967) *Phys. Rev. Lett.* **19**, 169.
FISK, S. and WIDOM, B. (1969) *J. Chem. Phys.* **50**, 3219.
FRENKEL, J. (1939) *J. Chem. Phys.* **7**, 200.
GAUNT, D. S. (1967) *Proc. Phys. Soc.* **92**, 150.
GAUNT, D. S. and BAKER, G. A. (1970) *Phys. Rev. B* **1**, 1184.
GAUNT, D. S. and DOMB, C. (1970) *J. Phys. C: Solid State Phys.* **3**, 1442.
GRIFFITHS, R. B. (1965) *Phys. Rev. Lett.* **14**, 623.
GRIFFITHS, R. B. (1967) *Phys. Rev.* **158**, 176.
GRIFFITHS, R. B. (1970) *Phys. Rev. Lett.* **24**, 1479.
HANKEY, A. and STANLEY, H. E. (1972) *Phys. Rev. B* **6**, 3515.
HILEY, B. J. and SYKES, M. F. (1961) *J. Chem. Phys.* **34**, 1531.
HO, J. T. and LITSTER, J. D. (1969) *Phys. Rev. Lett.* **22**, 603.
HOCKEN, R. and MOLDOVER, M. R. (1976) *Phys. Rev. Lett.* **37**, 29.
JOYCE, G. S. (1972) DG **2**, Ch. 10.
KADANOFF, L. (1966) *Physics* **2**, 263
KADANOFF, L. (1971) *Proc. Enrico Fermi School on Critical Phenomena*, ed. M. S. GREEN, London and New York: Academic Press, p. 100.
KAUFMAN, B. (1949) *Phys. Rev.* **76**, 1232.
KNOPS, H. J. F. (1973) *J. Math. Phys.* **14**, 1918.
KOSTERLITZ, J. M. and THOULESS, D. J. (1973) *J. Phys C: Solid State Phys.* **6**, 1181.
KOUVEL, J. M. and FISHER, M. E. (1964) *Phys. Rev. A* **136**, 1626.
LANFORD, D. E. and RUELLE, D. (1969) *Commun. Math. Phys.* **13**, 194.
LEVELT-SENGERS, J. M. H., GREEN, M. S. and VICENTINI-MISSONI (1969) *Phys. Rev. Lett.* **22**, 389.
LIU, L. L. and STANLEY, H. E. (1972) *Phys. Rev. Lett.* **29**, 927.
LIU, L. L. and STANLEY, H. E. (1973) *Phys. Rev. B* **8**, 2279.
MANDELBROT, B. (1977) *Fractals: Form, Chance & Dimension*, San Francisco: Freeman.
MILOSEVIC, S. and STANLEY, H. E. (1972) *Phys. Rev. B* **5**, 2526.
PATASHINSKII, A. Z. and POKROVSKII, V. L. (1966) *Sov. Phys. JETP* **23**, 292.
PFEUTY, P., JASNOW, J. and FISHER, M. E. (1974) *Phys. Rev. B* **10**, 2088.
PIPPARD, A. B. (1959) *Classical Thermodynamics*, Cambridge.
REHR, J. J and MERMIN, N. D. (1973) *Phys. Rev. A* **8**, 472.
RIEDEL, E. and WEGNER, F. (1969) *Z. Phys.* **225**, 195.
ROACH, P. R. (1968) *Phys. Rev.* **170**, 213.
RUSHBROOKE, G. S. (1963) *J. Chem. Phys.* **39**, 8942.
STANLEY, H. E. (1968) *Phys. Rev.* **176**, 718.
STANLEY, H. E. (1971) *Phase Transitions & Critical Phenomena*, Oxford.
VICENTINI-MISSONI, M. (1972) DG **2**, Ch. 2.
WATSON, P. G. (1968) PhD Thesis, London University.
WATSON, P. G. (1969) *J. Phys. C: Solid State Phys.* **2**, 2158.
WATSON, P. G. (1972) DG **2**, Ch. 4.
WEISS, P. and FORRER, R. (1926) *Ann. Phys., NY* **5**, 153.
WIDOM, B. (1965) *J. Chem. Phys.* **43**, 3898.
WIDOM, B. (1972) DG **2**, Ch. 3.

WILSON, K. G. (1971) *Phys. Rev. B* **4**, 3174, 3184.
WILSON, K. G. (1972) *Phys. Rev. Lett.* **28**, 548.
YANG, C. N. and LEE, T. D. (1952) *Phys. Rev.* **87**, 404, 410.

7

Renormalization Group

7.1 Introductory Review

The renormalization group (RG) does not produce exact solutions of the Onsager type, and its application involves quite drastic approximations. But unlike the closed-form approximations of the classical theory discussed in Chapters 2 and 3, which do not reflect physical behaviour near the critical temperature, T_c, except in a small number of specific cases, the RG endeavours to retain all the essential physical characteristics of the problem near T_c, whilst rejecting those features which are not significant. This requires continuous thought and attention – in Wilson's graphic language (1975a): 'One cannot write a renormalization cook book.' In his Nobel prize lecture, Wilson (1983) describes how hard it was in the initial stages of his work to find approximations which would be computable in practice.

No reliable method has been found of assessing directly the magnitude of the error at any stage of approximation of an RG treatment, and comparison with the results of exact solutions and series expansion estimates has played, and continues to play, a key role in the successful application of the RG. The methodology described in the previous chapters of this book has not become redundant as a result of the emergence of the RG – on the contrary its value has been enhanced.

The RG approach has been reviewed briefly in §1.7. Before amplifying this review we shall outline (following Wilson 1975b) the nature of the problem to be tackled in trying to account for critical behaviour. We shall then discuss qualitatively how the RG deals with it.

If we consider the behaviour of a fluid like water, far from the critical point there are microscopic density fluctuations on an atomic scale (wavelengths ~ 1 Å). If the temperature and pressure are increased towards the critical point, fluctuations become important at longer wavelengths. Sufficiently close to T_c and P_c there are fluctuations on the scale of 1000–10 000 Å which scatter ordinary light and give rise to critical opalescence, making the water look milky. However, the microscopic fluctuations have not decreased in size. Close to the critical point there are fluctuations at all wavelengths from 1 Å up to the correlation length ξ which tends to infinity at the critical point. The essence of the problem is how to deal with an assembly with so many scales of length.

The aim of the RG approach is to reduce in a systematic way the large number of degrees of freedom and the large number of length scales of the assembly. A transformation is set up in the parameter space of the Hamiltonian.

$$\mathscr{H}' = \mathbf{R}[\mathscr{H}] \tag{7.1}$$

which preserves the dimensionality and symmetries of the assembly, and reduces the correlation length by a factor $b > 1$ and the number of degrees of freedom from N to $N' = N/b^d$. The transformation is chosen so that the p.f. is preserved

$$Z_{N'}(\mathscr{H}') = Z_N(\mathscr{H}). \tag{7.2}$$

A wide choice of transformations is possible which can be characterized in *real space* or in *momentum space*. In real space the transformation usually follows the Kadanoff idea of introducing block spin variables, or eliminating some of the spins by partial summation (decimation transformation). However, the edge of the block is not long compared with the lattice spacing, but contains one or two lattice spacings; also the Kadanoff picture must be amplified, and the relation between the original spin and the block spin must be described precisely. In momentum space the aim is to eliminate high-momentum variables which correspond to short-wavelength fluctuations.

The transformation is iterated,

$$\mathscr{H}' = \mathbf{R}(\mathscr{H}), \quad \mathscr{H}'' = \mathbf{R}(\mathscr{H}'), \quad \ldots \tag{7.3}$$

and it is assumed that convergence takes place to a fixed-point Hamiltonian $\mathscr{H}^*$ satisfying

$$\mathscr{H}^* = \mathbf{R}[\mathscr{H}^*] \tag{7.4}$$

and that critical properties are determined by behaviour near the fixed point. Since iterative processes play a central role in the development of the theory, we shall devote the opening part of the next section to a discussion of their characteristic properties.

To give a specific illustration of the above ideas, for real-space renormalization, we take an example used by Wilson (1979) in relation to the simple Ising model on the two-dimensional SQ lattice of lattice spacing a. The block spin relates to a 2×2 cell of the lattice which contains nine spins. The value of the block spin is determined by a simple majority rule: if five or more of the nine spins point upwards, the block spin has value +1, otherwise it has value −1. It is a straightforward matter with the help of a computer to transform any configuration of the original lattice to one of the transformed lattice (Figure 7.1). The 2^9 possible configurations of a cell in the original lattice are reduced to two in the transformed lattice, the value of b above is 3, and the new lattice has spacing $3a$. Speaking roughly we can say that the transformation averages out fluctuations in spin direction whose scale is smaller than the block size $3a$. It is as if one looks at the lattice through an out-of-focus lens which blurs the smaller features but does not affect the larger ones.

When the transformation is applied for a second time, fluctuations in spin direction whose scale is smaller than 3^2a will have been averaged out. After n applications, fluctuations whose scale is smaller than $3^n a$ will have been averaged out. Eventually fluctuations at all scales up to the correlation length are averaged out.

Block spins do not have the same couplings as the spins in the original lattice. In the orignal formulation there is one nearest-neighbour interaction parameter. The first stage of the block spin transformation introduces a finite but substantial number of more distant neighbours and multispin couplings. If the second stage is carried out exactly a new round of coupling parameters will be introduced; if this is continued indefinitely an infinite number of coupling parameters will be required. An infinity of parameters is needed to furnish an exact solution. Wilson's crucial conjecture was that all but a finite and relatively small number of coupling parameters can be ignored without impairing the physical description of behaviour near the critical point. He later tested this conjecture by performing a calculation for the simple Ising model on the SQ lattice using a decimation transformation (developed by Kadanoff) and retaining 217 parameters (Wilson 1975b)! His conclusions:

> The dominant interaction at the fixed point is the nearest-neighbour coupling. For accurate calculations one must include many more; but for a qualitative picture the nearest-neighbour constant K plus perhaps the next-nearest neighbour coupling L should be enough. The remaining couplings are at least a factor 5 smaller than L .... Thus the old idea of Kadanoff that there would be effective nearest-neighbour Ising models for block spins is very close to the truth.

The term *group* is used for the set of transformations since they are generated by successive applications of a single transformation. In fact *semi-group* would be a more accurate description since the inverse of the transformations is not defined. One can go from spin to block spin, but not vice versa.

For the use of the RG in momentum space it proved convenient to employ a continuous spin adaptation of the Ising model. This enabled contact to be made with the Gaussian model (§5.4.2) which gives valid results for critical exponents when $d > 4$; replacing the lattice structure by a continuum (as for a fluid) the theory could be merged with the Landau–Ginzburg development (§4.9.1) and with the field theories with which the RG originated.

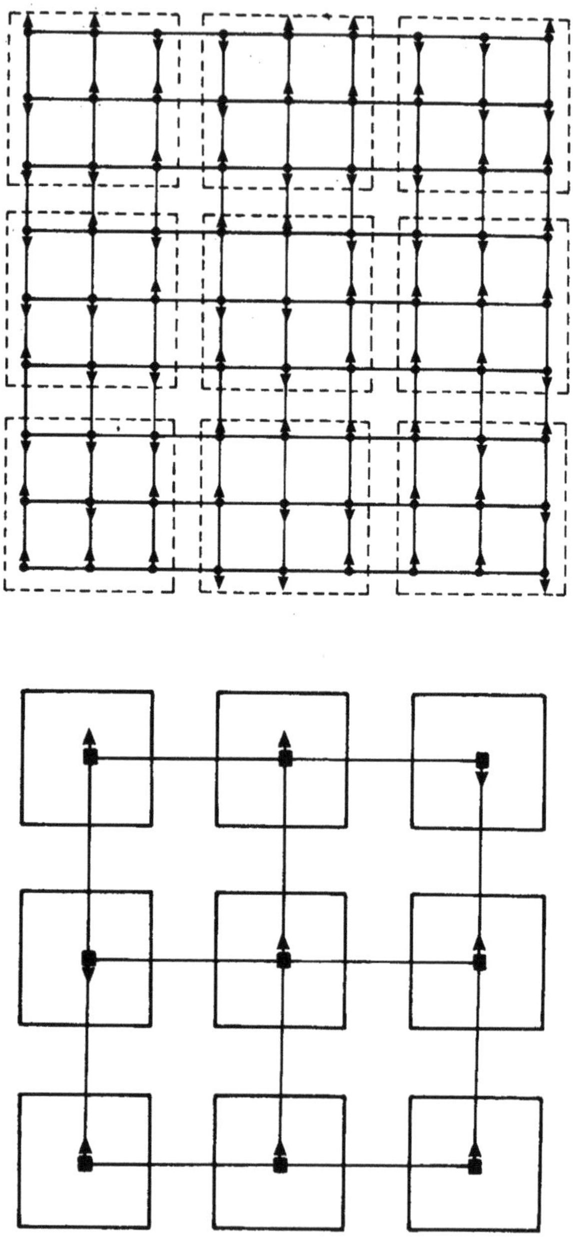

Figure 7.1 The block spin transformation. The top diagram represents a typical Ising model spin configuration. The lattice is subdivided into 2 × 2 blocks each containing nine spins. The sign of the block spin is determined by a majority rule, and the configuration of the transformed lattice of block spins is shown below. Finally the original length scale must be restored by reducing the dimensions by a factor of 3.

The p.f. for the $I(\frac{1}{2})$ model (equation (4.72)) can be written in the alternative form

$$Z_l = \int_{-\infty}^{\infty} d\sigma_l \, 2\delta(\sigma_l^2 - 1) \exp\left(K \sum_{\mathbf{l},\mathbf{m}} \sigma_l \sigma_\mathbf{m} + \beta\mu_0 H \sum_{\mathbf{l}} \sigma_l \right). \tag{7.5}$$

Here **l** and **m** are used to denote lattice points instead of **i** and **j** to indicate that σ_l is a continuous variable. Wilson replaced the function $2\delta(\sigma_l^2 - 1)$ by the continuous function

$$W(\sigma_l) = \exp(-\tfrac{1}{2}c\sigma_l^2 - u\sigma_l^4) \tag{7.6}$$

where u is a positive number. When $c = -4u$, $W(\sigma_l)$ has the form

$$W(\sigma_l) = \exp[-u(\sigma_l^2 - 1)^2 + u] \tag{7.7}$$

which reproduces the $I(\frac{1}{2})$ model when $u \to \infty$. For finite positive u, $W(\sigma_l)$ then has maxima at $\sigma_l = \pm 1$ (Figure 7.2), and if u is large the probability function is sharply peaked at $\sigma_l = \pm 1$. When $u = 0$ in (7.6) the Gaussian model is retrieved for positive c.

Following the idea of universality classes enunciated in §6.5 one might expect that any one weighting function of the form (7.7) would belong to the Ising universality class. Wilson made the more daring suggestion that as long as $u > 0$ *any* weighting function of the form (7.6) will correspond to the Ising universality class, i.e. that the only critical value of the parameters (c,u) in the sense of Griffiths (1970) is $u = 0$.

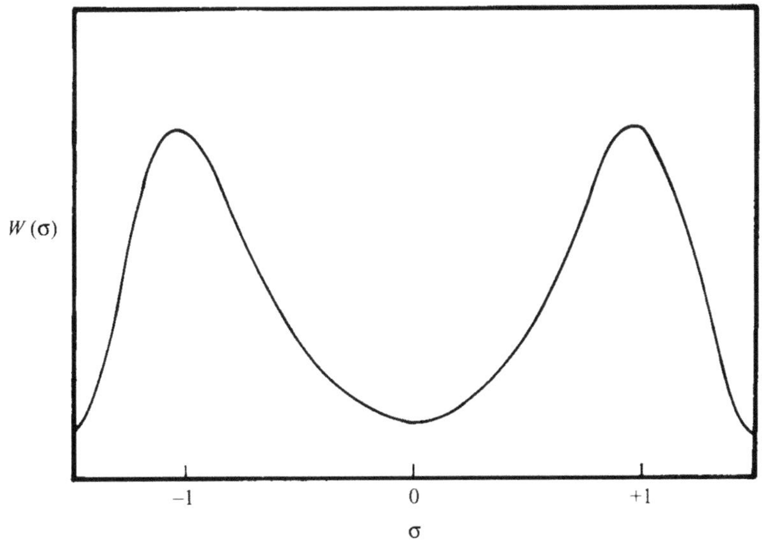

Figure 7.2 The weighting function $W(\sigma) = \exp[-u(\sigma^4 - 2\sigma^2)]$ for $u = 1.8$.

The transition to a continuum-space model can be achieved on replacing the $-\sigma_{\mathbf{l}}\sigma_{\mathbf{m}}$ in (7.5) by

$$\tfrac{1}{2}(\sigma_{\mathbf{l}} - \sigma_{\mathbf{m}})^2. \tag{7.8}$$

Since the lattice spacing is small and not expected to exercise a significant influence on critical behaviour, the discrete $\sigma_{\mathbf{l}}$ is replaced by the continuous function $\sigma(x,y,z)$, and (7.8) becomes

$$\tfrac{1}{2}(\nabla\sigma)^2. \tag{7.9}$$

To operate in momentum space it is convenient to introduce the Fourier-transformed spin variables (cf. equation (4.125))

$$\hat{\sigma}_{\mathbf{q}} = \sum \exp(i\mathbf{q}\cdot\mathbf{l})\sigma_{\mathbf{l}}. \tag{7.10}$$

For a lattice structure of spacing a, the momenta $\mathbf{q}$ are essentially restricted by $|q| \leq \pi/a$ (neglecting the shape of the Brillouin zone which, as one might expect, proves irrelevant). This cut-off is characteristic of the physics of condensed matter, and contrasts with quantum field theory where there is normally no natural cut-off and the momenta must be taken to infinity; the latter is therefore represented by zero lattice spacing. The RG transformation $\mathbf{R}$ of (7.1) is defined by averaging out all the $\hat{\sigma}_{\mathbf{q}}$ variables satisfying

$$b^{-1} < |q|a/\pi \leq 1. \tag{7.11}$$

This corresponds as before to a spatial rescaling factor b.

The first practical application of the RG was in momentum space taking $b = 2$ (Wilson 1971). Severe approximations were required in evaluating the integrals to put the results into a form from which iterative recursion relations could be derived. These relations were treated numerically to give the following values for the exponents of the three-dimensional $I(\tfrac{1}{2})$ model: $\nu \simeq 0.61$, $\gamma \simeq 1.22$, $\eta \simeq 0$. The exponent estimates were sufficiently close to the values derived from series expansions (§5.3.3) to indicate that the method was viable (the zero value obtained for η was approximate, and there were indications that neglected terms could give rise to a small non-zero value).

In a subsequent paper (Wilson and Fisher 1972) the method was streamlined, and the key suggestion advanced that the appropriate expansion parameter for the critical exponents is $\epsilon = 4 - d$, treated as a continuous parameter. Expansions to the first order in ϵ were derived for a number of models.

The following paper (Wilson 1972) generalized the treatment to the n-vector classical Heisenberg model, using for the exponent in (7.6)

$$-\tfrac{1}{2}c_0 \sum_{\alpha=1}^{n} \sigma_\alpha^2 - u_0\left(\sum_{\alpha=1}^{n} \sigma_\alpha^2\right)^2. \tag{7.12}$$

The continuum variables of field theory were used, an extra term being added to (7.9) to ensure a cut-off for high momenta. The Hamiltonian used for a

d-dimensional n-vector model was thus

$$-\beta\mathcal{H} = \int d\mathbf{x}\left[\frac{1}{2}c_0\sum_{i=1}^{n}\sigma_i^2(\mathbf{x}) + u_0\left(\sum_{i=1}^{m}\sigma_i^2(\mathbf{x})\right)^2 + \frac{1}{2}\sum_{i=1}^{n}[\nabla\sigma_i(\mathbf{x}) - \nabla\nabla^2\sigma_i(\mathbf{x})]^2\right], \tag{7.13}$$

$\mathbf{x}$ being a position vector in d dimensions.

The first two terms of the ϵ expansion were derived by the standard Feynman-graph methods of field theory. 'The Feynman-graph method of this paper is unrelated to the renormalization group methods .... The calculation of critical exponents in powers of ϵ is simpler in the graphical approach than in the exact renormalization-group approach'.

Feynman-graph methods and standard renormalization techniques had recently been applied to two problems in critical phenomena. Thouless (1969) had considered the Ising model with a long-range interaction, and Larkin and Khmelnitskii (1969) had treated the n-vector model in four dimensions. Wilson adapted their work diagram by diagram to non-integral d, and calculated the correlation function $\hat{\Gamma}(\mathbf{q})$ whose behaviour at T_c was known to be $\sim q^{-2+\eta}$ (equation (5.167)). He found that there is a unique choice $u_0 = u_0(\epsilon)$ for which the expansion for $\hat{\Gamma}(q)$ in terms of u_0 matches the expected critical behaviour. He was thus able to derive the first two terms of the ϵ expansion for the critical exponents. The results were as follows:

$$\gamma = 1 + \frac{(n+2)}{2(n+8)}\epsilon + \frac{(n+2)(n^2+22n+52)}{4(n+8)^3}\epsilon^2 + O(\epsilon^3); \tag{7.14}$$

$$\eta = \frac{(n+2)}{2(n+8)^2}\epsilon^2 + \frac{(n+2)}{2(n+8)^2}\left(\frac{6(3n+14)}{(n+8)^2} - \frac{1}{4}\right)\epsilon^3 + O(\epsilon^4); \tag{7.15}$$

$$\phi = 1 + \frac{n\epsilon}{2(n+8)} + \frac{\epsilon^2(n^3+24n^2+68n)}{4(n+8)^3}\epsilon^2 + O(\epsilon^3). \tag{7.16}$$

There are a number of interesting points about these expansions. The expansion (7.15) for η has no term of order ϵ, and this accounts for its small value. Even though ϵ is assumed to be a small parameter, substituting $\epsilon = 1$ gives results which compare surprisingly well with the series expansion estimates (for $n = 1$, $\gamma \simeq 1.244$ compared with 1.250 (§5.3.3.1); $\eta \simeq 0.037$ compared with 0.041 (equation (5.180′)). The values derived for $n = \infty$ correspond to the spherical model (Stanley 1968) for which exact results are available; the comparison is again surprisingly favourable (e.g. $\gamma = 1.75$ compared with 2).

This seminal paper initiated a new diagrammatic approach to the calculation of critical exponents, and powerful field theoretical techniques were used to extend the ϵ expansion to higher orders, and to improve the accuracy of the critical exponent estimates. The same techniques could also be used for the calculation of scaling and cross-over functions. In fact we shall see in §7.7.2

that it is these techniques that provided more accurate information than had been derived by series expansions.

Wilson (1976) regretted having to abandon the closed-form treatment of the RG for diagrammatic expansions.

> When I started studying the RG, my hope was that it would be useful for the problem of strong interactions of elementary particles where (at the time I began to study) Feynman diagram methods seemed hopeless. Again, in the case of critical phenomena I assumed that diagrammatic methods had and would fail. History has passed me by. Some of you who were struggling with field theoretical methods as applied to critical phenomena may wish that history had been kinder to me. To you I can only apologize for my part, in bringing the diagrammatic methods into the problem.

The similarity of (7.13) to the Landau–Ginzburg treatment mentioned at the end of §4.9.1 is apparent. But it is important to identify where the original Landau theory of §4.9 goes wrong, and how the RG approach corrects it.

In the rest of this chapter we shall endeavour to elaborate on the topics raised in this introduction. Our basic aim will be to formulate general principles and summarize what results have been achieved without entering into too many details of the technical aspects of the calculations.

7.2 Background Ideas

7.2.1 Iterative Processes

The simplest iterative process is the one-dimensional relation

$$x_{n+1} = f(x_n) \tag{7.17}$$

which leads to a solution of the equation

$$x = f(x). \tag{7.18}$$

Let ξ be a particular solution of this equation. Write

$$x_{n+1} = \xi + \theta_{n+1} \tag{7.19}$$

and assume that we are sufficiently close to ξ for a Taylor expansion to the first order about ξ to provide an adequate approximation. Substituting in (7.17), we find that

$$\theta_{n+1} \simeq f'(\xi)\theta_n. \tag{7.20}$$

Hence we see that if $|f'(\xi)| < 1$ successive deviations from ξ will decrease in a geometric progression, whereas if $|f'(\xi)| > 1$ they will increase in a geometric progression. Points of the first type are termed *stable fixed points*, since any deviation from the fixed point will be damped away. Points of the second type are termed *unstable fixed points.*

This is illustrated diagrammatically in Figure 7.3. The solutions of (7.18) are given by the intersection of the curve $y = f(x)$ with the line $y = x$. At fixed points A and B the slope $f'(\xi)$ is greater than 1, and they are therefore unstable; at fixed points A′ and B′ the slope $f'(\xi)$ is less than 1, and they are therefore stable. At points C and D $f'(\xi) = 1$, and the fixed point can be described as *marginal*; higher-order derivatives are needed to clarify the situation.

The behaviour of successive x_n in relation (7.17) is illustrated in Figure 7.4. For any initial value between x_A and x_B the x_n converge to $x_{A'}$; only for the isolated initial values x_A and x_B will the x_n remain at the unstable fixed point.

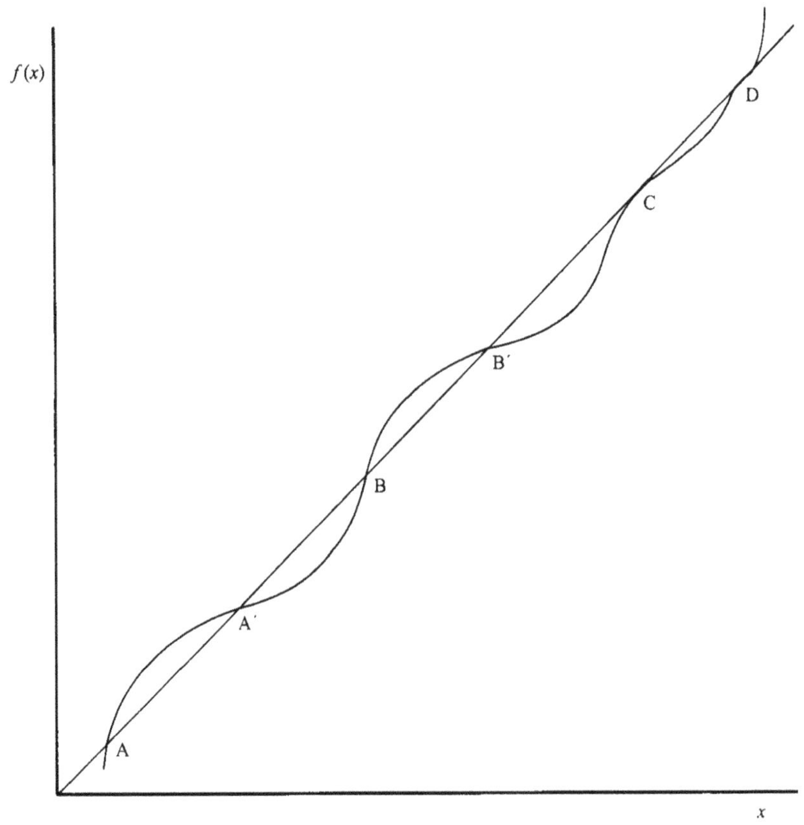

Figure 7.3 Fixed points in a one-dimensional iterative process: A and B are unstable, A′ and B′ are stable, C and D are marginal.

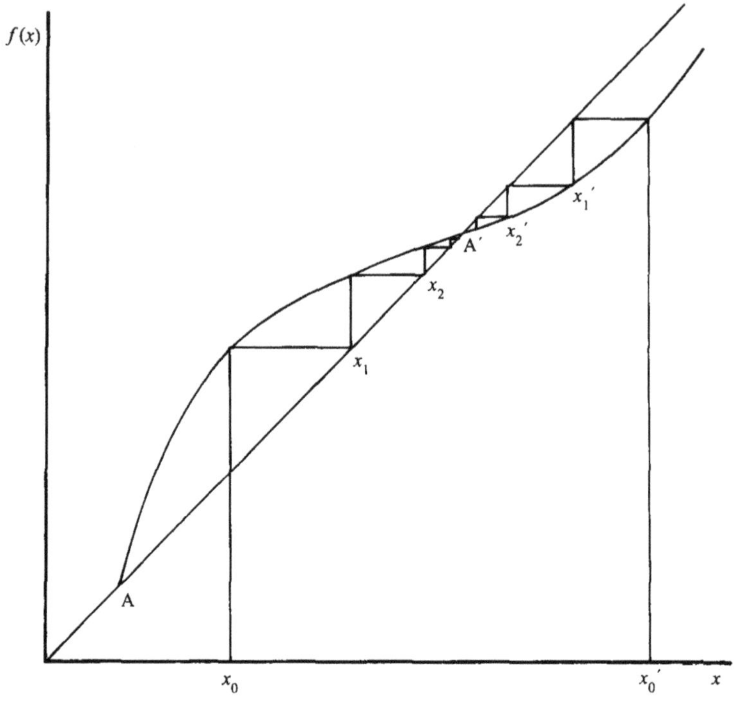

Figure 7.4 Convergence to a stable fixed point.

If we now move to a two-dimensional process,

$$\begin{aligned} x_{n+1} &= f_1(x_n, y_n) \\ y_{n+1} &= f_2(x_n, y_n), \end{aligned} \tag{7.21}$$

the fixed points are the solutions of the simultaneous equations

$$\begin{aligned} x &= f_1(x,y) \\ y &= f_2(x,y). \end{aligned} \tag{7.22}$$

Each of these equations represents a curve, and the fixed points are the points of intersection of the two curves.

Let (ξ,η) be a particular solution, and write

$$\begin{aligned} x_{n+1} &= \xi + \theta_{n+1} \\ y_{n+1} &= \eta + \theta_{n+1}. \end{aligned} \tag{7.23}$$

Substituting in (7.21) and expanding to first order, we find that

$$\begin{aligned} \theta_{n+1} &= (\partial f_1/\partial \xi)\theta_n + (\partial f_1/\partial \eta)\phi_n \\ \phi_{n+1} &= (\partial f_2/\partial \xi)\theta_n + (\partial f_2/\partial \eta)\phi_n \end{aligned} \tag{7.24}$$

or in matrix notation

$$\mathbf{\Theta}_{n+1} = \mathbf{\Delta}\mathbf{\Theta}_n \tag{7.25}$$

where

$$\mathbf{\Theta}_n = \begin{pmatrix} \theta_n \\ \phi_n \end{pmatrix}, \quad \mathbf{\Delta} = \begin{pmatrix} \partial f_1/\partial \xi & \partial f_1/\partial \eta \\ \partial f_2/\partial \xi & \partial f_2/\partial \eta \end{pmatrix}.$$

Suppose that the eigenvalues of the matrix $\mathbf{\Delta}$ are λ_1, λ_2, and the corresponding right-hand eigenvectors $\mathbf{t}_1$, $\mathbf{t}_2$. $\mathbf{\Theta}_0$ can then be expressed in the form

$$\mathbf{\Theta}_0 = a_1\mathbf{t}_1 + a_2\mathbf{t}_2 \tag{7.26}$$

and substituting in (7.25) we find that

$$\mathbf{\Theta}_n = a_1\lambda_1^n\mathbf{t}_1 + a_2\lambda_2^n\mathbf{t}_2. \tag{7.27}$$

If $\lambda_1 < 1$, $\lambda_2 < 1$, $\mathbf{\Theta}_n$ moves to zero, and the fixed point will be described as stable. If $\lambda_1 < 1$, $\lambda_2 > 1$, for all starting points with $a_2 = 0$, i.e. for all starting vectors orthogonal to $\mathbf{t}_2$, $\mathbf{\Theta}_n$ moves to zero, but for all other starting vectors it grows indefinitely; such a fixed point will be described as *1-unstable*. Finally, if $\lambda_1 > 1$, $\lambda_2 > 1$, $\mathbf{\Theta}_n$ always grows indefinitely, and the fixed point is *2-unstable*; only for points with $\mathbf{\Theta}_0 = 0$ will successive $\mathbf{x}_n$ stay at the fixed point. If either of λ_1 or λ_2 is equal to 1 we again have a *marginal* situation.

The generalization to an r-dimensional process is straightforward. We can now define stable, 1-unstable, ..., s-unstable fixed points if 0, 1, ..., s, ... of the eigenvalues are greater than 1. For an s-unstable fixed point, $\mathbf{\Theta}_n$ moves to zero for all vectors in the subspace of $(r - s)$ dimensions defined by the eigenvectors with eigenvalues less than 1.

In most applications of the RG to critical phenomena we are concerned with s-unstable fixed points with $s = 1$ to 3.

Proceeding now beyond the linear approximation in the neighbourhood of a particular fixed point, we can trace out a region of $(r - s)$ dimensions which is the *domain of attraction* of the fixed point; any starting value. $\mathbf{\Theta}_0$, in this region will lead to the fixed point. However, any starting point in the more general space of r dimensions will lead away from the fixed point, and its trajectory can be traced. It must end up either at another fixed point, or at zero or infinity, which are conventionally described as trivial fixed points.

The framework outlined above leads naturally to an explanation of universality. Different fixed points correspond to different universality classes, and we shall see how the properties of each particular universality class are related to analytical behaviour at the corresponding fixed point. The domain of attraction of a particular fixed point defines the values of parameters in the Hamiltonian for which critical exponents and other universal properties of critical behaviour which characterize the specific universality class are identical.

7.2.2 Amplifying the Kadanoff Treatment

In §6.3 we described Kadanoff's introduction of the idea of a block spin and many of the relations are taken over in the RG treatment. As we have noted in §7.1 the number of degrees of freedom is reduced from N to $N' = N/b^d$; the length scale is reduced by b, so that any length vector $\mathbf{r}$ is transformed to $\mathbf{r}' = \mathbf{r}/b$. A *block spin* must be related to the original spin in such a way that the transformation spin → block spin can be iterated. In §7.1 we mentioned a convenient definition for this purpose for an Ising system, the total spin positive corresponding to $+1$, and negative corresponding to -1. A spin rescaling factor must then be introduced, $\sigma'_\alpha = \sigma_\alpha/c$, corresponding to (6.77).

From (7.2) we deduce for the free energy per spin

$$f(h',\tau') = b^d f(h,\tau) \tag{7.28}$$

which is identical with (6.78), and for the correlation functions, using the same argument as in (5.80) and (5.81), we find that

$$\Gamma(\mathbf{r}',h',\tau') = c^{-2}\Gamma(\mathbf{r},h,\tau). \tag{7.29}$$

Kadanoff showed that all the critical exponents could be expressed in terms of two parameters x and y. The RG formulates a definite procedure for the calculation of the values of these parameters, and we shall see in §7.4 that for the simple two-dimensional Ising model the procedure can be followed through satisfactorily.

7.2.3 Landau–Ginzburg–Wilson Hamiltonian

The key to the successful application of the RG in momentum space, and the remarkable exploitation of the ideas and techniques of field theory, lay in the choice of a suitable continuum Hamiltonian. It is surely appropriate that the basic idea goes back to Landau. In his work *Key Problems of Physics and Astrophysics* Ginzburg (1978) says 'L.D. Landau told me once that his attempts to solve the problem of the second-order phase transitions had demanded greater effort than any other problem he had worked on'.

In §4.9 we discussed Landau's first paper in 1937 in which he introduced the concept of the *order parameter*, and emphasized the important role which symmetry plays in phase transitions. This symmetry of the physical system is reflected in the symmetry of the Hamiltonian and the free energy, and we shall see that it plays a central role in modern developments. We noted that Landau attempted to expand the free energy in the neighbourhood of T_c as a power series in the order parameter, which for a magnetic system is the magnetization. The form of the expansions is then (equation (4.131))

$$\Phi(M,T) = \Phi_0(T) + A(T)M^2 + B(T)M^4 + \cdots . \tag{7.30}$$

But we have also seen in Chapter 5 that, for two- and three-dimensional Ising systems, the free energy is singular at T_c, and no such expansion is possible.

In §4.9.1 we referred to a corresponding expansion of the *local free energy* which takes account of inhomogeneities and fluctuations, and which for magnetic systems takes the form

$$\Delta\Phi(M,T) = \tfrac{1}{2}pM^2 + vM^4 + \tfrac{1}{2}f(\nabla M)^2. \tag{7.31}$$

An expansion of type (7.31) can be developed rigorously provided that higher-order powers of M and its derivatives are included. In §4.9.1 we showed how Landau used formula (7.31) with the M^2 term only to derive an expression for the critical value of the correlations. Subsequently Ginzburg and Landau (1950), in dealing with inhomogeneities in superconducting systems, introduced a term involving the fourth power of the superconducting order parameter.

With remarkable insight Wilson appreciated that although expansion (7.30) for the bulk free energy with coefficients A, B, ... analytic is invalid as a description of critical behaviour, expansion (7.31) with f, p, v analytic can provide a legitimate starting basis for a Hamiltonian. He maintained that the fault in the Landau theory lies not in the starting point (7.31), but in the way it makes the transition from (7.31) to (7.30). (This will be analyzed in more detail in §7.2.5.) Wilson was led to an expression of the form (7.31) when he looked for a Hamiltonian which would represent the interaction of block spins in the Kadanoff picture, and in which the discontinuous effects of lattice and spin structure in the standard Ising and Heisenberg Hamiltonians would be smoothed out. Such a short-range smoothing should not affect the nature of the critical behaviour.

The basis which led him to this Hamiltonian has already been indicated in §7.1 (equations (7.5) and (7.6)). Following Aharony (1976) we shall give a derivation for a general n-component spin model.

Let $\boldsymbol{\sigma}$ as before represent the spin vector in n dimensions, so that

$$|\boldsymbol{\sigma}| = \left(\sum_{\alpha=1}^{n} \sigma_\alpha^2\right)^{1/2} \tag{7.32}$$

each σ being allowed to vary continuously. We introduce the weighting factor

$$W(\boldsymbol{\sigma}) = \exp(-\tfrac{1}{2}c|\boldsymbol{\sigma}|^2 - u|\boldsymbol{\sigma}|^4) \tag{7.33}$$

which serves to restrict the possible values which $\boldsymbol{\sigma}$ can take. If $\mathscr{H}$ is the microscopic Hamiltonian, the p.f. will be given by the following integral over all lattice spin vectors $\boldsymbol{\sigma}_l$:

$$Z = \int\left(\prod_{l} W(\boldsymbol{\sigma}_l)\, \mathrm{d}\boldsymbol{\sigma}_l\right) \exp(-\beta\mathscr{H}). \tag{7.34}$$

The interaction term $-J\boldsymbol{\sigma}_l \boldsymbol{\sigma}_m$ in $\mathscr{H}$ can be written in the form

$$\tfrac{1}{2}J(\boldsymbol{\sigma}_l - \boldsymbol{\sigma}_m)^2 - \tfrac{1}{2}J\boldsymbol{\sigma}_l^2 - \tfrac{1}{2}J\boldsymbol{\sigma}_m^2 \tag{7.35}$$

and if the lattice is now replaced by a continuum, $\boldsymbol{\sigma}_l$ being replaced by $\boldsymbol{\sigma}(\mathbf{x})$, the first term in (7.35) can reasonably be transformed into

$$\tfrac{1}{2}Ja^2 \sum_{\alpha=1}^{n} (\nabla\sigma_\alpha)^2 \quad (a = \text{lattice spacing}). \tag{7.36}$$

Using (7.33)–(7.36) the p.f. Z in (7.34) is now obtained by integrating over the whole of space the function

$$\int \mathrm{d}\boldsymbol{\sigma} \exp\left(-\tfrac{1}{2}(c - \beta qJ)|\boldsymbol{\sigma}|^2 - u|\boldsymbol{\sigma}|^4 - \tfrac{1}{2}\beta Ja^2 \sum_{\alpha=1}^{n} (\nabla\sigma_\alpha)^2 \right). \tag{7.37}$$

The integral in (7.37) can be identified with the local free energy (7.31); it is worth noting that the coefficient of $|\sigma|^2$ is temperature dependent and of the form originally assumed by Landau (§4.9.1).

One could not be sure a priori whether higher-order terms in the expansion (7.31) would affect the pattern of critical behaviour. But Wilson hoped that these terms would prove irrelevant, and this hope was dramatically fulfilled.

7.2.4 *Ginzburg Criterion*

Early experimental work on λ-point transitions seemed to indicate that there were two distinct types. For the superconducting and ferroelectric transitions a classical mean field treatment seemed to fit the experimental results, and there was no evidence of the presence of singularities at the critical point. But the liquid helium, order–disorder and ferromagnetic transitions seemed to contain a region around T_c in which the classical concepts break down.

Ginzburg accepted the Landau picture that all λ-point transitions are basically similar, but he appreciated, in view of the developments initiated by Onsager, that the Landau theory breaks down in a *non-classical region* around T_c, and needs to be modified. He assumed that for transitions like those which occur in superconductivity and ferroelectricity, the non-classical region is so narrow that it is never observed experimentally. In 1960 he endeavoured to formulate a criterion from which the width of the non-classical region could be estimated.

Ginzburg's basic idea was to differentiate between bulk and correlation energies. The bulk terms correspond to the Landau treatment of equation (7.30) as developed in §4.9, and near T_c we have from equation (4.132)

$$M^2 \simeq (e/2B)\Delta T \quad (\Delta T = T_c - T). \tag{7.38}$$

Note that this does not depend on dimension. The correlation or fluctuation energy is determined from (7.31) in the manner of §4.9.1, ignoring the term in M^4. By analogy with equation (4.144) we have

$$M_q^2 \simeq \frac{k_B T}{V(p + fq^2)} \quad (k_B = \text{Boltzmann's constant}) \tag{7.39}$$

and to calculate the total fluctuation contribution $(\Delta M)^2$, this must be integrated over all values of q from zero to q_{max} ($\sim 1/a$) with the appropriate weighting factor for the number of modes in the interval $(q, q + dq)$; this latter factor depends on dimension d. Ginzburg was concerned with three dimensions and wrote

$$(\Delta M)^2 = \frac{V}{2\pi^2} \int_0^{q_{max}} q^2 \, dq M_q^2 = \frac{k_B T}{2\pi^2} \int_0^{q_{max}} \frac{q^2 \, dq}{p + fq^2}. \tag{7.40}$$

Near T_c, p is approximately $e\Delta T$, and it is easy to show from (7.40) that

$$(\Delta M)^2 \sim (\Delta T)^{1/2}. \tag{7.41}$$

Comparing (7.38) with (7.41) we see that sufficiently near to T_c the fluctuations do indeed dominate. When Ginzburg substituted numerical values for all the constants, he found that for superconductivity the non-classical region is $\sim 10^{-16}$ K, whereas for liquid helium it is ~ 0.6 K, and this provided the explanation of the experimental data which he sought.

In the mid-1970s it was realized that the Ginzburg criterion could be used to determine the critical dimension above which fluctuations are unimportant, and a Landau treatment is adequate. In d dimensions the integral in (7.40) becomes

$$(\Delta M)^2 \sim \int_0^{q_{max}} \frac{q^{d-1} \, dq}{p + fq^2}. \tag{7.42}$$

Replacing the variable q in the integral by $q/(\Delta T)^{1/2}$ it is readily shown that

$$(\Delta M)^2 \sim (\Delta T)^{d/2 - 1} \tag{7.43}$$

with a logarithmic value for $d = 2$. We then deduce that for $d < 4$ the fluctuation contribution dominates in the non-classical region; when $d = 4$ the bulk contribution (7.38) and the fluctuation contribution (7.43) become comparable; when $d > 4$ the bulk contribution dominates, the fluctuation contribution is negligible, and a Landau treatment is satisfactory.

An alternative way of using the Ginzburg concept to determine the critical dimension is to note the difference of specific heat between the Landau treatment and the Gaussian model. The former uses only the bulk energy (7.30) and obtains a simple discontinuity ($\alpha = 0$, equation (4.137)). The latter, which corresponds to the use of (7.31) up to the M^2 term, gives rise to a singular specific heat of the form $\tau^{(d-4)/2}$ (equation (5.141)). The two results become unified when $d \geq 4$.

7.2.5 Correcting the Landau Picture

In a paper given in 1973 at the conference to celebrate the centenary of the publication of van der Waals' thesis, Wilson pinpointed what was wrong with the Landau treatment, and indicated how it could be corrected. To produce a continuum theory Landau had followed the standard procedures of hydrodynamics averaging over a sufficient number of spins in a region surrounding a point x to remove the discrete lattice or atomic structure. In this way a magnetization density $M(\mathbf{x})$ had been defined analogous to the fluid density $\rho(\mathbf{x})$ of hydrodynamics. Suppose the averaging takes place over a region of radius L, which satisfies $a \ll L \ll \xi$. ξ is the correlation range, which in the neighbourhood of T_c Landau showed to be of order $\tau^{-1/2}$. L is held fixed as $T \to T_c$ so that one can assume analyticity within the averaging region.

For the free energy Landau used the expansion (7.31), where the coefficients p and v are analytic functions of T even at T_c. But Landau failed to note that near T_c fluctuations with wavelength of order L must be taken into account, and hence that p and v are also dependent on L, i.e. $p(L,T)$, $v(L,T)$. The RG enables the form of this L dependence to be calculated, and Wilson gave a crude calculation of the exponents β and ν which contained the essence of the RG theory while avoiding mathematical complications.

The basic picture is that p and v are L dependent for $d \leq L \leq \xi$, but when L reaches the value ξ, they cease to be L dependent and remain effectively constant. If one wants to compute ξ or the spontaneous magnetization, the Landau theory can be used, but one must choose a large enough L so that all important fluctuations are included. This can be achieved by choosing $L = \xi$, and we shall adopt this procedure shortly.

In order to calculate the L dependence of p and v, the Landau free energy for two different averaging sizes L and $L + \delta L$ were compared. The averaging size was taken to mean that the statistical-mechanical average had been computed for all fluctuations with wavelengths $\lambda < L$, but not for wavelengths $\lambda > L$. To perform this averaging a volume V of magnet was studied ($L^d \ll V \ll \xi^d$), and the effect of changing the averaging length from L to $L + \delta L$ was calculated.

We refer the reader to the original papers (Wilson 1974, 1983) for details of the calculation, and the approximations involved. Wilson was led to the differential equations

$$\left.\begin{aligned} \mathrm{d}p/\mathrm{d}L &= -3L^{\epsilon-1}vp \\ \mathrm{d}v/\mathrm{d}L &= -9L^{\epsilon-1}v^2 \end{aligned}\right\} \quad (\epsilon = 4 - d) \tag{7.44}$$

which give rise to the simple solutions

$$p \sim L^{-\epsilon/3}\tau$$
$$v \sim L^{-\epsilon} \tag{7.45}$$

We saw in §6.3 that Kadanoff in his Ising model discussion had suggested a power-law dependence of type (7.45), but there was no prescription for calculating the exponents and they had been fitted empirically. The exponent estimates in (7.45) are not very accurate, but we shall see later in this chapter that they can be improved very substantially; they give useful qualitative insight into the nature of the inadequacies of the original Landau treatment.

We have already seen in §4.9.1 that the Landau fluctuation theory gives the same result for the pair correlation function as the Ornstein–Zernike theory. Wilson gave a direct demonstration of this as follows. In the presence of a magnetic field $H(x)$ the free energy corresponding to (7.31) becomes

$$\Phi(M,T) = \int d\mathbf{x}\{\tfrac{1}{2}pM^2(x) + vM^4(\mathbf{x}) + \tfrac{1}{2}f[\nabla M(\mathbf{x})]^2 - H(\mathbf{x})M(\mathbf{x})\}. \tag{7.46}$$

We introduce a local alignment field $H(\mathbf{x}) = H\delta(\mathbf{x})$ at $x = 0$, and study the magnetization $M(\mathbf{x})$ generated by this field. Take H to be small so that the M^4 term can be neglected. $M(\mathbf{x})$ is calculated by minimizing the free energy, and the standard calculus of variations treatment leads to the equation

$$-f\nabla^2 M(\mathbf{x}) + pM(\mathbf{x}) = H\delta(\mathbf{x}). \tag{7.47}$$

This has the solution

$$M(\mathbf{x}) = \frac{H}{f|\mathbf{x}|}\exp\left[-\left(\frac{r}{f}\right)^{1/2}|\mathbf{x}|\right] \tag{7.48}$$

so that the correlation range ξ is $vp^{-1/2}$.

Applying this argument to $p(L,T)$ with $L \sim \xi$, we obtain

$$\xi = 1/p(\xi,T)^{1/2} \tag{7.49}$$

and using (7.45)

$$\xi = \xi^{\epsilon/6}\tau^{-1/2} \tag{7.50}$$

which gives for the exponent ν the value

$$\nu = (2 - \epsilon/3)^{-1}. \tag{7.51}$$

For $d = 3$ this gives $\nu = 0.6$, which is an improvement on the Landau value and a reasonable first approximation to the true value.

The partition function based on the Hamiltonian (7.31) is obtained by averaging $\exp[-\beta\Delta\Phi(M,T)]$ over all possible functions $M(\mathbf{x})$, and is in conventional terminology a functional integral. If we replace the sum over all the functions $M(\mathbf{x})$ by the largest single contribution, i.e. that which maximizes the integrand, we must take $\nabla M = 0$, since the $(\nabla M)^2$ term is always positive. We then obtain the homogeneous system for which the original Landau theory of §4.9 applies, with p identified with A and v with B (equation (4.131)). The spontaneous magnetization is given by

$$M_0 = (-p/2v)^{1/2} \tag{7.52}$$

and, substituting $p = p(\xi,T)$ as before, we find

$$M_0 \sim \xi^{\epsilon/3}\tau'^{1/2} \tag{7.53}$$

so that the exponent β is given by

$$\beta = \tfrac{1}{2} - (\epsilon/6)[1 - (\epsilon/6)]^{-1}. \tag{7.54}$$

For $d = 3$, $\beta = 0.3$, which is again a reasonable first approximation.

The breakdown of the Landau theory arising from the L dependence is now clear. The function $p(\xi,T)$ from which ξ is computed behaves as $\tau\xi^{-\epsilon/3}$, i.e. $\tau^{1+\epsilon v/3}$, which is not analytic at T_c. If $p(\xi,T)$ were independent of L, $p(\xi,T)$ would be analytic in T, and the Landau theory would be valid. This is true when $d > 4$.

7.3 RG Derivation of Critical Behaviour

It is convenient to derive the general RG pattern of critical behaviour from real-space renormalization for discrete spins, and we follow the treatment of Thompson (1978). Consider the partition function

$$Z_N = \langle \exp(-\beta\mathscr{H})\rangle \tag{7.55}$$

which is characterized by a number of coupling constants K^1, K^2, K^3, ... which depend on T, H, ... and can be conveniently written as a vector $\mathbf{K}$. Suppose that the correlation length associated with this Hamiltonian is $\xi(\mathbf{K})$. We now apply a block spin or decimation transformation R_b to obtain a new set of parameters $\mathbf{K}'$ satisfying

$$\mathbf{K}' = f(\mathbf{K}) \tag{7.56}$$

where the function will usually be non-linear. Since we have reduced the unit of length by b we have the relation

$$\xi(\mathbf{K}') = b^{-1}\xi(\mathbf{K}) \tag{7.57}$$

and because of (7.2) we have for the partition function per particle $Z = (Z_N)^{1/N}$

$$[Z(\mathbf{K}')]^{N'} = [Z(\mathbf{K})]^N \tag{7.58}$$

and hence for the free energy per particle

$$\psi(\mathbf{K}') = b^d\psi(\mathbf{K}). \tag{7.59}$$

Applying the transformation successively as in (7.3) in the form

$$\mathbf{K}_{n+1} = f(\mathbf{K}_n) \tag{7.60}$$

we find that

$$\xi(\mathbf{K}_{n+1}) = b^{-1}\xi(\mathbf{K}_n) = \ldots\, b^{-n}\xi(\mathbf{K}_0). \tag{7.61}$$

Suppose now that $\mathbf{K}_n$ approaches a fixed point $\mathbf{K}^*$ satisfying

$$\mathbf{K}^* = f(\mathbf{K}^*). \tag{7.62}$$

It then follows from (7.61) that if we start at $\mathbf{K}_0$, then

$$\xi(\mathbf{K}_0) = b^n\xi(\mathbf{K}_{n+1}) \tag{7.63}$$

so that if $\xi(K^*)$ is not zero the correlation of the original system must be infinite, i.e. we must have been at the critical point. The locus of all points which after multiple iteration approach K^* is called the *surface of criticality* associated with K^*. For example, there will be exactly one such point on the surface corresponding to nearest-neighbour interactions. If, however, our starting point K_0 is slightly away from the critical point, then $\xi(\mathbf{K}_0)$ is finite, and we cannot be led to $\mathbf{K}^*$. That is why $\mathbf{K}^*$ must be unstable.

Each fixed point has its own range of attraction and all Hamiltonians within this range will have the same pattern of critical behaviour (universality). But different fixed points in the parameter space give rise to different universality classes.

To obtain critical behaviour near $\mathbf{K}^*$ we write

$$\mathbf{K}_n = \mathbf{K}^* + \mathbf{k}_n \tag{7.64}$$

where $\mathbf{k}_n$ is small. Applying this to (7.60) we can develop an expansion and retain only linear terms, obtaining

$$\mathbf{k}_{n+1} = \mathbf{L}\mathbf{k}_n \tag{7.65}$$

where relations (7.65) are linear so that $\mathbf{L}$ is a matrix. Let $\mathbf{l}_i$ be the eigenvectors and Λ_i the eigenvalues, satisfying

$$\mathbf{L}\mathbf{l}_i = \Lambda_i \mathbf{l}_i. \tag{7.66}$$

Expand $\mathbf{k}_{n0}$ in the form

$$\mathbf{k}_{n0} = \sum_i u_i^{(0)}\mathbf{l}_i \tag{7.67}$$

where n_0 is sufficiently large for (7.65) to be valid. We then write

$$\mathbf{k}_{n+n0} = \sum \Lambda_i^n u_i^{(0)}\mathbf{l}_i. \tag{7.68}$$

Any eigenvalue Λ_i whose modulus exceeds 1 will grow in importance (*relevant*); one whose modulus is less than 1 will be *irrelevant*; an eigenvalue for which $|\Lambda_i| = 1$ is called *marginal*. The $u_i^{(0)}$ are called *scaling fields* (and we can drop the superscript); for example, u_1 is the temperature τ, u_2 the magnetic field H, etc.

From (7.61) we have the relation for the correlation function

$$\xi(u_1, u_2, \ldots) = b^n \xi(\Lambda_1^n u_1, \Lambda_2^n u_2, \ldots) \tag{7.69}$$

whilst the corresponding relation for the free energy follows from (7.59),

$$\psi(u_1, u_2, \ldots) = b^{-n} \psi(\Lambda_1^n u_1, \Lambda_2^n u_2, \ldots). \tag{7.70}$$

When $H = 0$ and the only relevant field is τ we find from (7.69)

$$\xi(\tau, 0, \ldots) = b^n \xi(\Lambda_1^n \tau) \cong b^n(\Lambda_1^n \tau)^{-\nu}. \tag{7.71}$$

Hence we find that

$$b\Lambda_1^{-\nu} = 1 \quad \text{or} \quad \nu = \ln b / \ln \Lambda_1. \tag{7.72}$$

For the free energy we define scaling indices y_T and y_H by

$$\Lambda_1 = b^{dy_T}, \quad \Lambda_2 = b^{dy_H} \tag{7.73}$$

and we then have, for all λ,

$$\psi(\tau,H) = b^{-dn} \psi(b^{ndy_T}\, \tau,\, b^{ndy_H}\, H), \tag{7.74}$$

i.e.

$$\psi(\lambda^{y_T}\, \tau,\, \lambda^{y_H}\, H) = \lambda \psi(\tau, H) \tag{7.75}$$

where $\lambda = b^{dn}$.

The free energy is a *generalized homogeneous function* (equation (6.42)). Take λ to satisfy

$$\lambda^{y_T}\, \tau = 1, \quad \lambda = \tau^{-1/y_T} \tag{7.76}$$

and we have

$$\psi(\tau,H) = \tau^{2-\alpha} V(H\tau^{-\Delta}) \tag{7.77}$$

where

$$\begin{aligned} 2 - \alpha &= y_T^{-1} \\ \Delta &= y_H / y_T \end{aligned} \tag{7.78}$$

and

$$V(u) = \psi(1,u),$$

i.e. we have derived a relation analogous to (6.54). Finally we have the additional relation from (7.72) and (7.73)

$$\nu^{-1} = dy_T. \tag{7.79}$$

Hence *all* exponents are determined by y_T and y_H in agreement with Kadanoff, and in support of length scaling.

The extension to cross-over behaviour (§6.4) is immediate. One needs to use a scaling field g and to introduce a scaling index y_g satisfying a relation corresponding to (7.73), and this will lead to a free energy of the form (6.100). We identify the cross-over exponent ϕ with $\nu d y_g$.

7.4 Real-space Renormalization – Practical Examples

In the previous section the general theory of critical behaviour was developed in terms of the concepts of the RG using a real-space transformation. It is important to verify by means of suitably chosen numerical examples that the pattern of behaviour described is indeed fulfilled in practice. For a chosen lattice we wish to select a suitable set of parameters **K**, calculate the transformation $f(\mathbf{K})$ in (7.56), identify a fixed point $\mathbf{K}^*$ satisfying (7.62), use a Taylor expansion about K^* to determine the matrix $\mathbf{L}$ in (7.65), and calculate its eigenvalues Λ_i and eigenvalues $\mathbf{l}_i$ from (7.66). From these we can determine the critical exponents by means of (7.72), (7.73) and (7.78). We wish to investigate how the pattern of eigenvalues changes, and how the approximation to the critical exponents improves as the number of parameters in the set **K** is increased.

The most complete realization of the RG is provided by the block spin transformation, an example of which we shall discuss first. But the computational difficulties with this transformation are considerable, so that we shall then take an example of the less complete decimation transformation where the approximation can be pursued further.

7.4.1 *Block Spin Transformation*

The pioneering work on this transformation is due to Niemeijer and van Leeuwen (1973, 1974, 1976) for the triangular lattice. These authors noted that for this lattice there exists a simple and natural definition of an Ising block or cell spin as indicated in Figure 7.5. The shaded cells form a triangular lattice with an increased lattice spacing $\sqrt{3}\,a$ (so that $b = \sqrt{3}$), and each spin of the original lattice is associated with one of the shaded cells. Denote cell spins by σ', and the cells by $\mathbf{i}'$; the cell spin $\sigma'_{\mathbf{i}'}$ of a cell bounded by site spin $\sigma^1_{\mathbf{i}'}, \sigma^2_{\mathbf{i}'}, \sigma^3_{\mathbf{i}'}$, is defined by

$$\sigma'_{\mathbf{i}'} = \operatorname{sign}(\sigma^1_{\mathbf{i}'} + \sigma^2_{\mathbf{i}'} + \sigma^3_{\mathbf{i}'}). \tag{7.80}$$

For the four configurations (1,1,1), (−1,1,1), (1,−1,1), (1,1,−1) of $\sigma^1_{\mathbf{i}'}, \sigma^2_{\mathbf{i}'}, \sigma^3_{\mathbf{i}'}$, the cell spin $\sigma_{\mathbf{i}'}$ has the value 1, and for the four configurations (−1,−1,−1), (1,−1,−1), (−1,1,−1), (−1,−1,1), $\sigma_{\mathbf{i}'}$ has the value −1. For a given value of $\sigma_{\mathbf{i}'}$ the three neighbouring sites can be grouped into four configurations,

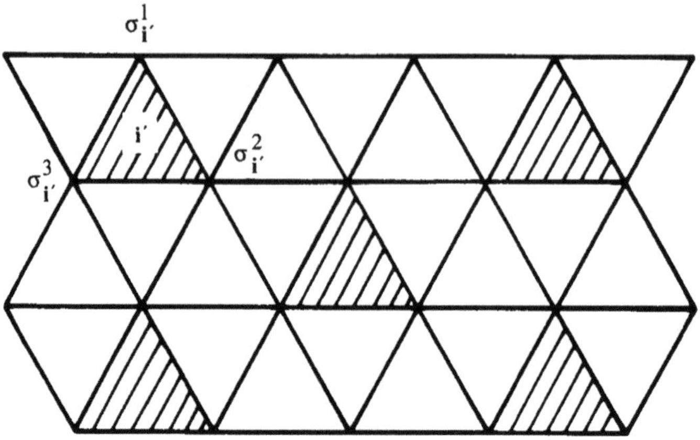

Figure 7.5 Block spin transformation on the triangular lattice.

labelled by t,

$$\begin{array}{lllll} t=0 & \sigma_{i'}^1 = & \sigma_{i'}^2 = & \sigma_{i'}^3 = \sigma_{i'} \\ t=1 & -\sigma_{i'}^1 = & \sigma_{i'}^2 = & \sigma_{i'}^3 = \sigma_{i'} \\ t=2 & \sigma_{i'}^1 = & -\sigma_{i'}^2 = & \sigma_{i'}^3 = \sigma_{i'} \\ t=3 & \sigma_{i'}^1 = & \sigma_{i'}^2 = & -\sigma_{i'}^3 = \sigma_{i'}. \end{array} \tag{7.81}$$

It is sometimes convenient to define the relation between site and cell spins by means of a 'sorting' factor $P(\sigma',\sigma)$

$$P(\sigma',\sigma) = \prod_{i'} \tfrac{1}{2}[1 + \sigma'_{i'}(\sigma_{i'}^1 + \sigma_{i'}^2 + \sigma_{i'}^3 - \sigma_{i'}^1\sigma_{i'}^2\sigma_{i'}^3)/2] \tag{7.82}$$

which is zero except when condition (7.80) is satisfied.

We now define a general Ising Hamiltonian which takes account of all neighbour and multispin Hamiltonians, as follows:

$$-\beta\mathscr{H}(\sigma) = \sum_a K_a \sigma_a \tag{7.83}$$

where a is any subset of sites, and σ_a is given by

$$\sigma_a = \prod_{i \in a} \sigma_i . \tag{7.84}$$

It is convenient to divide the subset a into even and odd numbers of spins, denoted by + and − superscripts. Thus, we can write (7.83) in a less formal manner as

$$K \sum_{n,n} \sigma_i\sigma_j + K_1^+ \sum_{n,n,n} \sigma_i\sigma_j + \cdots - \beta H \sum \sigma_i + K_1^- \sum_{\text{triangle}} \sigma_i\sigma_j\sigma_k \ldots . \tag{7.85}$$

Corresponding constants K', $K_1'^{+}$ ... H', $K_1'^{-}$ ... are defined for the cell-spin lattice and the Hamiltonian is denoted by $\mathscr{H}'(\sigma')$. The transformation (7.56) from **K** to **K**$'$ is obtained by summing $\exp(-\beta\mathscr{H})$ over the σ variables, using the relation

$$\sum_{\sigma} P(\sigma',\sigma) \exp[-\beta\mathscr{H}(\sigma)] = \exp\{-\beta[G + \mathscr{H}'(\sigma')]\}. \tag{7.86}$$

It is important to include an empty set to take account of terms independent of σ' which arise in the summation, and this gives rise to G; but G makes no direct contribution to the transformation.

The sum to be evaluated in (7.86) is much more complicated than the original Ising model, and an exact solution is not feasible. Three approximation methods have been introduced, which parallel approximations used for the original Ising model, as follows:

A Evaluation for finite lattices.
B Cumulant expansions (analogous to series expansions).
C Evaluation for finite clusters (cf. Domb 1974).

For details of the computations the reader is referred to the review article of Niemeijer and van Leeuwen (1976) and to the original papers mentioned. However, we shall quote some numerical results drawing particular attention to method C since this has been pursued furthest.

Regarding the calculations, the Hamiltonian in (7.85) has been divided into an even part which is invariant under a flip of all spins, and an odd part which changes sign under such a spin flip, and $P(\sigma',\sigma)$ is invariant under a simultaneous flip of σ' and σ. As a result of this symmetry fixed points are usually found in the subspace of even interactions, for which the matrix $\mathbf{L}$ in (7.65) breaks up into an even–even and odd–odd part. This is a useful simplification since the two submatrices can be treated separately. We shall use the superscripts T, H to differentiate between the even and odd submatrices, i.e. $\mathbf{L}^{\mathrm{T}}$, $\mathbf{L}^{\mathrm{H}}$, Λ^{T}, Λ^{H}, u_1^{T}, u_1^{H}, $\mathbf{l}_1^{\mathrm{T}}$, $\mathbf{l}_1^{\mathrm{H}}$.

In the linear approximation the critical surface is perpendicular to $\mathbf{l}_1^{\mathrm{T}}$ through $\mathbf{K}^*$ (§7.2.1). If $\mathbf{l}_1^{\mathrm{T}}$ is the left-hand eigenvector, this is given by the equation

$$\mathbf{l}_1^{\mathrm{T}} \cdot (\mathbf{K} - \mathbf{K}^*) = 0. \tag{7.87}$$

The critical temperature K_c of the simple Ising model is obtained by taking the vector **K** with all elements K_1^+, K_2^+, ... in (7.85) equal to zero, the only non-zero element being K.

Let us first use the known results for the simple Ising model on the triangular lattice to determine the exact values of the parameters we are seeking to calculate. Since $\nu = 1$ and $b = \sqrt{3}$, we find from (7.72) that

$$\Lambda_1^{\mathrm{T}} = \sqrt{3} \simeq 1.732. \tag{7.88}$$

Also $\alpha = 0$, $\Delta = 15/8$, so that from (7.78)

$$y_T = 1/2, \quad y_H = 15/16. \tag{7.89}$$

Using (7.73) we find that

$$\Lambda_1^{\mathrm{H}} = 3^{15/16} \simeq 2.801. \tag{7.90}$$

Finally from the exact value for the critical temperature given in (5.20), we deduce that

$$K_c = \tfrac{1}{4}(\ln 3) \simeq 0.2744. \tag{7.91}$$

We now quote some results obtained in various approximations. Using method B, Niemeijer and van Leeuwen (1974) obtained for the fixed point

$$K^* = 0.2789, \quad K_1^* = -0.0143, \quad K_2^{*+} = -0.0152 \tag{7.92}$$

and for the matrix $\mathbf{L}$

$$\mathbf{L}^{\mathrm{T}} = \begin{pmatrix} 1.8966 & 1.3446 & 0.8964 \\ -0.0403 & 0.0 & 0.4482 \\ -0.0782 & 0.0 & 0.0 \end{pmatrix}. \tag{7.93}$$

This has eigenvalues

$$\Lambda_1^{\mathrm{T}} = 1.7835, \quad \Lambda_2^{\mathrm{T}} = 0.2286, \quad \Lambda_3^{\mathrm{T}} = -0.1156. \tag{7.94}$$

The left-hand eigenvector $\tilde{\mathbf{l}}_1^{\mathrm{T}}$ is

$$(1, 0.7539, 1.0961) \tag{7.95}$$

giving an estimate for critical temperature

$$K_c = 0.2514. \tag{7.96}$$

The odd terms were treated independently, but with fewer parameters, and yielded for the eigenvalue

$$\Lambda_1^{\mathrm{H}} = 3.306. \tag{7.97}$$

These estimates are not very striking in comparison with the exact values (7.88), (7.90) and (7.91). But the impressive feature of the calculation is the complete agreement of the pattern with that anticipated in the previous section, i.e. the existence of a fixed point, the distribution of eigenvalues with exactly two greater than 1 and relevant, and the remainder less than 1 and irrelevant.

Results of a similar kind were obtained for the cluster approximation, but it is possible to see the improvement in numerical estimates as the size of the cluster increases. This is clearly shown in Table 7.1 which lists the values of y_T, y_H and K_c for clusters containing from two to seven spins. The eigenvalues

Table 7.1

Cluster	y_T	y_H	K_c
	0.7908	2.0217	0.365
	0.7394	1.6688	0.255
	0.8177	1.7562	0.253
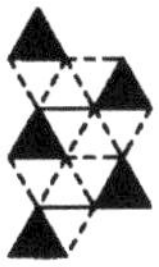	1.0518	1.9232	0.281
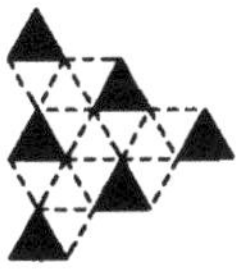	1.0518	1.9219	0.281
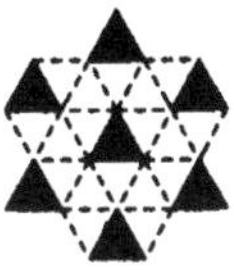	1.0281	1.873 75	0.274 16
Exact	1.000	1.875 00	0.274 65

for the five cluster approximation are as follows:

$$\Lambda_1^{\mathrm{T}} = 1.784, \quad \Lambda_2^{\mathrm{T}} = 0.585, \quad \Lambda_3^{\mathrm{T}} = 0.195, \quad \Lambda_4^{\mathrm{T}} = 0.112,$$
$$\Lambda_5^{\mathrm{T}} = -0.000\,03, \quad \Lambda_6^{\mathrm{T}} = -0.040, \quad \Lambda_7^{\mathrm{T}} = -0.150; \tag{7.98}$$

$$\Lambda_1^{\mathrm{H}} = 3.186, \quad \Lambda_2^{\mathrm{H}} = 0.716, \quad \Lambda_3^{\mathrm{H}} = 0.271, \quad \Lambda_4^{\mathrm{H}} = 0.199,$$
$$\Lambda_5^{\mathrm{H}} = -0.078, \quad \Lambda_6^{\mathrm{H}} = 0.002, \quad \Lambda_7^{\mathrm{H}} = -0.0007, \quad \Lambda_8^{\mathrm{H}} = -0.027. \tag{7.99}$$

The agreement of this distribution with the description of the previous section is particularly striking.

It is a common feature of all the approximations that, in the fixed-point Hamiltonian, the nearest-neighbour coupling is larger by a factor of 10 than the second and third neighbour couplings, which in turn are three or four times larger than the four-spin interactions. Such approximations which use only a few parameters already show all the characteristic features expected of the RG transformation.

7.4.2 *Decimation Transformation* (based on Thompson (1978))

The aim of the decimation transformation is to eliminate a fraction of the spins by partial summation, these spins being selected in such a manner that the remaining spins constitute a lattice identical with the original lattice. The most elementary application of the transformation is to the one-dimensional Ising chain with Hamiltonian

$$\mathscr{H}(\sigma) = -J \sum_{i=1}^{2N} \sigma_i \sigma_{i+1} \quad (\sigma_1 \equiv \sigma_{2N+1}). \tag{7.100}$$

The number of degrees of freedom can be decreased by a factor of 2, and the spacing increased by a factor of 2 ($b = 2$) by summing the even spins σ_2, $\sigma_4, \ldots, \sigma_{2N}$ in the p.f.

$$Z_{2N}(K) = \exp\left(K \sum_{i=1}^{2N} \sigma_i \sigma_{i+1} \right). \tag{7.101}$$

We first derive an elementary transformation by summing over the crossed spin σ_2 in Figure 7.6. We have

$$\begin{aligned} \sum_{\sigma_2 = \pm 1} \exp[K(\sigma_1\sigma_2 + \sigma_2\sigma_3)] &= 2 \cosh K(\sigma_1 + \sigma_3) \\ &= \exp(K\sigma_1)\exp(K\sigma_3) + \exp(-K\sigma_1)\exp(-K\sigma_3) \\ &= (\cosh K + \sigma_1 \sinh K)(\cosh K + \sigma_3 \sinh K) \\ &\quad + (\cosh K - \sigma_1 \sinh K)(\cosh K - \sigma_3 \sinh K) \\ &= 2\cosh^2 K(1 + \sigma_1\sigma_3 \tanh^2 K). \end{aligned} \tag{7.102}$$

If we now write this in the form

$$f(K') \exp(K'\sigma_1\sigma_3) \tag{7.103}$$

we derive relations from which K' and $f(K')$ can be determined in terms of K. A particularly simple way of calculating these relations is to substitute any two independent sets of values of σ_1 and σ_3 and these are sufficient to determine K'

Figure 7.6 Decimation transformation in one dimension.

and $f(K')$. For example, using

$$\sigma_1 = \sigma_3 = 1, \quad 2\cosh 2K = f(K')\exp(-K')$$
$$\sigma_1 = 1, \quad \sigma_3 = -1, \quad 2 = f(K')\exp(-K') \tag{7.104}$$

we find that

$$K' = (\ln\cosh 2K)/2 \tag{7.105}$$

$$f(K') = 2(\cosh 2K)^{1/2}. \tag{7.106}$$

Applying this identity to all the even spins in (7.101), we find that

$$Z_{2N}(K) = 2^N(\cosh 2K)^{N/2} Z_N(K'). \tag{7.107}$$

The RG transformation is given by (7.105), which defines successive coupling constants K_n satisfying

$$K_{n+1} = [\ln(\cosh 2K_n)]/2. \tag{7.108}$$

It is easy to see that the equation

$$K^* = [\ln(\cosh 2K^*)]/2 \tag{7.109}$$

has no non-trivial solution (the only solutions are $K^* = 0, \infty$). This is what we expect for a one-dimensional chain.

For the two-dimensional SQ lattice the spins to be eliminated in the transformation are shown by crosses in Figure 7.7, and it will be seen that the new lattice is also SQ, and that $b = \sqrt{2}$. In order to perform the sum of the crossed spins we make use of the *star–square transformation* shown in Figure 7.8,

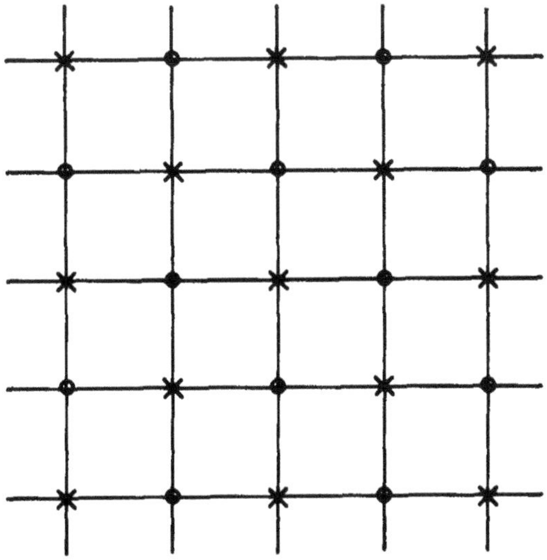

Figure 7.7 Decimation transformation for the SQ lattice.

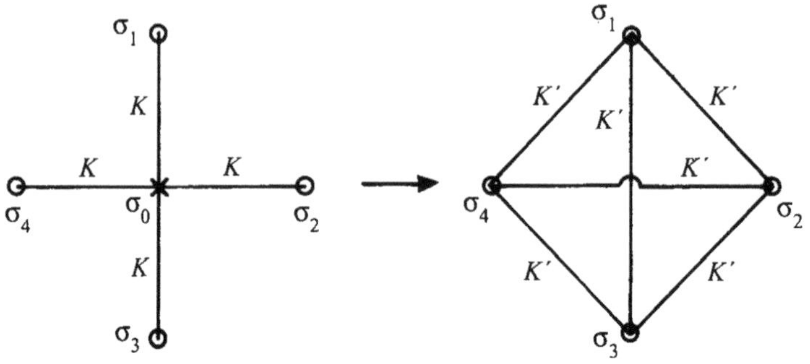

Figure 7.8 The star–square transformation.

which is analogous to the star–triangle transformation of §5.1.3. We now have

$$\sum_{\sigma_0=\pm 1} \exp[K\sigma_0(\sigma_1+\sigma_2+\sigma_3+\sigma_4)] = 2\cosh K(\sigma_1+\sigma_2+\sigma_3+\sigma_4) \tag{7.110}$$

and if we expand the RHS of (7.110) in the same manner as (7.102), we find that the only terms which enter are

$$\sum_{\substack{i,j=1\\ i\neq j}} \sigma_i\sigma_j \quad \text{and} \quad \sigma_1\sigma_2\sigma_3\sigma_4 .$$

We can therefore write the identity

$$2\cosh K(\sigma_1+\sigma_2+\sigma_3+\sigma_4) \equiv A(K)\exp(K'\textstyle\sum \sigma_i\sigma_j + U(K)\sigma_1\sigma_2\sigma_3\sigma_4) \tag{7.111}$$

and the functions $A(K)$, $K'(K)$ and $U(K)$ can readily be determined from three independent sets of values of σ_1, σ_2, σ_3, σ_4. Taking $\sigma_1=\sigma_2=\sigma_3=\sigma_4=1$, we obtain

$$2\cosh 4K = A\exp(6K'+U) \tag{7.112}$$

whilst from $\sigma_1=\sigma_2=1$, $\sigma_3=\sigma_4=-1$, we have

$$2 = A\exp(-2K'+U). \tag{7.113}$$

Dividing (7.112) by (7.113) we deduce that

$$\begin{aligned} \cosh 4K &= \exp(8K') \\ K' &= (\ln\cosh 4K)/8. \end{aligned} \tag{7.114}$$

Taking $\sigma_1=\sigma_2=\sigma_3=1$, $\sigma_4=-1$, we obtain

$$2\cosh 2K = A\exp(-U). \tag{7.115}$$

From (7.113) and (7.115) we deduce that

$$\begin{aligned} \cosh 2K &= \exp(2K')\exp(-2U) \\ \exp(2U) &= \frac{\exp(2K')}{\cosh 2K} = \frac{(\cosh 4K)^{1/4}}{\cosh 2K} \end{aligned} \tag{7.116}$$

$$U = [(\ln \cosh 4K) - (\ln \cosh 2K)/2]/8 = [\ln(1 - \tanh^4 2K)]/8.$$

Finally, from (7.113) we derive the value of A,

$$A = 2 \exp(U) \cosh 2K = 2(\cosh 4K)^{1/8}(\cosh 2K)^{1/2}. \tag{7.117}$$

When we sum over all the crossed spins for the lattice we derive the transformed Hamiltonian

$$\mathscr{H}'(\sigma) = \frac{N^2}{2} A(K) + 2K' \sum_{nn} \sigma_i \sigma_j + K' \sum_{nnn} \sigma_i \sigma_j + U(K) \sum_{\text{square}} \sigma_i \sigma_j \sigma_k \sigma_l. \tag{7.118}$$

The factor 2 in the near-neighbour interaction arises since two independent crossed spins contribute; but this is not the case for the next-near-neighbour interactions.

We now examine the numerical values of the various constants in (7.118). The critical value of K_c is known to be 0.440 887 from Onsager's solution. From (7.114), (7.116) and (7.117) we find that

$$A(K_c) = 1.003\,76 \tag{7.119}$$

$$K'_c = 0.137\,327 \tag{7.120}$$

$$U(K_c) = -0.035\,960. \tag{7.121}$$

It will be noted that $U(K_c)$ is small, and it is reasonable to ignore this term in a first approximation.

$\mathscr{H}'(\sigma)$ has been derived from the particular system consisting of near-neighbour interactions only. To derive the RG transformation (7.56) we must now iterate in accordance with (7.3). Let us examine (following Wilson 1975b) what is involved in the second stage of the iteration. We first note that the diagonal next-near-neighbour (n.n.n.) spin couples different crossed spins (e.g. 0, 5; 0, 7) and likewise the four-coupled spins (e.g. 0, 5, 6, 7). Hence when we sum over those crossed spins, we can no longer isolate them from another, and even at the second stage, interactions of arbitrary type and range will be introduced. However, the strengths of these couplings are small; even at T_c, K'_c is only 0.14, and U_c is -0.04 compared with the value 0.44 of K_c. Hence it is reasonable to treat them by perturbation series expansions.

Wilson pursued this transformation retaining 217 different interaction constants. The decimation transformation has a fault in comparison with the block spin transformation described previously that only the lattice *spacing* is scaled and not the value of the lattice *spin*. Hence only values of y_T will be obtained with any accuracy, and not those of y_H. The transformation can be modified to correct this fault, and by using such a modification Wilson was able to estimate critical exponents with an accuracy of 0.2%.

Our aim in the rest of this section will be to derive an approximation based on only two parameters: the n.n. interaction K and the n.n.n. interaction $\bar{K}$.

Expanding K' to the second order, we find from (7.118)

$$K_1 = 2K_0^2 \qquad \bar{K}_1 = K_0^2. \tag{7.122}$$

These terms will continue to the ith order of iteration, but an extra term will be introduced into the first equation from the n.n.n. coupling which now transforms to the n.n. coupling. (In Figure 7.7, for example, 2–4 is the n.n. interaction at the second stage.) Hence instead of (7.122) we have

$$K_{n+1} = 2K_n^2 + \bar{K}_n \qquad \bar{K}_{n+1} = K_n^2. \tag{7.123}$$

The fixed points are determined by the equations

$$K^* = 2K^{*2} + \bar{K}^* \qquad \bar{K}^* = K^{*2} \tag{7.124}$$

and other than the trivial values $K^*, \bar{K}^* = 0, \infty$, the non-trivial fixed point is given by $K^* = 1/3$, $\bar{K}^* = 1/9$.

Taking our initial value as K_0 we find that for $0 < K_0 < 0.3921$ the fixed point is (0,0), for $K_0 > 0.3921$ the fixed point is (∞,∞), and for the unique value $K_0 = 0.3921$ it is (1/3,1/9). The value 0.3921 therefore gives the approximation to K_c (0.4407).

Writing $\mathbf{K}_n = \mathbf{K}^* + \mathbf{k}_n$ as in (7.64) we find that the matrix $\mathbf{L}$ is given by

$$\mathbf{L} = \begin{pmatrix} 4/3 & 1 \\ 2/3 & 0 \end{pmatrix} \tag{7.125}$$

with eigenvalues

$$\Lambda_1 = \tfrac{1}{3}(2 + \sqrt{10}) \quad \Lambda_2 = \tfrac{1}{3}(2 - \sqrt{10}).$$

Clearly Λ_1 is relevant and Λ_2 irrelevant.

Since $b = \sqrt{2}$, we find using (7.72) that

$$\nu = \ln\sqrt{2}/\ln[(1/3)(2 + \sqrt{10})] = 0.638 \tag{7.126}$$

to be compared with the exact value $\nu = 1$.

The eigenvectors corresponding to Λ_1 are

$$\tilde{\mathbf{l}}_1 = \begin{pmatrix} 2 \\ \sqrt{10} - 2 \end{pmatrix} \quad \mathbf{l}_1 = \begin{pmatrix} 3 \\ \sqrt{10} - 2 \end{pmatrix}. \tag{7.127}$$

Hence the direction of the critical surface through the fixed point is

$$2k + (\sqrt{10} - 2)\bar{k} = 0. \tag{7.128}$$

The pattern of critical behaviour in relation to the fixed point is illustrated by Figure 7.9. If we start with a $\bar{K}_0 \neq 0$, for each starting value there corresponds a unique $K_0 = K_c(\bar{K}_0)$ such that starting from $(K_c,\bar{K}_0)$ the iterates

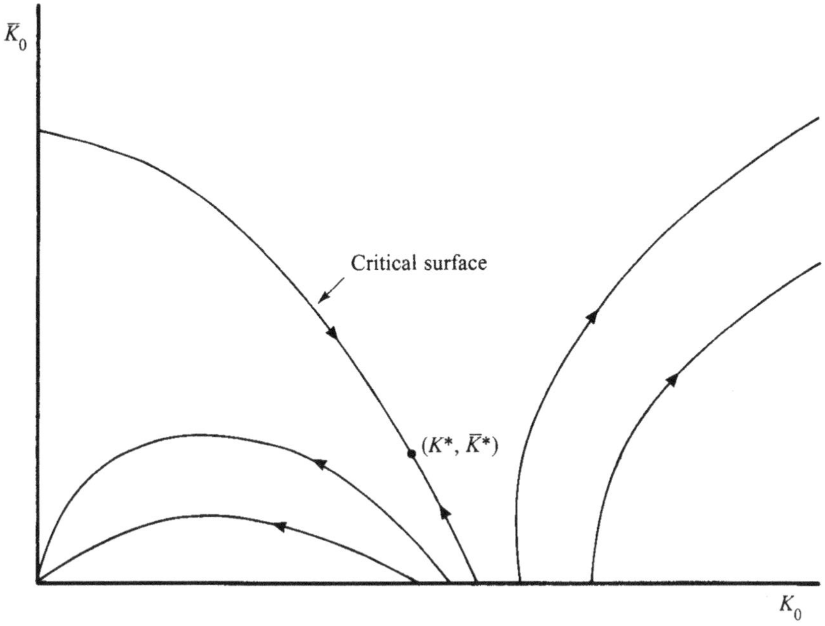

Figure 7.9 Trajectories illustrating the flow to the fixed points from different starting conditions.

approach the fixed point $(K^*,\bar{K}^*)$. The critical exponents are identical (universality). It will be seen that this simple example shows the characteristic features described in §7.3.

Despite considerable effort and much ingenuity the accuracy achieved by real-space renormalization for three-dimensional models has not proved comparable with that obtained by momentum-space renormalization. The important contribution of real-space renormalization has been to demonstrate in detail the validity of the general theory outlined in §7.3, and to show by two-dimensional examples that with sufficient computational effort accurate estimates of critical behaviour could be derived.

7.5 Momentum-space Renormalization

In the following sections we shall modify our previous notation as follows: an Ising spin on a lattice in real space in d dimensions will be denoted by $s_{\mathbf{l}}(\mathbf{l} = \mathbf{l}_1\mathbf{a}_1 + \mathbf{l}_2\mathbf{a}_2 + \cdots + \mathbf{l}_d\mathbf{a}_d)$. Its Fourier transform (equation (7.10)) will be denoted by

$$\sigma_{\mathbf{q}} = \sum_{\mathbf{l}} s_{\mathbf{l}} \exp(i\mathbf{q} \times \mathbf{l}) \tag{7.129}$$

where the sum is taken over all lattice points $\mathbf{l}$. We thus dispense with the 'hat' notation which has previously characterized Fourier transforms. The components of a classical n-vector spin will be denoted by $s_{\mathbf{l}}^{\alpha}$, where α goes from 1 to n; the corresponding Fourier transform will be denoted by $\sigma_{\mathbf{q}}^{\alpha}$. Finally we shall follow the conventional practice of absorbing the constant β $(=1/kT)$ into the Hamiltonian $\mathscr{H}$, so that the p.f. Z will be given by $\exp(-\mathscr{H})$.

As in §7.1 the $s_{\mathbf{l}}$ will be continuous variables taking all real values, with a weighting factor of type (7.6) to limit the effect of high values. It is possible to derive an exact transformation from a model with discrete spin values to a model with continuous spin values with a weighting factor. This is called the Kac–Hubbard–Stratonovitch transformation (see e.g. Fisher 1983).

We first discuss the Fourier transformation (7.129) in more detail. From the definition it is easy to see that the $\sigma_{\mathbf{q}}$ represent a periodic function in the continuum $\mathbf{q}$ space of the reciprocal lattice (§4.8.4). The $s_{\mathbf{l}}$ thus correspond to the Fourier coefficients of a conventional Fourier series, and are given by (Ziman 1964, Ch. 1)

$$s_l = \frac{(2\pi)^d}{v_0} \int \sigma_{\mathbf{q}} \exp(-\mathbf{l}\mathbf{q} \times \mathbf{l})\, d\mathbf{q} \tag{7.130}$$

the integral being taken over the first Brillouin zone of the reciprocal lattice; v_0 is the volume of unit cell of the original lattice.

We note that when $s_{\mathbf{l}} = 1$ in equation (7.129), $\sigma_{\mathbf{q}}$ is zero when $\mathbf{q} \neq 0$ and infinity when $\mathbf{q} = 0$, and therefore $\sigma_{\mathbf{q}}$ is a delta function. From the inverse relation (7.130) we deduce that

$$\sigma_{\mathbf{q}} = \frac{v_0}{(2\pi)^d}\, \delta^{(d)}(\mathbf{q}) \tag{7.131}$$

where $\delta^{(d)}(\mathbf{q})$ is the d-dimensional delta function.

From (7.130) we see that

$$s_{\mathbf{l}}^2 = \left(\frac{(2\pi)^d}{v_0}\right)^2 \int d\mathbf{q} \int d\mathbf{q}' \sigma_{\mathbf{q}} \sigma_{\mathbf{q}'} \exp[-\mathrm{i}(\mathbf{q} + \mathbf{q}')] \tag{7.132}$$

and if we sum over all lattice points $\mathbf{l}$, and use (7.131), we find that

$$\sum_{\mathbf{l}} s_{\mathbf{l}}^2 = \frac{(2\pi)^d}{v_0} \int d\mathbf{q} \sigma_{\mathbf{q}} \sigma_{-\mathbf{q}}. \tag{7.133}$$

In a similar manner

$$\sum_{\mathbf{l}} s_{\mathbf{l}}^4 = \left(\frac{(2\pi)^d}{v_0}\right)^3 \int d\mathbf{q} \int d\mathbf{q}' \int d\mathbf{q}'' \sigma_{\mathbf{q}} \sigma_{\mathbf{q}'} \sigma_{\mathbf{q}''} \sigma_{-\mathbf{q}-\mathbf{q}'-\mathbf{q}''}. \tag{7.134}$$

Integrals such as (7.133) and (7.134) are to be taken over all possible paths $\sigma_{\mathbf{q}}$ in the continuum set of points in $\mathbf{q}$ space, and are thus *functional integrals*. The above treatment applies to an infinite system. Fisher (1974, 1983) has pointed out that for a finite system with periodic boundary conditions, equation (7.129) will be summed over N lattice points, and the reciprocal relation (7.130) then applies to a finite set of N points spread appropriately through the first Brillouin zone; the integral $\int d\mathbf{q}$ in (7.130) must now be replaced by the sum $(1/N)\Sigma$ over these points, the integral being attained in the limit $N \to \infty$. This observation may be conceptually useful in leading to a natural definition of the functional integrals.

7.5.1 The Gaussian Model

The Gaussian model was discussed in §5.4.2, and an exact solution was derived by direct integration over the $\sigma_{\mathbf{l}}$ variables and summation over $\mathbf{l}$. We shall now outline the treatment of Wilson and Kogut (1974) based on the RG in which they demonstrate that for the Gaussian model an RG transformation can be formulated exactly in momentum space which leads to a *Gaussian fixed point.* When an $s_{\mathbf{l}}^4$ term is introduced as in (7.6), they use a perturbation treatment to show that for dimension d less than 4 a new *Ising fixed point* arises; changes in critical behaviour are then calculated to the first order in $\epsilon = 4 - d$.

Restricting the interactions to n.n. in a d-dimensional SC lattice ($v_0 = a^d$) the integrand in equation (5.129) will be modified to

$$\exp\left(-\tfrac{1}{2}c\sum_{\mathbf{l}} s_{\mathbf{l}}^2 + K\sum_{\mathbf{j},\mathbf{l}} s_{\mathbf{l}} s_{\mathbf{l}+\mathbf{j}}\right) \tag{7.135}$$

where $\mathbf{j}$ runs over positive values so as to include each near neighbour once when summing over $\mathbf{l}$. Replacing $s_{\mathbf{l}} s_{\mathbf{l}+\mathbf{j}}$ by $-\frac{1}{2}(s_{\mathbf{l}} - s_{\mathbf{l}+\mathbf{j}})^2 + \frac{1}{2}s_{\mathbf{l}}^2 + \frac{1}{2}s_{\mathbf{l}+\mathbf{j}}^2$, we obtain

$$\exp\left[-\tfrac{1}{2}K\sum_{\mathbf{j},\mathbf{l}}(s_{\mathbf{l}} - s_{\mathbf{l}+\mathbf{j}})^2 - (\tfrac{1}{2}c - dK)\sum_{\mathbf{l}} s_{\mathbf{l}}^2\right]. \tag{7.136}$$

Applying a Fourier transformation as given by (7.129) and (7.130) this becomes

$$\exp\left[-\frac{1}{2}\left(\frac{2\pi}{a}\right)\right]^d \int_{\mathbf{q}} d\mathbf{q}\left(K\sum_{j}[1 - \exp(-iq_j)][1 - \exp(iq_j)] + \tilde{p}\right)\sigma_{\mathbf{q}}\sigma_{-\mathbf{q}} \tag{7.137}$$

where $\tilde{p} = c - 2dK$, and q_j is the jth component of $\mathbf{q}$.

We have already pointed out that the integral in (7.137) is a functional integral. We shall introduce the simplification of changing the hypercube representing the region of integration $-\pi/a < q_i < \pi/a$ into a hypersphere of equivalent volume. Moreover, we shall define our units of momentum in such a manner that this hypersphere becomes $0 < |q| < 1$. Critical behaviour arises from long-wave fluctuations, i.e. small q, and the deviations from spherical symmetry are related to lattice structure, which is of no significance for long waves. The term $|1 - \exp(-iq_j)|^2$ will therefore be replaced by its form for small q, namely q_j^2. Because of spherical symmetry the angular terms can be integrated out, and using polar coordinates in d dimensions (equation (4.52)) we find that $\int d\mathbf{q}$ is to be replaced by

$$\frac{2\pi^{d/2}}{\Gamma(d/2)} \int q^{d-1}\, dq = C_d \int q^{d-1}\, dq. \tag{7.138}$$

Subsequently it will be assumed that this formula remains valid for non-integral d.

We may note that the RG allows the approximation of taking into account the terms in q^2 in (7.137) and ignoring higher powers of q to be tested; the parameter representing the coefficient of q^4 is found to be irrelevant.

The integral (7.137) now reduces to a standard path integral (Feynman and Hibbs 1965) to which normal changes of variable can be applied.

For convenience the spins will be redefined so that $\sigma_\mathbf{q}$ is replaced by $\sigma_\mathbf{q}/K$. Equation (7.137) reduces finally to

$$\exp\left(-\frac{1}{2}\int_\mathbf{q} (q^2 + p)\sigma_\mathbf{q}\sigma_{-\mathbf{q}}\right) \quad \left(p = \frac{\tilde{p}}{K}\right). \tag{7.139}$$

We use the notation $\int_\mathbf{q}$ to incorporate the $(2\pi/a)^d$ factor from (7.137) and to indicate that when there are $\sigma_\mathbf{q}$ factors, they must be taken over all possible spherically symmetric paths. In the terminology of Feynman and Hibbs (7.139) would be written as

$$\exp\left(-\frac{1}{2}\int (q^2 + p)\delta(\mathbf{q} + \mathbf{q}_1)\right)\mathscr{D}\sigma_\mathbf{q}\mathscr{D}\sigma_{\mathbf{q}_1}. \tag{7.139$'$}$$

By integrating over all possible spherically symmetric paths of the $\sigma_\mathbf{q}$ in (7.139) we obtain the p.f. Z.

To apply an RG transformation in momentum space the spin components $\sigma_\mathbf{q}$ with $1/b < |\mathbf{q}| < 1$ are integrated out leaving unintegrated the components $\sigma_\mathbf{q}$ with $0 < |\mathbf{q}| < 1/b$. Thus the rapidly fluctuating parts of the spin field are integrated out leaving unintegrated the more slowly fluctuating parts. Originally Wilson and Kogut took $b = 2$, but a general b is no more difficult to deal with, and this fits the general theory of §7.3. It is easier to follow the structure of equations and formulae for general b, and the disappearance of b from the final description of critical behaviour provides a useful check.

This calculation is quite trivial since there is no coupling between $\sigma_\mathbf{q}$ for $|\mathbf{q}| > 1/b$ and $\sigma_\mathbf{q}$ for $|\mathbf{q}| < 1/b$. Hence the only result of the path integral over $\sigma_\mathbf{q}$ with $|\mathbf{q}| > 1/b$ is a constant which multiplies

$$\exp\left(-\frac{1}{2}\int_\mathbf{q} (q^2 + p)\sigma_\mathbf{q}\sigma_{-\mathbf{q}}\right). \tag{7.140}$$

Although this contributes to the p.f. it cancels when averages are taken (e.g. for correlations) and plays no part in our treatment; the RG transformation thus replaces (7.139) by (7.140), in which the range of $|\mathbf{q}|$ is from 0 to $1/b$.

To reduce $\mathscr{H}'$ to the same form as $\mathscr{H}$ transformations of scale are needed. First a momentum $\mathbf{q}' = b\mathbf{q}$ is introduced, so that the new momentum variable $\mathbf{q}'$ has the range $0 < |\mathbf{q}'| < 1$ like the original momentum $\mathbf{q}$. Second the spin variable is scaled

$$\sigma_{\mathbf{q}} = \zeta\sigma_{\mathbf{q}'} = \zeta\sigma_{b\mathbf{q}} \tag{7.141}$$

where the scale factor ζ will be determined shortly. In terms of the new spin variable σ' the Hamiltonian $\mathscr{H}'$ is given from (7.139) by

$$\mathscr{H}' = \tfrac{1}{2}(\zeta^2 b^{-d}) \int_{\mathbf{q}'} \left(\frac{q'^2}{b^2} + p\right)\sigma_{\mathbf{q}'}\sigma_{-\mathbf{q}'}. \tag{7.142}$$

Note that each σ in the integrand of (7.139) gives rise to a factor ζ, and each variable of integration $\mathbf{q}$ to a factor b^{-d}. ζ will now be chosen so that (7.142) is of the same form as the integrand in (7.139)

$$\mathscr{H}' = \frac{1}{2}\int_{\mathbf{q}'} (q'^2 + p')\sigma_{\mathbf{q}'}\sigma_{-\mathbf{q}'}. \tag{7.143}$$

Here the coefficient of q'^2 is arranged to be unity, which requires

$$\zeta = b^{1+d/2}. \tag{7.144}$$

We then have

$$p' = b^2 p. \tag{7.145}$$

The transformation (7.145) can be iterated and leads to a critical value $p_c = 0$ to satisfy (7.62). Using the same procedure as in §7.3 we find that $\Lambda_1 = b^2$, and

$$\nu = \ln b/\ln \Lambda_1 = \tfrac{1}{2}. \tag{7.146}$$

7.5.2 The s^4 Model

A small quartic term $-u\sum_l s_l^4$ is now added, as in equation (7.6), and the modifications to the previous result are calculated by perturbation theory, using diagrammatic methods to specify the various terms. From (7.134) the new Hamiltonian in momentum space is given by

$$\mathscr{H}[\sigma] = \mathscr{H}^{G}[\sigma] + u \int_{\mathbf{q}_1}\int_{\mathbf{q}_2}\int_{\mathbf{q}_3} \sigma_{\mathbf{q}_1}\sigma_{\mathbf{q}_2}\sigma_{\mathbf{q}_3}\sigma_{-\mathbf{q}_1-\mathbf{q}_2-\mathbf{q}_3} \tag{7.147}$$

where the first term represents the Gaussian Hamiltonian

$$\mathscr{H}^{G}[\sigma] = \frac{1}{2}\int_{\mathbf{q}} (q^2 + p)\sigma_{\mathbf{q}}\sigma_{-\mathbf{q}} \tag{7.148}$$

and the second interaction term is denoted by $\mathscr{H}_{I}[\sigma]$.

Following the ideas of the previous section an attempt will be made to define a new physical system in which the high-frequency modes of the present system are integrated out. The effective Hamiltonian governing the new physical system will be defined to be as similar as possible to (7.147). In place of the constants p and u, new constants p' and u' will appear in the effective Hamiltonian, which will be determined in terms of p and u via simple formulae.

Write the function $\sigma_{\mathbf{q}}$ in the form

$$\sigma_{\mathbf{q}} = \sigma_{0\mathbf{q}} + \sigma_{1\mathbf{q}} \tag{7.149}$$

where

$$\sigma_{0\mathbf{q}} = \sigma_{\mathbf{q}} \quad \text{if } 0 < |\mathbf{q}| < 1/b, \text{ zero otherwise;}$$

$$\sigma_{1\mathbf{q}} = \sigma_{\mathbf{q}} \quad \text{if } 1/b < |\mathbf{q}| < 1, \text{ zero otherwise.}$$

Our task is then to integrate out the $\sigma_{1\mathbf{q}}$ terms and rescale the residual $\sigma_{0\mathbf{q}}$ terms,

$$\sigma_{0\mathbf{q}} = \zeta\sigma'_{b\mathbf{q}} \quad (0 < |\mathbf{q}| < 1/b). \tag{7.150}$$

The p.f. is given by

$$\begin{aligned} Z &= \int_{\sigma} \exp(-\{\mathscr{H}[\sigma]\}) = \int_{\sigma 0}\left(\int_{\sigma 1} \exp\{-(\mathscr{H}[\sigma_0 + \sigma_1])\}\right) \\ &= \int_{\sigma'} \exp(-\{\mathscr{H}'[\sigma']\}). \end{aligned} \tag{7.151}$$

The precise relation between $\mathscr{H}$ and $\mathscr{H}'$ can be calculated perturbatively if u is small; the most general form that we can expect for $\mathscr{H}'$ can be written

$$\begin{aligned} \mathscr{H}' &= \frac{1}{2}\int_{\mathbf{q}} u'_2(\mathbf{q})\sigma_{\mathbf{q}'}\sigma_{-\mathbf{q}'} + \int_{\mathbf{q}1'}\int_{\mathbf{q}2'}\int_{\mathbf{q}3'} u'_4 \\ &\quad \times (q'_1, q'_2, q'_3, -q'_1 - q'_2 - q'_3)\sigma_{\mathbf{q}1'}\sigma_{\mathbf{q}2'}\sigma_{\mathbf{q}3'}\sigma_{-\mathbf{q}1', -\mathbf{q}2', -\mathbf{q}3'} \\ &\quad + \text{higher-order terms in } \sigma'. \end{aligned} \tag{7.152}$$

To calculate u'_2, u'_4 we write

$$\begin{aligned} \exp(-\{\mathscr{H}'[\sigma']\}) &= \int_{\sigma 1} \exp(-\{\mathscr{H}^{G}[\sigma] + \mathscr{H}_{I}[\sigma]\}) \\ &= \exp(-\{\mathscr{H}^{G}[\sigma_0]\}) \int_{\sigma 1} \exp(-\{\mathscr{H}^{G}[\sigma_1]\})\exp(-\{\mathscr{H}_{I}[\sigma]\}). \end{aligned} \tag{7.153}$$

The term $\exp(-\{\mathscr{H}^G[\sigma_0]\})$ can be written in terms of σ' as before, from (7.142)

$$\mathscr{H}^G[\sigma_0] = \tfrac{1}{2}(\zeta^2 b^{-d-2})_{\mathbf{q}}(q'^2 + b^2 p)\sigma_{\mathbf{q}'}\sigma_{-\mathbf{q}'}, \tag{7.154}$$

whilst the factor $\exp(-\{\mathscr{H}_I[\sigma]\})$ can be expanded as

$$1 - \mathscr{H}_I + \frac{\mathscr{H}_I^2}{2!} - \frac{\mathscr{H}_I^3}{3!} \cdots \tag{7.155}$$

The latter can be written diagrammatically in the form

$$1 - \mathsf{X} + \tfrac{1}{2}(\mathsf{X}, \mathsf{X}) - \tfrac{1}{6}(\mathsf{X}, \mathsf{X}, \mathsf{X}) \ldots. \tag{7.156}$$

In this graphical expansion each end point of each X represents a variable $\sigma_{\mathbf{q}} = \sigma_{0\mathbf{q}} + \sigma_{1\mathbf{q}}$.

For each diagram we must consider all possible products which arise from the choice of $\sigma_{0\mathbf{q}}$ or $\sigma_{1\mathbf{q}}$ for the different end points. The $\sigma_{0\mathbf{q}}$ are then free variables in a path integral which constitutes the renormalized Hamiltonian. The variable $\mathbf{q}$ ranges from 0 to $1/b$ and will be rescaled to range from 0 to 1. The $\sigma_{1\mathbf{q}}$ represent paths over which integration must be performed, $\mathbf{q}$ ranging from $1/b$ to 1. It is not difficult to see that this involves integrals of the form

$$I(\mathbf{q}_1, \mathbf{q}_2, \ldots, \mathbf{q}_k) = \int_{\sigma_1} \sigma_{1\mathbf{q}_1}\sigma_{1\mathbf{q}_2} \cdots \sigma_{1\mathbf{q}_k} \exp(-\{\mathscr{H}^G[\sigma_1]\}). \tag{7.157}$$

The standard Gaussian integral

$$\int_{-\infty}^{\infty} \exp\left[-\left(\sum_{i,j=1}^{N} \tfrac{1}{2}a_{ij}t_i t_j + \sum_{k=1}^{N} h_k t_k\right)\right] dt_1\, dt_2 \ldots dt_N \tag{7.158}$$

can be evaluated by transforming the matrix $\mathbf{A}$ of coefficients a_{ij} to diagonal form, and completing the square in the new variables. Its value is then found to be

$$C \exp(\tfrac{1}{2}\mathbf{h}'\mathbf{A}^{-1}\mathbf{h}) \quad [C = (2\pi)^{N/2}(\det \mathbf{A})^{1/2}]. \tag{7.159}$$

C is the value of the integral when $\mathbf{h} = 0$. Integrals of the form

$$K(a, b, \ldots, c) = \int_{-\infty}^{\infty} t_a t_b \ldots t_c \exp\left[-\left(\sum_{i,j=1}^{N} a_{ij}t_i t_j\right)\right] dt_1 \ldots dt_N \tag{7.160}$$

can be derived by calculating

$$\frac{\partial}{\partial h_a}\frac{\partial}{\partial h_b} \cdots \frac{\partial}{\partial h_c}[C \exp(\tfrac{1}{2}\mathbf{h}'\mathbf{A}^{-1}\mathbf{h})]$$

and putting $h_a, h_b, \ldots, h_c$ equal to zero. If we expand the exponential as a power series it will be seen that the integrals are non-zero only when the number of $t_a, t_b, \ldots, t_c$ is even $(=2s)$, and that the only relevant term is then $(\frac{1}{2}\mathbf{h}'\mathbf{A}^{-1}\mathbf{h})^s/s!$.

The method is discussed in detail by Wilson and Kogut (1974, pp. 95–6) but since it is of central importance in the following sections we shall provide some illustrative examples. When $s = 2$, in order to determine $K(i,j,k,l)$ we must calculate

$$\frac{(\partial/\partial h_i)(\partial/\partial h_j)(\partial/\partial h_k)(\partial/\partial h_l)(\frac{1}{2}\mathbf{h}'\mathbf{A}^{-1}\mathbf{h})(\frac{1}{2}\mathbf{h}'\mathbf{A}^{-1}\mathbf{h})}{2!}.$$

There will be three terms arising, $(\mathbf{A}^{-1})_{ij}(\mathbf{A}^{-1})_{kl}$, $(\mathbf{A}^{-1})_{ik}$, $(\mathbf{A}^{-1})_{jl}$ and $(\mathbf{A}^{-1})_{il}(\mathbf{A}^{-1})_{jk}$, and there is a factor of 2 from each bracket, and a factor of 2 from the ordering of the pair of brackets. Hence the combinatorial factor 8 is exactly balanced by $(\frac{1}{2})^2\frac{1}{2}!$ and we have

$$K(i,j,k,l) = C[(\mathbf{A}^{-1})_{ij}(\mathbf{A}^{-1})_{kl} + (\mathbf{A}^{-1})_{ik}(\mathbf{A}^{-1})_{jl} + (\mathbf{A}^{-1})_{il}(\mathbf{A}^{-1})_{jk}]. \tag{7.161}$$

It should be noted that the formula remains valid when any of i,j,k,l are equal, but some of the terms may then be identical.

When $s = 3$ the number of terms in $K(i,j,k,l,m,n)$ is the number of ways of choosing three different pairs from six objects, i.e. $6!/2!2!2!3!$ or 15. Again, the $(\frac{1}{2})^2\frac{1}{3}!$ cancels with the combinatorial factor, and we obtain $C[(\mathbf{A}^{-1})_{ij}(\mathbf{A}^{-1})_{kl}(\mathbf{A}^{-1})_{mn} + 14 \text{ similar terms}]$.

The integral (7.157) can be evaluated by analogy with (7.160) and reduced to products of integrals of pairs

$$I(\mathbf{q}_1, \mathbf{q}_2) = \int_{\sigma_1} \sigma_{1\mathbf{q}_1}\sigma_{1\mathbf{q}_2} \exp(-\{\mathscr{H}^{\mathrm{G}}[\sigma_1]\}). \tag{7.162}$$

(This can be seen readily by using the Fisher approach for a finite system described at the end of §7.5 in which the functional integrals reduce to standard integrals.) Using the analogue of the coefficients of the inverse matrix it can be shown that the appropriate coefficient is equal to

$$\frac{\delta^{[d]}(\mathbf{q}_1 + \mathbf{q}_2)}{(q_1^2 + p)}. \tag{7.163}$$

Let us now consider the terms which arise from the simplest diagram X whose vertices will be labelled q_1, q_2, q_3, q_4. We must enumerate the terms in the product

$$(\sigma_{0\mathbf{q}_1} + \sigma_{1\mathbf{q}_1})(\sigma_{0\mathbf{q}_2} + \sigma_{1\mathbf{q}_2})(\sigma_{0\mathbf{q}_3} + \sigma_{1\mathbf{q}_3})(\sigma_{0\mathbf{q}_4} + \sigma_{1\mathbf{q}_4})$$

and from the above discussion only terms with even numbers of the $\sigma_{1\mathbf{q}}$ give rise to non-zero contributions. There are three types of terms:

1 $\sigma_{0\mathbf{q}_1}\sigma_{0\mathbf{q}_2}\sigma_{0\mathbf{q}_3}\sigma_{0\mathbf{q}_4}$. This contains four free variables and contributes to u_4 in (7.152). It will be denoted diagrammatically by X.

2 $\sigma_{0\mathbf{q}_1}\sigma_{0\mathbf{q}_2}\sigma_{1\mathbf{q}_3}\sigma_{1\mathbf{q}_4}$. This contains two free variables $\sigma_{0\mathbf{q}_1},\sigma_{0\mathbf{q}_2}$, and the factors $\sigma_{1\mathbf{q}_3},\sigma_{1\mathbf{q}_4}$ give rise to an integral of the form (7.162). It contributes to u_2' in (7.152) and will be denoted diagrammatically by .

3 $\sigma_{1\mathbf{q}_1}\sigma_{1\mathbf{q}_2}\sigma_{1\mathbf{q}_3}\sigma_{1\mathbf{q}_4}$. This contains no free vertices and gives rise to a product of integrals of the form (7.162). Hence it is a constant, and has no relevance to our development. It will be denoted diagrammatically by .

We may summarize diagrammatically by writing

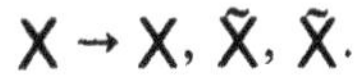 (7.164)

Here the wavy line represents the factor (7.163). Finally when the logarithm of expansion (7.155) is taken to obtain $\mathscr{H}'$ all disconnected diagrams are eliminated, as in §5.2, and we can confine our attention to connected diagrams.

We shall develop the expansions to the first order in ϵ, and for this the only diagrams required are the diagrams in (7.164), and the following diagram from the $\frac{1}{2}\mathscr{H}_1^2$ term in (7.155):

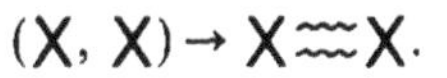 (7.165)

Diagrams with two free vertices relate to the renormalization of the parameter p, and those with four free vertices to the renormalization of the parameter u.

Each diagram carries a combinatorial factor corresponding to the number of independent ways in which it can be realized, remembering that the free vertices of the diagram from which it is derived have different labels. For example, the second diagram on the RHS in (7.164) can be realized from the LHS by joining together any pair of vertices chosen from (q_1,q_2,q_3,q_4), and there are six such pairs; this diagram therefore carries a factor 6. Because of symmetry, the second diagram on the RHS carries a factor 3.

For the diagram in (7.165), if the vertices of the disjoint graphs on the LHS are labelled (1,2,3,4) and (1′,2′,3′,4′), we can choose any one of six independent pairs from the first set to connect with any one of six independent pairs from the second set; there then remain two ways of making the connection (e.g. 1,1′ and 2,2′ or 1,2′ and 2,1′). This diagram therefore carries a factor 72.

Using (7.163) we find that the second diagram  makes a contribution

$$\int_{\mathbf{q}_1}\sigma_{0\mathbf{q}_1}\int_{\mathbf{q}_2}\sigma_{0\mathbf{q}_2}\int_{\mathbf{q}_3}\int_{\mathbf{q}_4}\delta^{(d)}(\mathbf{q}_3+\mathbf{q}_4)\delta^{(d)}(\mathbf{q}_1+\mathbf{q}_2+\mathbf{q}_3+\mathbf{q}_4)(q_3^2+p)^{-1}$$

$$=6u\int_{\mathbf{q}_1}\sigma_{0\mathbf{q}_1}\sigma_{0-\mathbf{q}_1}\int_{\mathbf{q}_3}\frac{1}{q_3^2+p}\left(0<|\mathbf{q}_1|<\frac{1}{b},\frac{1}{b}<|\mathbf{q}_3|<1\right). \quad (7.166)$$

When the scale is changed $q'\to bq$, $\sigma_\mathbf{q}\to\zeta\sigma'_{\mathbf{q}'}$ this is transformed to

$$6u\zeta^2 b^{-d}A\left(\frac{1}{b},p\right)\int_{\mathbf{q}'}\sigma_{\mathbf{q}'}\sigma_{-\mathbf{q}'} \quad (7.167)$$

where, from (7.138),

$$A\left(\frac{1}{b}, p\right) = C_d \int_{1/b}^{1} \frac{q^{d-1}\,\mathrm{d}q}{q^2 + p}. \tag{7.168}$$

The term (7.167) is clearly of the same form in $\mathbf{q}'$ as the corresponding term in the original Hamiltonian (7.147), (7.148) in q.

We can now write down the equation for $u_2'(q')$ in (7.152) using (7.154) and (7.167),

$$u_2'(q) = \zeta^2 b^{-d}\left[\frac{q'^2}{b^2} + p' + 12uA\left(\frac{1}{b}, p\right)\right] \tag{7.169}$$

and using relation (7.144) as before, we find for the renormalized coefficient p',

$$p' = b^2\left[p + 12uA\left(\frac{1}{b}, p\right)\right]. \tag{7.170}$$

In order to deal with diagrams like (7.165) it is useful to summarize the results we have derived so far in the following rules (Wilson and Kogut 1974):

1 Label the four-point vertices of the initial term in the expansion (7.156) from which the connected diagram is derived.
2 With each free vertex associate a spin variable $\sigma_{0\mathbf{q}} = \zeta\sigma'_{b\mathbf{q}} = \zeta\sigma_{\mathbf{q}'}$.
3 With each wavy line connecting $\mathbf{q}_1$ and $\mathbf{q}_2$ associate a 'propagator' $\delta^d(\mathbf{q}_1 + \mathbf{q}_2)(q_1^2 + p)^{-1}$.
4 Momenta on free vertices range from 0 to $1/b$, and on connected vertices from $1/b$ to 1.
5 Associate a factor $u\delta^{(d)}(\mathbf{q}_1 + \mathbf{q}_2 + \mathbf{q}_3 + \mathbf{q}_4)$ with each four-point vertex.
6 Integrate $\int_{\mathbf{q}}$ over the momenta according to rule 4.

Let us now apply these rules to (7.165) using the labelling scheme of Figure 7.10. We obtain the result

$$u^2 \int_{\mathbf{q}1} \cdots \int_{\mathbf{q}2} \delta^{(d)}(\mathbf{q}_1 + \mathbf{q}_2 + \mathbf{q}_5 + \mathbf{q}_7)\delta^{(d)}(\mathbf{q}_3 + \mathbf{q}_4 + \mathbf{q}_6 + \mathbf{q}_8)\delta^{(d)}(\mathbf{q}_5 + \mathbf{q}_6)$$
$$\times\, \delta^{(d)}(\mathbf{q}_7 + \mathbf{q}_8)\sigma_{0_{\mathbf{q}1}}\sigma_{0_{\mathbf{q}2}}\sigma_{0_{\mathbf{q}3}}\sigma_{0_{\mathbf{q}4}}(q_5^2 + p)^{-1}(q_7^2 + p)^{-1} \tag{7.171}$$

which can readily be simplified to

$$u^2 \int_{\mathbf{q}1}\int_{\mathbf{q}2}\int_{\mathbf{q}3} \sigma_{0_{\mathbf{q}1}}\sigma_{0_{\mathbf{q}2}}\sigma_{0_{\mathbf{q}3}}\sigma_{0-(\mathbf{q}_1+\mathbf{q}_2+\mathbf{q}_3)} \int_{\mathbf{q}5} (q_5^2 + p)^{-1}$$
$$\times\, [(q_5 + q_1 + q_2)^2 + p]^{-1}. \tag{7.172}$$

This is of the form required in (7.152), but we need a simplification to make further progress. Our aim is to renormalize the parameters p and u which are not $\mathbf{q}$ dependent, whereas the coefficient in (7.172) is $\mathbf{q}$ dependent. Since we are

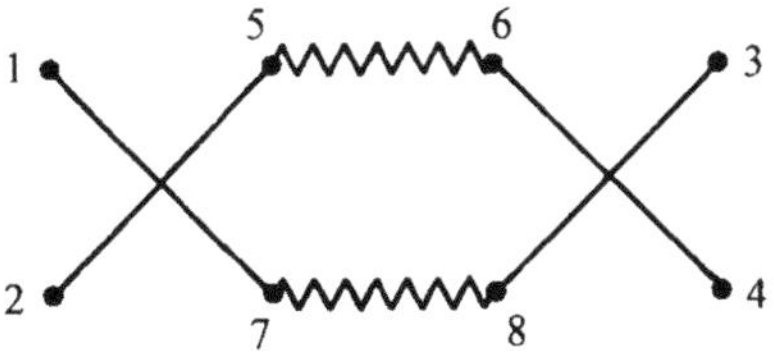

Figure 7.10 A diagram arising in the second-order perturbation term.

in the first place interested in the recursion relation for a non-q-dependent u, we can put $q_1 = q_2 = 0$ in (7.172) to determine this relation. As in real-space renormalization application of the RG introduces new parameters, coefficients of u^2q^2, u^2q^4, ... which we ignore at this stage. The significance of such q-dependent terms will be discussed at the end of the next section. When we rescale and introduce the combinatorial factor 72 and the factor $-\frac{1}{2}$ from the coefficient of $\mathscr{H}_I^2$ we obtain

$$-36u^2\zeta^4 b^{-3d} C\left(\frac{1}{b}, p\right) \int_{\mathbf{q}_1'} \int_{\mathbf{q}_2'} \int_{\mathbf{q}_3'} \sigma_{\mathbf{q}_1'} \sigma_{\mathbf{q}_2'} \sigma_{\mathbf{q}_3'} \sigma_{-(\mathbf{q}_1'+\mathbf{q}_2'+\mathbf{q}_3')} \tag{7.173}$$

where

$$C\left(\frac{1}{b}, p\right) = C_d \int_{1/b}^{1} q^{d-1}\, \mathrm{d}q/(q^2+p)^2. \tag{7.174}$$

There are no convergence problems with integrals like $A(1/b, p)$ in (7.168), or $C(1/b, p)$ in (7.174), since the range of integration does not include the origin. In fact they can be expanded in powers of p,

$$\begin{aligned} A\left(\frac{1}{b}, p\right) &= A\left(\frac{1}{b}, 0\right) + pA_1\left(\frac{1}{b}, 0\right) + O(p^2) \\ C\left(\frac{1}{b}, p\right) &= C\left(\frac{1}{b}, 0\right) + pC\left(\frac{1}{b}, 0\right) + O(p^2) \end{aligned} \tag{7.175}$$

and it will readily be seen, on differentiating (7.168) w.r.t. p, that

$$A_1\left(\frac{1}{b}, 0\right) = -C_1\left(\frac{1}{b}, 0\right). \tag{7.176}$$

This identity will be important in the subsequent calculations of the critical exponents.

From (7.147) and (7.173), using (7.145), we find for renormalized coefficient u',

$$u' = b^{4-d}\left[u - 36u^2 C\left(\frac{1}{b}, p\right)\right]. \tag{7.177}$$

We now use equations (7.170) and (7.177) to determine the fixed point (p^*,u^*). If $\epsilon < 0$ (i.e. $d > 4$) it is clear from the second equation that successive iterations must lead to $u^* = 0$ because of the damping factor b^ϵ, and hence we are left at the Gaussian fixed point $u^* = 0$, $p^* = 0$. However, if $\epsilon > 0$, the second equation ensures the growth of u, and it should be possible to get away from the Gaussian fixed point. This is illustrated by assuming that p and u are small, and seeking a solution to the first order in ϵ. Replacing p,p' by p^* and u,u', by u^*, we derive the equations

$$\begin{aligned}(b^2 - 1)p^* &= -12b^2 A\left(\frac{1}{b}, 0\right)u^* \\ b^\epsilon - 1 &= 36C\left(\frac{1}{b}, 0\right)u^*.\end{aligned} \tag{7.178}$$

To the first order in ϵ the solution is

$$\begin{aligned}u^* &= \frac{\epsilon \ln b}{36C(1/b, 0)} \\ p^* &= -\frac{b^2}{3(b^2 - 1)}\frac{A(1/b, 0)}{C(1/b, 0)}\epsilon \ln b.\end{aligned} \tag{7.179}$$

The formalism of §7.3 can now be used to obtain the matrix **L**, the eigenvalues Λ_1, Λ_2, and hence the critical exponent ν. The matrix **L** is given by the derivatives of the RHS of (7.170) and (7.177) w.r.t. (p,u) at (p^*,u^*),

$$\mathbf{L} = \begin{vmatrix} b^2 + 12b^2u^*A_1\left(\frac{1}{b}, 0\right) & 12b^2\left[A\left(\frac{1}{b}, 0\right) + A_1\left(\frac{1}{b}, 0\right)p^*\right] \\ 0 & b^\epsilon\left[1 - 72u^*C\left(\frac{1}{b}, 0\right)\right]\end{vmatrix} \tag{7.180}$$

To the first order in ϵ this is

$$\begin{vmatrix} b^2\left(1 - \frac{1}{3}\epsilon \ln b\right) & 12b^2A\left(\frac{1}{b}, 0\right)\left[1 - \frac{b^2}{3(b^2 - 1)}\epsilon \ln b\right] \\ 0 & 1 - \epsilon \ln b\end{vmatrix} \tag{7.181}$$

The two eigenvalues are

$$\begin{aligned}\Lambda_1 &= b^2\left(1 - \frac{\epsilon}{3}\ln b\right) \\ \Lambda_2 &= 1 - \epsilon \ln b\end{aligned} \tag{7.182}$$

the first being relevant, and the second irrelevant. From (7.72)

$$\nu = \frac{\ln b}{\ln \Lambda_1} = \frac{\ln b}{\ln b^2 - \frac{1}{3}\epsilon \ln b} = \frac{1}{2} + \frac{\epsilon}{12}. \tag{7.183}$$

We have thus derived the exponent ν for the Ising model to the first order in ϵ. Putting $\epsilon = 1$ we obtain the numerical estimate 0.583 which is a substantial step in the right direction. We note that as anticipated the critical properties do not depend on b.

7.5.3 *The n-vector Model*

To generalize the above treatment to the n-vector model $s_{\mathbf{l}}^2$ will be replaced by $\left(\sum_{\alpha=1}^{n} s_{\mathbf{l}}^{\alpha} s_{\mathbf{l}}^{\alpha}\right)$ and $s_{\mathbf{l}}^4$ by $\left(\sum_{\alpha=1}^{n} s_{\mathbf{l}}^{\alpha} s_{\mathbf{l}}^{\alpha}\right)^2$, and the interaction in (7.135) by

$$K \sum_{\mathbf{j},\mathbf{l}} \sum_{\alpha=1}^{n} s_{\mathbf{l}}^{\alpha} s_{\mathbf{l}+\mathbf{j}}^{\alpha}.$$

Proceeding as in §7.5.1 from (7.135) to (7.137) we will be led to a relation analogous to (7.139)

$$\exp\left(-\frac{1}{2}\int_{\mathbf{q}} (q^2 + p) \sum_{\alpha=1}^{n} \sigma_{\mathbf{q}}^{\alpha} \sigma_{-\mathbf{q}}^{\alpha}\right). \tag{7.184}$$

The Gaussian model takes only the square terms into account, and can be treated as before. It is instructive to look at the anisotropic Gaussian model with additional interaction

$$\sum_{\alpha=1}^{n} g_{\alpha} s_{\mathbf{l}}^{\alpha} s_{\mathbf{l}+\mathbf{j}}^{\alpha} \tag{7.185}$$

and to trace with relative ease the transition from an isotropic to an anisotropic fixed point and the corresponding cross-over behaviour (for a summary see Aharony (1976, pp. 366–70)).

The fourth-order term $-u(\sum_{\alpha=1}^{n} s_{\mathbf{l}}^{\alpha} s_{\mathbf{l}}^{\alpha})^2$ is dealt with as in §7.5.2. Fourier transformation now leads to a term

$$-u \int_{\mathbf{q}_1} \int_{\mathbf{q}_2} \int_{\mathbf{q}_3} \sum_{\alpha,\beta} \sigma_{\mathbf{q}_1}^{\alpha} \sigma_{\mathbf{q}_2}^{\alpha} \sigma_{\mathbf{q}_3}^{\beta} \sigma_{-\mathbf{q}_1-\mathbf{q}_2-\mathbf{q}_3}^{\beta}. \tag{7.186}$$

The treatment closely parallels that of §7.5.2, the basic new feature arising in the calculation of the combinatorial factor when the n different values of α, β are taken into account. As before only two diagrams contribute. To deal with X as in (7.164) we write the original term in full as

$$\sum_{\alpha,\beta=1}^{n} (\sigma_{0\mathbf{q}_1}^{\alpha} + \sigma_{1\mathbf{q}_1}^{\alpha})(\sigma_{0\mathbf{q}_2}^{\alpha} + \sigma_{1\mathbf{q}_2}^{\alpha})(\sigma_{0\mathbf{q}_3}^{\beta} + \sigma_{1\mathbf{q}_3}^{\beta})(\sigma_{0\mathbf{q}_4}^{\beta} + \sigma_{1\mathbf{q}_4}^{\beta}) \tag{7.187}$$

from which we must calculate the number of terms of the form

$$\sum_{\alpha=1}^{n} \sigma^{\alpha}_{0\mathbf{q}_i}\sigma^{\alpha}_{0\mathbf{q}_j} \tag{7.188}$$

which will reproduce the original form of (7.184). When $\alpha = \beta$ six different pairs can be chosen from (q_1, q_2, q_3, q_4) as before. When $\alpha \neq \beta$ there are $(n-1)$ different $\sigma^{\beta}_{1\mathbf{q}_3}\sigma^{\beta}_{1\mathbf{q}_4}$ corresponding to $\sigma^{\alpha}_{0\mathbf{q}_1}\sigma^{\alpha}_{0\mathbf{q}_2}$ and $(n-1)\sigma^{\beta}_{1\mathbf{q}_1}\sigma^{\beta}_{1\mathbf{q}_2}$ corresponding to $\sigma^{\alpha}_{0\mathbf{q}_3}\sigma^{\alpha}_{0\mathbf{q}_4}$. This gives a total of $2(n+2)$. The contribution of this diagram (see (7.167)) is thus

$$2(n+2)u\zeta^2 b^{-d} A\left(\frac{1}{b}, p\right)\int_{\mathbf{q}'}\sum_{\alpha=1}^{n}\sigma^{\alpha}_{\mathbf{q}'}\sigma^{\alpha}_{-\mathbf{q}'} \tag{7.189}$$

where as before

$$A\left(\frac{1}{b}, p\right) = C_d\int_{1/b}^{1}\frac{q^{d-1}\,\mathrm{d}q}{(q^2+p)}. \tag{7.190}$$

The second diagram (7.165) is longer and more tricky. We first note that any pairing of the form $\sigma^{\alpha}_{1\mathbf{q}_1}\sigma^{\beta}_{1\mathbf{q}_{2'}}$ will give zero unless $\alpha = \beta$. This follows from the discussion of Gaussian integrals in (7.161), since the corresponding element of the inverse of the generalized matrix is zero. The original diagram on the LHS is represented by terms of the form

$$\sigma^{\gamma}_{\mathbf{q}_1}\sigma^{\gamma}_{\mathbf{q}_2}\sigma^{\delta}_{\mathbf{q}_3}\sigma^{\delta}_{\mathbf{q}_4}\sigma^{\theta}_{\mathbf{q}'_1}\sigma^{\theta}_{\mathbf{q}'_2}\sigma^{\phi}_{\mathbf{q}'_3}\sigma^{\phi}_{\mathbf{q}'_4} \tag{7.191}$$

where the indices γ, δ, θ, ϕ go from 1 to n. For specific α and β we must compute the number of terms of the form

$$\sigma^{\alpha}_{0\mathbf{q}_i}\sigma^{\alpha}_{0\mathbf{q}_j}\sigma^{\beta}_{0\mathbf{q}_{k'}}\sigma^{\beta}_{0\mathbf{q}_{l'}} \quad\text{or}\quad \sigma^{\alpha}_{0\mathbf{q}_i}\sigma^{\beta}_{0\mathbf{q}_j}\sigma^{\alpha}_{0\mathbf{q}_{k'}}\sigma^{\beta}_{0\mathbf{q}_{l'}}.$$

We shall first deal with the more difficult general case when $\alpha \neq \beta$ and shall list the different types of configuration with their appropriate combinatorial factors.

(i) 1 2 3 4 1′ 2′ 3′ 4′

$\alpha\ \alpha\ \gamma\ \gamma\quad \beta\ \beta\ \gamma\ \gamma\quad (\gamma \neq \alpha \text{ or } \beta)$.

Combinatorial factor $8(n-2)$.

(The γ can be connected in two ways 33′, 44′ or 34′, 43′, and the α and the γ and the β and the γ can be interchanged.)

(ii) 1 2 3 4 1′ 2′ 3′ 4′

$\alpha\ \alpha\ \alpha\ \alpha\quad \beta\ \beta\ \alpha\ \alpha$

Combinatorial factor $6 \times 2 \times 2 \times 2 = 48$.

(There are six ways of choosing a pair of α on the LHS, two ways of connecting the remaining α with those on the RHS, the α and the β on the RHS can be interchanged, and the LHS and RHS can be interchanged.)

(iii) 1 2 3 4 1′ 2′ 3′ 4′

$\alpha\ \alpha\ \beta\ \beta\quad \alpha\ \alpha\ \beta\ \beta.$

We must select one α and one β on the LHS and one α and one β on the RHS.

Combinatorial factor $2 \times 2 \times 2 \times 2 \times 2 = 32$.

(Each α and each β can be selected in two ways, the connections of the remainder then being uniquely determined. Also the α and the β can be interchanged on one of the sides.) Hence, we find for the total combinatorial factor $8(n + 8)$.

When $\alpha = \beta$, the problem is much simpler and there are only two possibilities:

(i) 1 2 3 4 1′ 2′ 3′ 4′

$\alpha\ \alpha\ \gamma\ \gamma\quad \alpha\ \alpha\ \gamma\ \gamma\quad (\gamma \neq \alpha)$

Combinatorial factor $8(n - 1)$.

(ii) 1 2 3 4 1′ 2′ 3′ 4′

$\alpha\ \alpha\ \alpha\ \alpha\quad \alpha\ \alpha\ \alpha\ \alpha$

Combinatorial factor 72.

Total combinatorial factor again $8(n + 8)$.

Proceeding as in the previous section we find, instead of (7.173), the contribution

$$-4(n + 8)u^2\zeta^4 b^{-3d} C\left(\frac{1}{b}, p\right) \int_{\mathbf{q}'_1} \int_{\mathbf{q}'_2} \int_{\mathbf{q}'_3} \sum_{\alpha,\beta} \sigma^{\alpha}_{\mathbf{q}'_1} \sigma^{\alpha}_{\mathbf{q}'_2} \sigma^{\beta}_{\mathbf{q}'_3} \sigma^{\beta}_{-\mathbf{q}'_1-\mathbf{q}'_2-\mathbf{q}'_3} \tag{7.192}$$

where as before

$$C\left(\frac{1}{b}, p\right) = C_d \int_{1/b}^{1} \frac{q^{d-1}\,dq}{(q^2 + p)^2}. \tag{7.193}$$

To ensure that the coefficient of q^2 in (7.184) remains unchanged by the RG transformation we take, as before,

$$\zeta = b^{1+d/2}. \tag{7.194}$$

Equations (7.170) and (7.177) generalize to

$$\begin{aligned} p' &= b^2[p + 4(n + 2)uA(1/b, p)] \\ u' &= b^{\epsilon}[u - 4(n + 8)u^2 C(1/b, p)]. \end{aligned} \tag{7.195}$$

The fixed point is given, instead of (7.179), by

$$u^* = \frac{\epsilon \ln b}{4(n+8)C(1/b,0)}$$
$$p^* = -\frac{b^2}{b^2-1}\frac{n+2}{n+8}\frac{A(1/b,0)}{C(1/b,0)}\epsilon \ln b. \tag{7.196}$$

The matrix $\mathbf{L}$ which determines the eigenvalues is, to the first order in ϵ,

$$\mathbf{L} = \begin{vmatrix} b^2\left(1 - \frac{n+2}{n+8}\epsilon \ln b\right) & 4(n+2)b^2 A\left(\frac{1}{b}, 0\right)\left(1 - \frac{b^2}{b^2-1}\frac{n+2}{n+8}\epsilon \ln b\right) \\ 0 & 1 - \epsilon \ln b \end{vmatrix}$$

with two eigenvalues (7.197)

$$\Lambda_1 = b^2\left(1 - \frac{n+2}{n+8}\epsilon \ln b\right)$$
$$\Lambda_2 = 1 - \epsilon \ln b. \tag{7.198}$$

The critical exponent ν is given by

$$\nu = \frac{\ln b}{\ln \Lambda_1} = \frac{\ln b}{2 \ln b - [(n+2)/(n+8)]\epsilon \ln b} = \frac{1}{2} + \frac{\epsilon(n+2)}{4(n+8)} \tag{7.199}$$

which is again independent of b as expected. We note the importance of the combinatorial factors for the key diagrams in determining the $(n+2)$ and $(n+8)$ in (7.199).

Although the calculation of expansions like (7.199) to higher orders in ϵ is a matter largely for specialists, we feel that it is appropriate to indicate the lines along which such expansions can be pursued. New diagrams must be included which correspond to higher powers of u. The procedure for evaluating their contribution has already been described, and they do not give rise to any difficulty.

Integrals like (7.190) and (7.193) depend on d, and replacing d by $4 - \epsilon$, an expansion in powers of ϵ must be developed. C_d defined in (7.138) is also d dependent and therefore ϵ dependent. Either we can retain it in our equations remembering its ϵ dependence, or, more neatly, the parameters p and u can be redefined after absorbing the C_d factor, and recursion relations can be developed in the new parameters.

Finally the q-dependent terms in the diagrams which enter at the second order in u will affect the coefficient of q'^2 in (7.169), and the value of ζ will have to be modified to make the coefficient of q'^2 unity as before. The diagrams which are involved at second order in u are shown in Figure 7.11.

There is a simple scaling argument which shows the most convenient way in which to modify ζ to take the q^2-dependent terms into account. We know from equation (5.163) that the spin pair correlation funtion, $\Gamma(\mathbf{r},0)$, has the asymptotic form $1/r^{d-2+\eta}$ at the critical temperature. Hence, under the RG

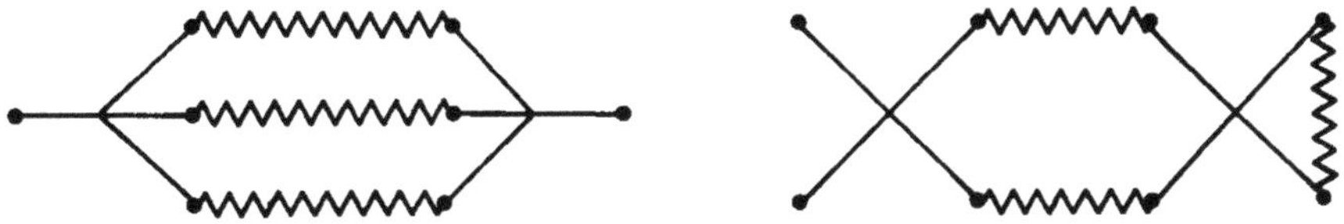

Figure 7.11 Diagrams arising for the second-order term in the parameter u.

transformation $r' \to r/b$,

$$\Gamma(\mathbf{r},0) = b^{2-d-\eta}\Gamma(\mathbf{r}',0). \tag{7.200}$$

But $\Gamma(\mathbf{r},0)$ is defined as $\langle s_0 s_\mathbf{r} \rangle$, and the scaling factor $\sigma_\mathbf{q} \to \zeta\sigma'_{\mathbf{q}'}$ is accompanied, from the Fourier transform relation (7.130), by the scaling factor $s_\mathbf{l} \to (\zeta/b^d)s'_{\mathbf{l}'}$. Hence we find from (7.200) that

$$\frac{\zeta^2}{b^{2d}} = b^{2-d-\eta} \quad \text{or} \quad \zeta^2 = b^{2+d-\eta}. \tag{7.201}$$

Away from the critical point we define η in (7.201) as a function of (p,u) of the second order in u which can be determined from the q^2 terms in the recursion relation for p.

It is found (Aharony 1976) that

$$\eta = 8(n+2)(C_d u)^2 + O(u^3). \tag{7.202}$$

When the fixed-point values (p^*,u^*) are substituted into the expansion for η, an ϵ expansion for the critical exponent η is derived. The first term in this expansion is

$$\eta = \frac{(n+2)}{2(n+8)^2}\epsilon^2 + O(\epsilon^3). \tag{7.203}$$

The corresponding ϵ expansion for the critical exponent ν is obtained from the leading eigenvalue of the matrix $\mathbf{L}$. To the second order this is

$$\nu = \frac{1}{2} + \frac{(n+2)}{4(n+8)}\epsilon + \frac{(n+2)(n^2+23n+60)}{8(n+8)^3}\epsilon^2 + O(\epsilon^3). \tag{7.204}$$

From ν and η all other critical exponents can be derived using the scaling relations of Table 6.1.

7.5.4 Competition between Fixed Points of Different Symmetry

We have referred in §4.9 to Landau's emphasis on the important role which is played by symmetry in the theory of the critical point. His ideas were completely vindicated by RG theory. Landau expanded the free energy as a power series in the order parameter around the critical point, grouping together terms in the expansion corresponding to different types of symmetry. This expansion is invalid because of the singularity in the free energy at the critical point. But there is a corresponding valid expansion for the original Hamiltonian, and RG recursion relations can be developed for the parameters in this expansion. These recursion relations give rise to fixed points of different symmetry whose regions of influence can then be delineated.

Following Aharony (1973) we shall introduce the fourth-order terms

$$u\left(\sum_{\alpha=1}^{n} s_{\mathbf{l}}^{\alpha} s_{\mathbf{l}}^{\alpha}\right)^{2} + v\left(\sum_{\alpha=1}^{n} (s_{\mathbf{l}}^{\alpha})^{4}\right). \tag{7.205}$$

When $u = v = 0$ we revert to the Gaussian fixed point; $v = 0$ corresponds to a Heisenberg fixed point with spherical symmetry; when $u = 0, v \neq 0$ the system splits up into independent Ising models, but there is also a fixed point with $u \neq 0$, $v \neq 0$ corresponding to cubic symmetry.

The calculations are similar to those described above, and the result up to second-order terms in u and v is (Aharony 1976)

$$\begin{aligned} u' &= b^{\epsilon-2\eta}\{u - 4C_4 \ln b[(n+8)u^2 + 6uv]\} \\ v' &= b^{\epsilon-2\eta}[v - 4C_4 \ln b(12uv + 9v^2)] \\ \eta &= 8C_4^2[(n+2)u^2 + 6uv + 3v^2]. \end{aligned} \tag{7.206}$$

If we write these equations in the form

$$\begin{aligned} u' &= a_{10}u + a_{20}u^2 + a_{11}uv \\ v' &= b_{01}v + b_{02}v^2 + b_{11}uv \end{aligned} \tag{7.207}$$

and put $u = u' = u^*$, $v = v' = v^*$ to determine the fixed points, we find the following four solutions to the first order in ϵ:

$$\text{Gaussian} \quad u^{\mathrm{G}} = v^{\mathrm{G}} = 0 \tag{7.208}$$

$$\text{Ising} \quad u^{\mathrm{I}} = 0, \quad v^{\mathrm{I}} = \frac{(1-b_{01})}{b_{02}} = \frac{\epsilon}{36C_4} \tag{7.209}$$

$$\text{Heisenberg} \quad v^{\mathrm{H}} = 0, \quad u^{\mathrm{H}} = \frac{(1-a_{10})}{a_{20}} = \frac{\epsilon}{4(n+8)C_4} \tag{7.210}$$

$$\begin{aligned} \text{Cubic} \quad u^{\mathrm{c}} &= \frac{b_{02}(1-a_{10}) - a_{11}(1-b_{02})}{a_{20}b_{02} - a_{11}b_{11}} = \frac{\epsilon}{12C_4 n} \\ v^{\mathrm{c}} &= \frac{a_{20}(1-b_{02}) - b_1(1-a_{10})}{a_{20}b_{02} - a_{12}b_{12}} = \frac{\epsilon}{12C_4 n}. \end{aligned} \tag{7.211}$$

To determine the eigenvalues, and hence the stability, we must linearize about the fixed points. Writing $\Lambda_1 = b^{\lambda 1}$ and $\Lambda_2 = b^{\lambda 2}$, the calculations yield

$$\lambda_1^{\mathrm{G}} = \lambda_2^{\mathrm{G}} = \epsilon \tag{7.212}$$

$$\lambda_1^{\mathrm{I}} = \tfrac{1}{3}\epsilon, \quad \lambda_2^{\mathrm{I}} = -\epsilon \tag{7.213}$$

$$\lambda_1^{\mathrm{H}} = -\epsilon, \quad \lambda_2^{\mathrm{H}} = \frac{n-4}{n+8}\,\epsilon. \tag{7.214}$$

$$\lambda_1^c = -\epsilon, \quad \lambda_2^c = \frac{4-n}{3n}\epsilon \tag{7.215}$$

Hence the Gaussian fixed point is 2-unstable and the Ising fixed point is 1-unstable; for conventional values of n (≤ 3) the Heisenberg fixed point is stable, and the cubic fixed point is 1-unstable. At a critical value of n the Heisenberg fixed point becomes 1-unstable, and the cubic fixed point becomes stable.

The schematic flow diagram for this iterative pattern showing the regions of attraction of the different fixed points is shown in Figure 7.12 for the two cases (a) $n < n_c$, (b) $n > n_c$ (following Aharony 1976).

7.5.5 Long-range Spherically Symmetrical Interactions

In §5.4.2.2 we discussed exact calculations for the spherical model, and summarized the results of Joyce's work on long-range interactions of the form

$$J(r) \sim r^{-(d+\sigma)}.$$

The main conclusions are to be found in equations (5.158)–(5.161); for sufficiently strong forces ($\sigma < d/2$) mean field thermodynamic critical exponents are attained, whilst for sufficiently large σ (>2) the critical exponents are those of short-range forces. We also referred in §5.4 to Stanley's important discovery for short-range forces that the spherical model can be regarded as an n-vector model in the limit $n \to \infty$.

The RG treatment introduced by Fisher, Ma and Nickel (1972) applies to a general n-vector model, and rapidly focuses attention on the above specific values of the parameter σ.

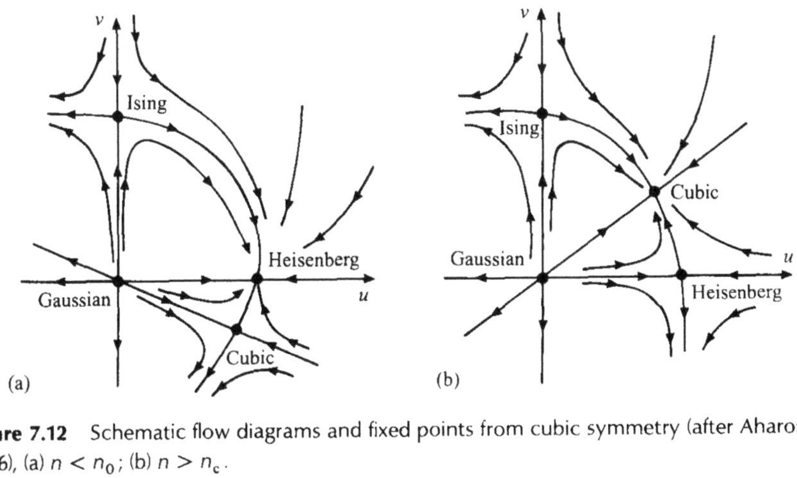

Figure 7.12 Schematic flow diagrams and fixed points from cubic symmetry (after Aharony 1976), (a) $n < n_0$; (b) $n > n_c$.

Following the treatment of §7.5.3, we shall use the interaction

$$\sum_{\mathbf{l},k} K(\mathbf{k}) \sum_{\alpha} s_{\mathbf{l}}^{\alpha} s_{\mathbf{l}+k}^{\alpha} \tag{7.216}$$

where $K(\mathbf{k})$ is of the form $k^{-(d+\sigma)}$ and $\mathbf{k}$ must be summed over all lattice points to infinity.

On applying the Fourier transformation (7.130) to the $s_{\mathbf{l}}^{\alpha}$ we will be led to a contribution to the integral

$$\int_{\mathbf{q}} d\mathbf{q} \hat{K}(\mathbf{q}) \sum_{\alpha} \sigma_{\mathbf{q}}^{\alpha} \sigma_{-\mathbf{q}}^{\alpha} \tag{7.217}$$

in (7.184), with

$$\hat{K}(\mathbf{q}) = \sum_{\mathbf{k}} \exp[\mathrm{i}(\mathbf{q} \cdot \mathbf{k}) K(\mathbf{k})]. \tag{7.218}$$

The sum (7.218) can be split into two contributions: those due to near neighbours which give a term $\sim q^2$ as before; and those due to distant neighbours whose contribution can be determined by the integral

$$\int \frac{r^{d-1} \exp(\mathrm{i}qr)\, \mathrm{d}r}{r^{d+\sigma}} \sim q^{\sigma}. \tag{7.219}$$

We therefore replace the integral in (7.184) by

$$\int_{\mathbf{q}} (p + A_2 q^2 + A_{\sigma} q^{\sigma}) \sum_{\alpha=1}^{n} \sigma_{\mathbf{q}}^{\alpha} \sigma_{-\mathbf{q}}^{\alpha}. \tag{7.220}$$

When we apply an RG transformation dividing the length scale by b, and rescaling the $\sigma_{\mathbf{q}}$ by a factor ζ, (7.220) transforms into

$$\zeta^2 b^{-d} \int_{\mathbf{q}'} \left[p + A_2 \left(\frac{q'}{b} \right)^2 + A_{\sigma} \left(\frac{q'}{b} \right)^{\sigma} \right] \sum_{\alpha=1}^{n} \sigma_{\mathbf{q}'}^{\alpha} \sigma_{-\mathbf{q}'}^{\alpha}. \tag{7.221}$$

If as before we keep the coefficient of q^2 unchanged by taking $\zeta^2 = b^{d+2}$ we will have

$$A'_{\sigma} = b^{2-\sigma} A_{\sigma}. \tag{7.222}$$

Thus A_{σ} is irrelevant if $\sigma > 2$, marginal if $\sigma = 2$, and relevant if $\sigma < 2$; we therefore see immediately the significance of the particular value $\sigma = 2$. In the first case the system behaves like a short-range force model.

But if A_{σ} is relevant it is appropriate to arrange the rescaling factor ζ so as to keep the coefficient of q^{σ} unchanged. For this purpose we choose

$$\zeta^2 = b^{d+\sigma} \tag{7.223}$$

and we then find that

$$A'_2 = b^{\sigma-2} A_2 \tag{7.224}$$

so that A_2 is an irrelevant parameter. We then have the recursion relation

$$p' = b^{\sigma} p \tag{7.225}$$

so that $\Lambda_1 = b^{\sigma}$, and

$$\nu = 1/\sigma. \tag{7.226}$$

But we know the value of ζ^2 at the critical point from (7.200). Hence we conclude from (7.223) that

$$\eta = 2 - \sigma. \tag{7.227}$$

We may also note that relation (7.222) enables us to identify the cross-over exponent between A_2 and A_σ, following the discussion at the end of §7.3, as

$$\phi = \nu(2 - \sigma) = \tfrac{1}{2}(2 - \sigma). \tag{7.228}$$

We must now investigate the effect of the s_1^4 terms in the latter case when A_σ is relevant and A_2 irrelevant. Proceeding as before the following recursion relation for u can be derived, instead of (7.195):

$$u' = b^{2\sigma - d}\left[u - 4(n + 8)u^2 C^{(\sigma)}\left(\frac{1}{b}, p\right)\right] \tag{7.229}$$

where $C^{(\sigma)}$ is given by (7.193) with q^{σ} replacing q^2 in the denominator. Thus u is irrelevant if $d > 2\sigma$, which is the second specific value mentioned in the opening paragraph. When this condition is satisfied the critical exponents have the exact values outlined above. But if $d < 2\sigma$, the critical exponents depend on $\epsilon' = (2\sigma - d)$; expansions for the exponents can then be derived in powers of ϵ' (for details see Fisher, Ma and Nickel (1972)).

7.5.6 Long-range Dipolar Forces

In Chapter 3 we learned of Heisenberg's important identification of forces giving rise to ferromagnetism as exchange interactions. They are strong interactions which decay exponentially with distance, and they are therefore short ranged. But from elementary magnetostatic considerations we know of another *dipolar* interaction between any pair of magnetic dipoles, which must also be present in any ferromagnet. Usually the dipolar forces are much weaker than the exchange forces and play little direct part in critical behaviour, although they play a significant role in characterizing domain structure. But there are substances in which the dipolar interaction is dominant; in any case, even weak forces will be expected to exercise an influence in a region sufficiently close to the critical point.

The form of the dipolar interaction can be derived in an elementary manner from first principles. The potential due to a magnetic monopole is a Coulomb potential of the type $1/r$. If we put together at the origin two monopoles of opposite sign to form a dipole, they will give rise to a potential

$$\mu_0 \frac{\partial}{\partial x_i}\left(\frac{1}{r}\right) = \frac{\boldsymbol{\mu}_i \cdot \mathbf{r}}{r^3} \tag{7.230}$$

where $\boldsymbol{\mu}_i = \mu_0 \mathbf{i}$ is the dipole moment, and $\mathbf{i}$ is a unit vector in the direction of the dipole. By similar reasoning a second dipole at a distance $\mathbf{r}$ from the first dipole will have interaction energy

$$\mu_0 \frac{\partial^2}{\partial x_i \partial x_j}\left(\frac{1}{r}\right) = \mu_0 \frac{\partial}{\partial x_j}\left(\frac{\boldsymbol{\mu}_i \cdot \mathbf{r}}{r^3}\right) = \frac{1}{r^3}\left(\boldsymbol{\mu}_i \cdot \boldsymbol{\mu}_j - \frac{3(\boldsymbol{\mu}_i \cdot \mathbf{r})(\boldsymbol{\mu}_j \cdot \mathbf{r})}{r^2}\right) \tag{7.231}$$

where $\boldsymbol{\mu}_j$ is the moment of the second dipole.

The interaction (7.231) decays only as $1/r^3$ and is therefore very long range; it also has a strong dependence on direction. Its macroscopic consequences are familiar to us from the shape-dependent 'demagnetizing factor' which arises in the determination of the internal magnetic field in a magnetic sample. Similar considerations arising from electric dipolar interactions determine the internal electric field in a dielectric.

Because of the conditional convergence of lattice sums involving the dipolar interaction (7.231), the microscopic theory gives rise to considerable difficulties. Even the problem of establishing the nature and energy of the ground state for a given lattice and a given sample shape, and whether it is ferromagnetic or antiferromagnetic, are quite sophisticated problems, and engaged the attention of early researchers (e.g. Onsager 1939, Sauer 1940, Luttinger and Tisza 1946, Nijboer and de Wette 1958). It is possible to avoid such considerations by approaching the critical point from the high-temperature side. Series expansions were pioneered by Hiley and Joyce (1965) and Levy and collaborators (e.g. Levy and Stinchcombe 1968) who succeeded in identifying the shape-dependent terms, and developing a formalism for the exploration of critical behaviour. But the practical difficulties of computation were formidable, and no results of significance emerged.

Real progress in the treatment of dipolar forces came with the application of the RG by Aharony and Fisher (1973) (for more detailed references see Aharony (1976); preliminary results had been obtained previously using field theory by Vaks, Larkin and Pikin (1966) and Larkin and Khmelnitskii (1969)). The details of the calculations are involved, and we shall therefore content ourselves with a brief summary of the main features and conclusions as outlined in the review article by Aharony (1976).

In order to use the RG we must first generalize the dipolar interaction (7.231) to d dimensions. The analogue of the Coulomb potential decays as $1/r^{d-2}$, and using the same method as for three dimensions we finalize the interaction energy

$$\frac{1}{r^d}\left(\boldsymbol{\mu}_i \boldsymbol{\mu}_j - \frac{d(\boldsymbol{\mu}_i \cdot \mathbf{r})(\boldsymbol{\mu}_j \cdot \mathbf{r})}{r^2}\right) \tag{7.232}$$

instead of (7.231), where $\boldsymbol{\mu}_i$, $\boldsymbol{\mu}_j$, $\mathbf{r}$ are d-dimensional vectors. Since the interaction involves scalar products of the form $(\boldsymbol{\mu}_i \cdot \mathbf{r})$, the spin dimension n must be equal to d. In §3.2 we pointed out that the magnetic moment is proportional to the spin

$$\boldsymbol{\mu}_i = g\beta_B s_i . \tag{7.233}$$

We shall therefore write the dipolar interaction in the form

$$(g\beta_B)^2 \sum_{\mathbf{l},\mathbf{k}} \sum_{\alpha,\beta} \frac{\partial^2}{\partial l^\alpha \partial l^\beta} \left(\frac{1}{|\mathbf{l}' - \mathbf{l}|}\right)^{d-2} s_\mathbf{l}^\alpha s_\mathbf{l}^\beta \; (\mathbf{l}' = \mathbf{l} + \mathbf{k}). \tag{7.234}$$

The short-range interaction will be as in §7.5.3, the square term giving a contribution (7.184).

Before focusing attention on the interaction (7.234) we shall generalize our previous treatment to an anisotropic interaction

$$\sum_{\mathbf{l},\mathbf{k}} K^{\alpha\beta}(\mathbf{k}) s_\mathbf{l}^\alpha s_{\mathbf{l}+\mathbf{k}}^\beta . \tag{7.235}$$

Applying the Fourier transformation (7.130) again to the $s_\mathbf{l}^\alpha$, we will be led to a contribution to the integral

$$\int_\mathbf{q} \mathrm{d}\mathbf{q} \hat{K}^{\alpha\beta}(\mathbf{q}) \sigma_\mathbf{q}^\alpha \sigma_{-\mathbf{q}}^\beta \tag{7.236}$$

in (7.184) with

$$\hat{K}^{\alpha\beta}(\mathbf{q}) = \sum_\mathbf{k} \exp[\mathrm{i}(\mathbf{q} \cdot \mathbf{k}) K^{\alpha\beta}(\mathbf{k})]. \tag{7.237}$$

The first important obstacle which Aharony and Fisher had to overcome was the evaluation of the sum (7.237) for the dipolar interaction (7.234). More than 50 years previously Ewald (1921) had introduced a transformation to deal with the three-dimensional problem, and this could be readily adapted to d dimensions. For a cubic hyperlattice they found that

$$\begin{aligned}\hat{K}^{\alpha\beta}(\mathbf{q}) = a_1 \frac{q^\alpha q^\beta}{q^2} - a_2 q^\alpha q^\beta - [a_3 + a_4 q^2 + a_5 (q^\alpha)^2]\delta_{\alpha\beta} \\ + O(q^4, (q^\alpha)^4, (q^\alpha)^2 (q^\beta)^2)\end{aligned} \tag{7.238}$$

for $\mathbf{q} \neq 0$, where the a_i are of order unity. For $q = 0$ a shape-dependent term is obtained.

Hence it will be seen that (7.184) must be replaced by

$$\exp\left(-\frac{1}{2}\int \sum_{\alpha,\beta} V_2^{\alpha\beta}(\mathbf{q}) \sigma_\mathbf{q}^\alpha \sigma_{-\mathbf{q}}^\beta\right) \tag{7.239}$$

with

$$V_2^{\alpha\beta}(\mathbf{q}) = [p + q^2 + f(q^\alpha)^2]\delta_{\alpha\beta} + g \frac{q^\alpha q^\beta}{q^2} - h q^\alpha q^\beta + \text{terms of order } q^4. \tag{7.240}$$

Aharony and Fisher noted that with the exception of the $f(q^\alpha)^2$ term the integrand in (7.239) is a scalar and invariant under a rotational transformation. They therefore termed the interaction

$$(p + q^2)\delta_{\alpha\beta} + g\frac{q^\alpha q^\beta}{q^2} - hq^\alpha q^\beta \tag{7.241}$$

the 'isotropic dipolar interaction' which is lattice independent; the term $f(q^\alpha)^2$ arises because of the cubic symmetry of the lattice, and one might reasonably expect it to have minor significance.

Following our previous treatment we will need to calculate integrals of the form (7.162), and we must therefore determine the reciprocal of the $d \times d$ matrix (7.240) with $f = 0$. Writing such a matrix in the general form

$$\mathbf{m} = \alpha\delta_{\alpha\beta} + cq_\alpha q_\beta \tag{7.242a}$$

it is reasonable to try to find a reciprocal $\mathbf{M}$ of the form

$$\mathbf{M} = A\delta_{\alpha\beta} + Cq_\alpha q_\beta . \tag{7.242b}$$

By standard matrix multiplication, we find

$$(\mathbf{mM})_{\alpha\beta} = aA\delta_{\alpha\beta} + (aC + Ac + Cc)q_\alpha q_\beta . \tag{7.243}$$

Hence, to satisfy $\mathbf{mM} = 1$, we must have

$$aA = 1, \quad aC + Ac + Cc = 0. \tag{7.244}$$

These equations have a simple solution for A and C. Applying it to our case, with

$$a = p + q^2, \quad c = g - hq^2 \tag{7.245}$$

we find, for the analogue of $(q^2 + p)^{-1}$ in (7.163),

$$G^{\alpha\beta}(\mathbf{q}) = \frac{1}{p + q^2}\left(\delta_{\alpha\beta} - \frac{q_\alpha q_\beta}{q^2}\right) + \frac{1}{p + g + (1 - h)q^2}\frac{q^\alpha q^\beta}{q^2} . \tag{7.246}$$

Incidentally $G^{\alpha\beta}(\mathbf{q})$ represents the correlation function for the Gaussian model of isotropic dipolar interactions.

Aharony and Fisher then added the fourth-order term (7.186) which has the same isotropic symmetry, and proceeded to determine the recursion relations for the parameters in the Hamiltonian. For g they were led to the simple recursion relation

$$g' = b^{2-\eta}g \tag{7.247}$$

for which the fixed points are zero or infinity. The first corresponds to antiferromagnetism, which requires the expansion of $K^{\alpha\beta}(\mathbf{q})$ about the ordering wave vector $\mathbf{k}_0$ (Aharony 1973). It is then found that h is usually irrelevant, and the behaviour corresponds to an isotropic short-ranged Heisenberg interaction. The second fixed point, $g = \infty$, represents dipolar behaviour, and because of (7.247) the cross-over exponent from $g = 0$ is given by

$$\phi = \nu(2 - \eta) = \gamma \tag{7.248}$$

following the discussion at the end of §7.3. The detailed calculations of the recursion relations undertaken by Aharony and Fisher are complicated, but their conclusions are surprisingly simple. The critical exponents are very close to those of a short-ranged isotropic interaction, as shown in Table 7.2 (taken from Aharony 1976). It seems that the long-ranged aspect of the dipolar forces which might be expected to yield mean field exponents is almost exactly counterbalanced by the directional dependence of these forces.

7.5.7 Corrections to Scaling

In the discussions of the series expansion method in Chapter 5, attention was drawn to the importance of including correction terms to the dominant singularity in the assessment of the critical behaviour of a thermodynamic quantity. All of the known exact solutions gave rise to a Darboux form of singularity, equation (5.97), with analytic correction terms, and this form was therefore generally adopted. Similar considerations applied to the correction terms in

Table 7.2

	Isotropic short range		Isotropic dipolar	
Exponent	ϵ expansion	$\epsilon = 1$	ϵ expansion	$\epsilon = 1$
η	$\frac{1}{48}\epsilon^2$	0.0208	$\frac{20}{867}\epsilon^2$	0.0231
2ν	$1 + \frac{1}{4}\epsilon + \frac{1}{8}\epsilon^2$	1.375	$1 + \frac{9}{34}\epsilon + \frac{7013}{58\,956}\epsilon^2$	1.384
γ	$1 + \frac{1}{4}\epsilon + \frac{11}{96}\epsilon^2$	1.365	$1 + \frac{9}{34}\epsilon + \frac{2111}{19\,652}\epsilon^2$	1.372
α_s	$-\frac{1}{8}\epsilon^2$	−0.125	$-\frac{1}{34}\epsilon - \frac{6223}{58\,956}\epsilon^2$	−0.135
δ	$3 + \epsilon + \frac{11}{24}\epsilon^2$	4.458	$3 + \epsilon + \frac{787}{1734}\epsilon^2$	4.454
β	$\frac{1}{2} - \frac{1}{8}\epsilon + \frac{1}{192}\epsilon^2$	0.380	$\frac{1}{2} - \frac{2}{17}\epsilon - \frac{55}{58\,956}\epsilon^2$	0.381

the scaling hypothesis, as discussed in §6.2.5. Using an RG treatment Wegner (1972) demonstrated that this assumption is incorrect, and that in general secondary *non-integral exponents* must be taken into account in the correction terms.

The origin of the correction terms lies in the irrelevant eigenvalues of the matrix **L** of §7.3. Let us introduce a scaling field g, as at the end of §7.3, which modifies equation (7.75) to

$$\psi(\lambda^{y_T}\tau,\lambda^{y_H}H,\lambda^{y_g}g) = \lambda\psi(\tau,H,g) \tag{7.249}$$

and equation (7.77) to

$$\psi(\tau,H) = -\tau^{2-\alpha}V(H\tau^{-\Delta},g\tau^{-\phi}) \tag{7.250}$$

with $\phi = \nu dy_g$ (the minus is used for consistency with (6.54)). When ϕ is positive, equation (7.250) represents normal cross-over behaviour in the parameter g. But when ϕ is negative, the parameter g ceases to be relevant. However, we have seen that the function V is usually analytic, and we can therefore expand around $\tau = 0$ in powers of the small parameter $g\tau^{\theta}$ ($\theta = -\phi$). We then find

$$\psi(\tau,H) = -\tau^{2-\alpha}[V(H\tau^{-\Delta}) + g\tau^{\theta}V_1(H\tau^{-\Delta}) + \ldots]. \tag{7.251}$$

This simple form leads to the conclusion that a single exponent, θ, is needed to provide a correction term for the critical behaviour of *all* thermodynamic quantities (see Table 5.4). For example, for the initial magnetic susceptibility

$$\chi_0 \sim a_0\tau^{-\gamma} + a_1\tau^{-\gamma+\theta}; \tag{7.252}$$

for the spontaneous magnetization

$$m_0 \sim b_0\tau^{\beta} + b_1\tau^{\beta+\theta}. \tag{7.253}$$

The magnetization at $\tau = 0$ after a little more manipulation gives

$$H \sim c_0 m^{\delta} + c_1 m^{\delta(1+\theta/\Delta)}. \tag{7.254}$$

Moreover, from (7.69) by a similar argument we can deduce that the same correction exponent applies to the behaviour of the correlation function near T_c,

$$\xi(\tau) \sim d_0\tau^{-\nu} + d_1\tau^{-\nu+\theta}. \tag{7.255}$$

Equations like (7.252)–(7.255) were of major importance for a re-analysis of series expansion data.

In addition to introducing the new parameter θ the RG was able to provide good estimates of its value. We can illustrate the method of calculation from the treatment of §7.5.3. From equation (7.198) we see that the eigenvalue Λ_2 is irrelevant, and has the value $1 - \epsilon \ln b \sim b^{-\epsilon}$. Hence from the discussion in §7.3, dy_g is equal to $-\epsilon$, and to first order θ is equal to $\nu\epsilon = \epsilon/2$. Putting $\epsilon = 1$ for three dimensions we find that $\theta = 1/2$ which is close to the value which has been derived finally. Aharony (1976) gives the value to second order:

$$dy_g = -\epsilon + \frac{3(3n + 14)}{(n + 8)^2}\epsilon^2. \tag{7.256}$$

7.6 Perturbation Expansions

In the previous section we have shown how Wilson used the RG in momentum space to develop an ϵ expansion for critical exponents. However, once he had established the basic principles governing the behaviour in the critical region, he discovered that the terms of the expansion could be calculated more simply using the standard perturbation expansions of quantum field theory. We will give a brief outline of this approach, and will derive the first term of the expansion for the exponents γ and η. The reader interested in a more detailed treatment is referred to the book by Ma (1976a), and to the review articles by Ma (1976b) on the $1/n$ expansion, and by Wallace (1976) on the ϵ expansion.

Most of the literature dealing with this topic assumes that the reader is familiar with the ideas and terminology of quantum field theory. Since the present author lacked such a familiarity, and had to struggle to understand what was going on, he will endeavour to tackle the subject from first principles, explaining the diagrammatics in some detail. Fortunately, most of the material needed has already been introduced in the previous section.

We will use the n-vector model Hamiltonian of §7.5.3, starting with the Gaussian square terms (7.184), which we will write as

$$\exp(-\mathscr{H}^{\mathrm{G}}) = \exp\left(-\frac{1}{2}\int_{\mathbf{q}} (q^2 + p_0) \sum_{\alpha=1}^{n} \sigma_{\mathbf{q}}^{\alpha}\sigma_{-\mathbf{q}}^{\alpha}\right). \tag{7.257}$$

The reason for writing p_0 will become apparent later. The fourth-order term represented by (7.186) will be treated as a perturbation

$$\mathscr{H}_1 = u \int_{\mathbf{q}_1}\int_{\mathbf{q}_2}\int_{\mathbf{q}_3} \sigma_{\mathbf{q}_1}^{\alpha}\sigma_{\mathbf{q}_2}^{\alpha}\sigma_{\mathbf{q}_3}^{\beta}\sigma_{-\mathbf{q}_1-\mathbf{q}_2-\mathbf{q}_3}^{\beta}; \tag{7.258}$$

$\exp(-\mathscr{H}_1)$ will be expanded, as before in (7.155), with the diagrammatic representation (7.156). However, we will now write down an appropriate integral to be calculated for each term of the perturbation expansion. Our use of a lattice model has automatically placed a finite upper limit of integration to the $\mathbf{q}$ as explained in §7.5.1 (in quantum field theory an artificial cut-off Λ must be introduced to prevent ultraviolet divergencies as $q \to \infty$). The divergencies with which we are concerned occur at the $q = 0$ limit and are usually termed 'infrared divergencies'.

If we now try to evaluate the p.f. we will be led, as before, to Gaussian integrals of the form (7.157), the term in u involving a product of four $\sigma_{\mathbf{q}_i}$

variables, the term in u^2 involving a product of eight $\sigma_{\mathbf{q}_i}$ variables, etc. The key to the calculation of all the integrals which arise in the perturbation expansion lies in the result, (7.160) and (7.161), that all possible pair connections among the t_i must be specified, and each pair replaced by the corresponding coefficient of the inverse matrix $\mathbf{A}^{-1}$.

The corresponding diagram is constructed by joining the vertices in pairs, no vertex being free, and each vertex having one joining line attached to it. In the previous section we denoted such connections by wavy lines (Figures 7.10, 7.11). We shall now adopt the conventional practice of replacing wavy lines by standard lines, which, since they have no free vertices, are usually called *internal legs*; but we shall identify the vertices from which the diagram is constructed so that its origin remains clear. As before, each diagram will have an associated combinatorial factor arising from the different possible labellings of the original vertices, and an n-dependent factor arising from the spin components $\alpha, \beta \ldots = 1, 2, \ldots, n$.

To evaluate the p.f. Z all diagrams (connected and disjoint) must be taken into account, but for $\ln Z$ only connected diagrams enter (as in §5.2). There is one first-order diagram which is shown in Figure 7.13(a), and there are two second-order diagrams which are shown in Figures 7.14(a) and 7.14(c).

The rules for associating an integral with a given diagram follow a similar pattern to those listed after equation (7.170), as follows:

1 Label each vertex with a momentum q.
2 The momenta at the ends of a joining line must be equal and opposite.
3 The sum of the four momenta at a four-point vertex must be zero.
4 To each internal leg assign a propagator $(q_i^2 + p_0)^{-1}$, denoted by $G_0(q_i)$.

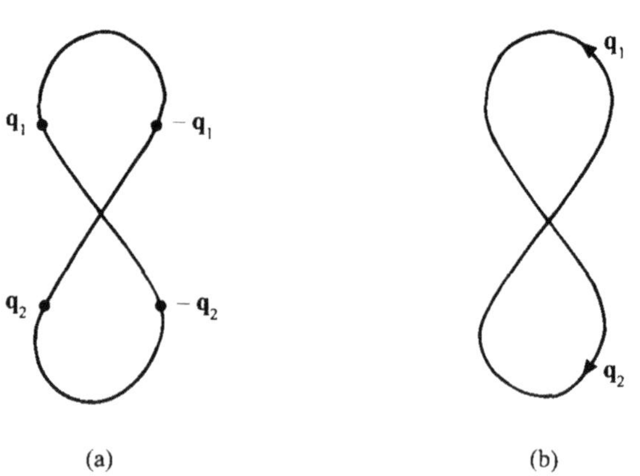

Figure 7.13 First-order diagram in perturbation theory for the p.f.: (a) vertex labelling, (b) leg labelling.

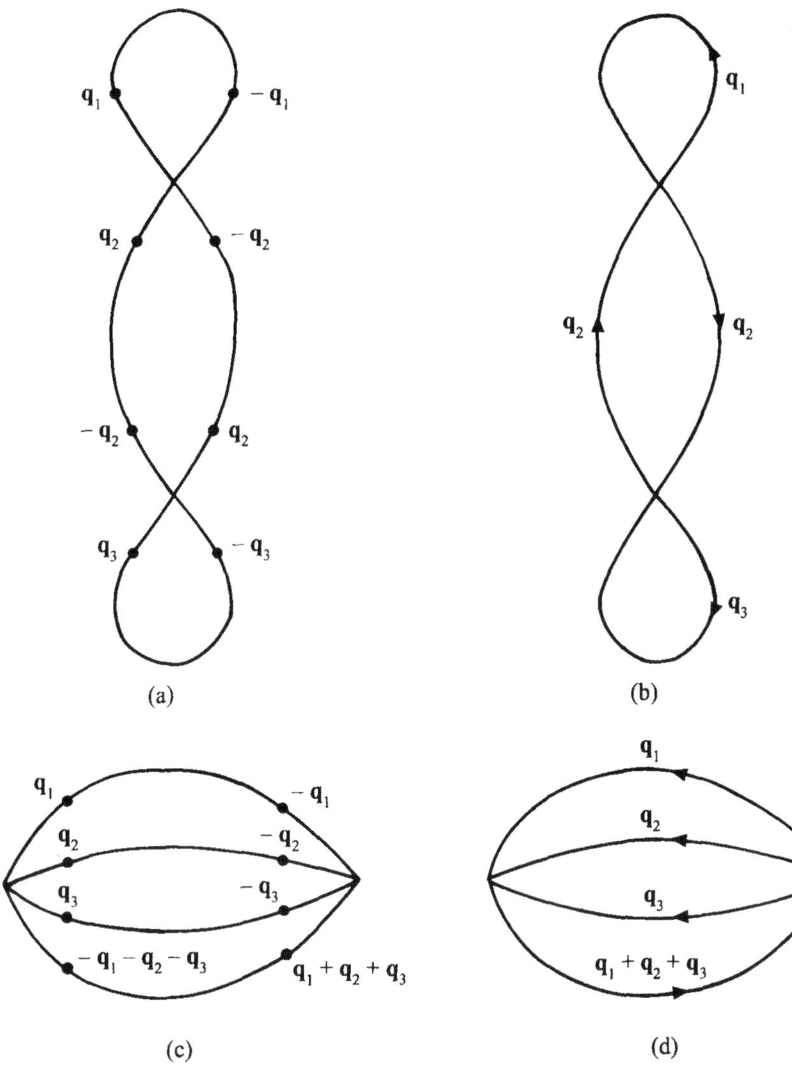

Figure 7.14 Second-order diagrams in perturbation theory for the p.f.: (a) and (c) vertex labelling, (b) and (d) leg labelling.

One must then calculate the combinatorial and spin factor relevant to the diagram.

The vertices in Figures 7.13(a) and 7.14(a, c) have been labelled with appropriate momenta in accordance with this scheme. For some purposes it is convenient to consider momentum as directed along the legs from negative to positive, and relabel the diagrams as in Figures 7.13(b) and 7.14(b, d); momentum is conserved at the vertices. As an example we write down the integral associated with the diagram of Figure 7.14(a):

$$\int_{\mathbf{q}1}\int_{\mathbf{q}2}\int_{\mathbf{q}3}(q_1^2+p_0)^{-1}(q_2^2+p_0)^{-2}(q_3^2+p_0)^{-1}$$

$$=\left(\int_{\mathbf{q}}(q^2+p_0)^{-1}\right)^2\left(\int_{\mathbf{q}}(q^2+p_0)^{-2}\right). \tag{7.259}$$

Our main concern, however, will not be with the calculation of the free energy, but of the pair correlation, for which important simplifications arise in the diagrammatics. In real space the pair correlation is defined by

$$\langle s_{\mathbf{l}}^1 s_{\mathbf{l}'}^1\rangle = \frac{\int s_{\mathbf{l}}^1 s_{\mathbf{l}'}^1 \exp(-\mathscr{H})}{\exp(-\mathscr{H})} \tag{7.260}$$

where the 1-component of the spin has been selected arbitrarily for a specified direction. Since we use a cyclic boundary condition for the lattice spins, this is translationally invariant, i.e.

$$\langle s_{\mathbf{l}}^1 s_{\mathbf{l}'}^1\rangle = \langle s_0^1 s_{\mathbf{l}'-\mathbf{l}}^1\rangle \tag{7.261}$$

which depends only on $(\mathbf{l}'-\mathbf{l})$. When we Fourier-transform to momentum space to the corresponding function $\langle\sigma_{\mathbf{q}}^1\sigma_{\mathbf{q}'}^1\rangle$, we find that the condition of translational invariance leads to the momentum condition

$$\mathbf{q}+\mathbf{q}'=0.$$

The symbol $\mathbf{q}$ has been used conventionally for a varying momentum, which then becomes a variable of integration. In considering the pair correlation function we will also be concerned with a fixed momentum which we will denote by $\mathbf{k}$. Our aim will be to evaluate $\langle\sigma_{\mathbf{k}}^1\sigma_{-\mathbf{k}}^1\rangle$, the Fourier transform of $\langle s_0^1 s_{\mathbf{l}'-\mathbf{l}}^1\rangle$, using the basic formula

$$\Gamma(k)=\langle\sigma_{\mathbf{k}}^1\sigma_{-\mathbf{k}}^1\rangle=\frac{\int_{\mathbf{q}}\sigma_{\mathbf{k}}^1\sigma_{-\mathbf{k}}^1\exp[-(\mathscr{H}^{\mathrm{G}}+\mathscr{H}_{\mathrm{I}})]}{\int_{\mathbf{q}}\exp[-(\mathscr{H}^{\mathrm{G}}+\mathscr{H}_{\mathrm{I}})]}. \tag{7.262}$$

Proceeding as before with the expansion of $\exp(-\mathscr{H}_{\mathrm{I}})$, we must now distinguish two special vertices with momenta $\mathbf{k}$ and $-\mathbf{k}$ which we denote by ⊙. The first term of our expansion will involve σ_k and σ_{-k}, the second term of order u will involve σ_k, σ_{-k} and four q_i variables, the second term of order u^2 will involve σ_k, σ_{-k} and eight q_i variables, etc. As indicated in §5.2, here again only connected diagrams enter, and the rules for associating an integral with a diagram are as outlined above. However, there is no integration associated with the variable $\mathbf{k}$, and a line connecting two vertices with momenta $\mathbf{k}$, $-\mathbf{k}$ gives rise to a factor

$$G_0(k)=(p_0+k^2)^{-1}. \tag{7.263}$$

Such a connection will be denoted by a dashed line.

Diagrams corresponding to the first three terms are listed in Figure 7.15, vertices being labelled by appropriate momenta. The values of the terms are

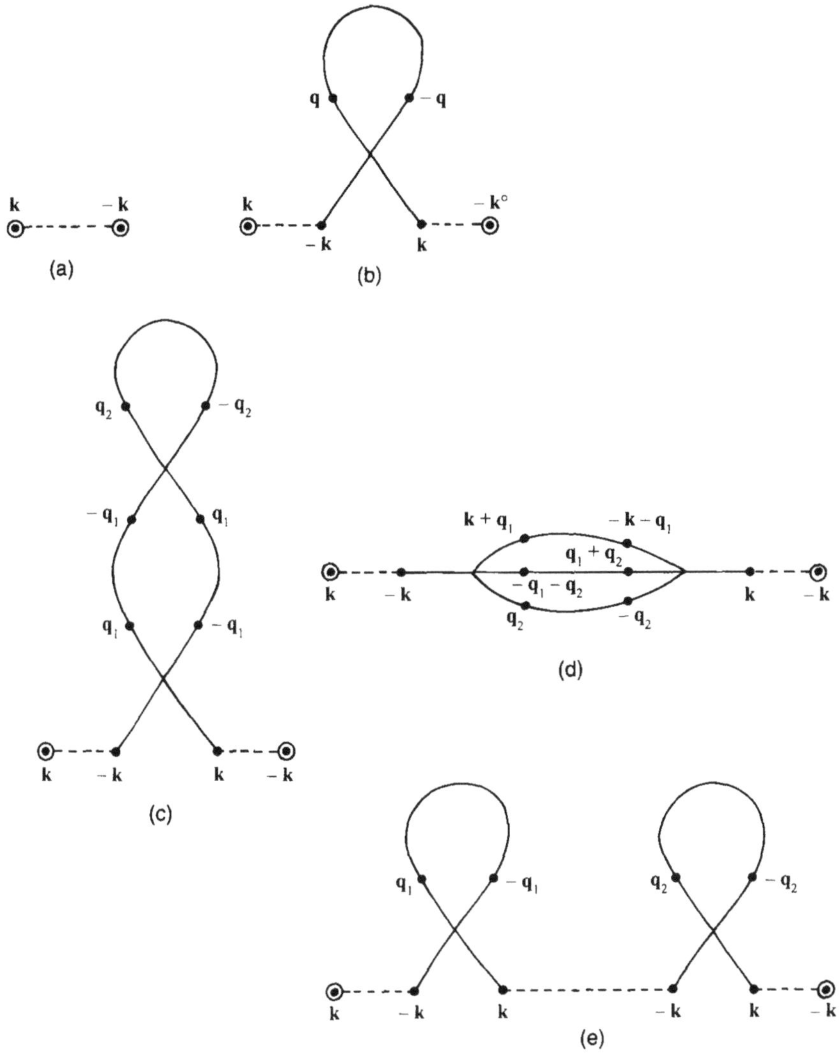

Figure 7.15 Early terms in the expansion of the correlation function represented diagrammatically.

readily calculated. Figure 7.15(a) gives $G_0(k)$. Figures 7.15(b)–(d) all contain a factor $[G_0(k)]^2$; the individual integrals are

$$\int_{\mathbf{q}} G_0(\mathbf{q}), \quad \int_{\mathbf{q}_1} [G_0(\mathbf{q}_1)]^2 \int_{\mathbf{q}_2} G_0(\mathbf{q}_2) \tag{7.264}$$

$$\int_{\mathbf{q}_1}\int_{\mathbf{q}_2} G_0(\mathbf{q}_2)G_0(\mathbf{q}_1 + \mathbf{q}_2)G_0(\mathbf{k} - \mathbf{q}_2). \tag{7.265}$$

Only in the last case of (7.265) does the value of the integral depend on **k**. The final diagram in Figure 7.15(e) illustrates a specific property of correlation diagrams which enables great progress to be made in summing them. Its value is

$$[G_0(k)]^3\left(\int_{\mathbf{q}} G_0(\mathbf{q})\right)^2 \tag{7.266}$$

and it is formed by stringing together two of the diagrams of Figure 7.15(b) in the manner shown. But the stringing process can be continued to give terms

$$[G_0(k)]^4\left(\int_{\mathbf{q}} G_0(q)\right)^3, \quad [G_0(k)]^5\left(\int_{\mathbf{q}} G_0(\mathbf{q})\right)^4, \ldots$$

and other diagrams like Figures 7.15(c) and 7.15(d) can be strung together similarly. In fact if the diagrams are defined without the dashed lines as in Figure 7.16 (called self-energy diagrams in quantum field theory) all the terms in the stringing process are as depicted in Figure 7.17.

We now observe that all diagrams formed from t sets of 4-vertices are taken from the tth term of the Taylor expansion of $\exp(-\mathscr{H}_{\mathrm{I}})$, and therefore carry a coefficient $1/t!$. But the permutation of these t sets gives rise to a combinatorial factor $t!$ which exactly cancels the coefficient (we still have to deal with the 4! permutations of the 4-vertices among themselves).

Denoting the sum of diagrams in Figure 7.16 by $\Sigma(k,p_0)$, we can write, from Figure 7.17,

$$\begin{aligned}\Gamma(k) &= G_0(k) + [G_0(k)]^2\Sigma(k,p_0) + [G_0(k)]^3[\Sigma(k,p_0)]^2 + \ldots \\ &= [G_0(k)]^{-1}[1 - G_0(k)\Sigma(k,p_0)]^{-1} = [p_0 + k^2 - \Sigma(k,p_0)]^{-1}.\end{aligned} \tag{7.267}$$

Following Dyson (1949), this is usually termed 'Dyson's equation' in the literature. If we now refer back to the discussion of correlations in Chapter 5 and

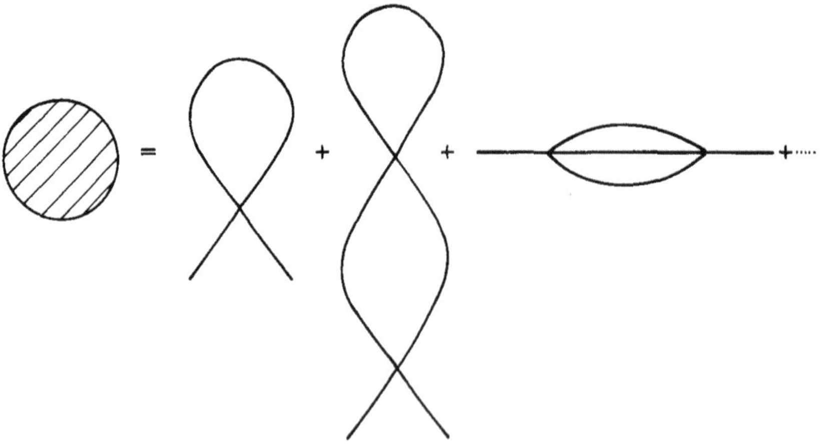

Figure 7.16 Self-energy diagrams in quantum field theory.

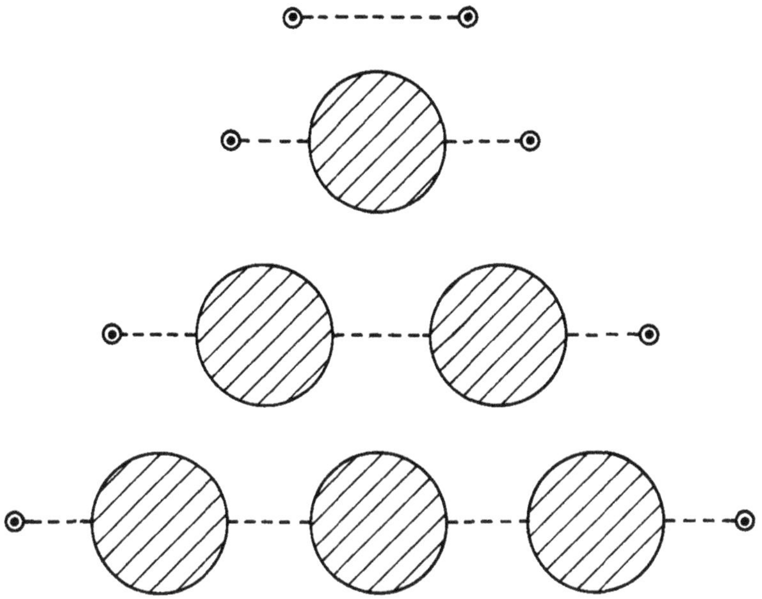

Figure 7.17 Diagrammatic representation of all terms in the expansion of the correlation function.

in particular to equation (5.176) (which is readily generalized to the Heisenberg model), we can identify $\Gamma(k)$ with $\hat{\chi}(k,w)$, and putting $k = 0$, we find that

$$\chi_0^{-1} = p_0 - \Sigma(0,p_0) = p. \tag{7.268}$$

The critical temperature is given by $\chi_0^{-1} = 0$, from which we can determine $p_{\sigma c}$. The variable p_0 is linear in temperature, and is not directly related to critical behaviour. By contrast, the variable p has direct physical significance, and it is more convenient to explore the critical region with a variable which takes the value zero at the critical point. We therefore change $G_0(k)$ to

$$G(k) = p + k^2 \tag{7.269}$$

and try to derive an expansion for $\Gamma(k)$ using $G(k)$ as propagator and p as the *renormalized* variable. In quantum field theory this procedure is termed *mass renormalization.*

7.6.1 *Expansions in the Renormalized Variable*

Let us first choose an arbitrary p, and take as our unperturbed Hamiltonian, instead of (7.257),

$$\mathscr{H}_G = \frac{1}{2}\int_{\mathbf{q}} (q^2 + p) \sum_{\alpha=1}^{n} \sigma_{\mathbf{q}}^{\alpha}\sigma_{-\mathbf{q}}^{\alpha}. \tag{7.270}$$

In order to obtain the same total Hamiltonian as before we now require two terms in the interaction Hamiltonian,

$$\mathscr{H}_{\mathrm{I}} = u \int_{\mathbf{q}_1} \int_{\mathbf{q}_2} \int_{\mathbf{q}_3} \sum_{\alpha,\beta=1}^{n} \sigma^{\alpha}_{\mathbf{q}_1} \sigma^{\alpha}_{\mathbf{q}_2} \sigma^{\beta}_{\mathbf{q}_3} \sigma^{\beta}_{-\mathbf{q}_1-\mathbf{q}_2-\mathbf{q}_3} + \frac{1}{2}(p_0 - p) \int_{\mathbf{q}} \sum_{\alpha=1}^{n} \sigma^{\alpha}_{\mathbf{q}} \sigma^{\alpha}_{-\mathbf{q}}. \tag{7.271}$$

The perturbation theory is developed diagrammatically; the first term in u gives rise to the same diagrams as before with p replacing p_0, and $G(k)$ replacing $G_0(k)$ in the expressions derived previously. The second term gives rise to 2-vertex sets which must be joined with the 4-vector sites to form connected graphs. The new diagrams can be conveniently represented as in Figure 7.18, each cross representing the insertion of a $(p_0 - p)$ 2-vertex set.

We now proceed exactly as before for the evaluation of the correlation function. The set of diagrams in Figure 7.16 is extended to include diagrams with crosses, and the sum of these diagrams is denoted by $\bar{\Sigma}(k,p)$. The single cross representing $(p_0 - p)$ is kept separately, and we now write for the pair correlation, instead of (7.267),

$$\Gamma(k,p) = \{p + k^2 - [\bar{\Sigma}(k,p) + p_0 - p]\}^{-1}. \tag{7.272}$$

So far the treatment applies for arbitrary p. But if we now particularize to the value of p in (7.268), we find putting $k = 0$ the simple relation

$$p_0 - p = -\bar{\Sigma}(0,p). \tag{7.273}$$

The substitution of $(p_0 - p)$ from (7.273) into (7.272) gives

$$\Gamma(k,p) = \{p + k^2 - [\bar{\Sigma}(k,p) - \bar{\Sigma}(0,p)]\}^{-1}. \tag{7.274}$$

For each diagram we must take off the value at $k = 0$, and diagrams which are independent of k make no contribution.

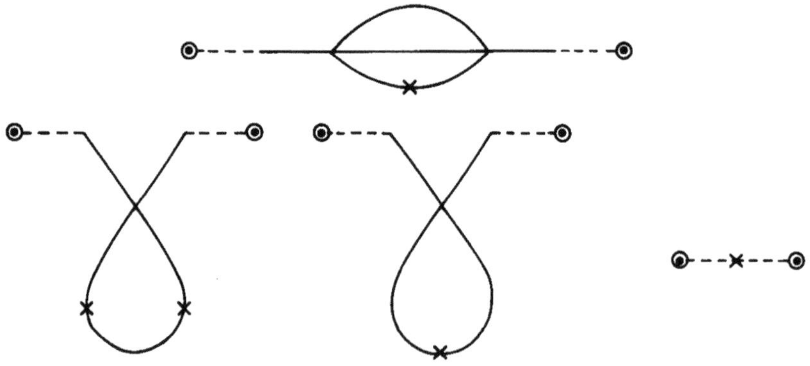

Figure 7.18 Additional diagrams which enter the expansion in terms of the renormalized variable.

Relation (7.273) determines $(p_0 - p)$. Although the RHS contains diagrams with crosses, i.e. terms in $(p_0 - p)$, these can be eliminated perturbatively since $(p_0 - p)$ is of order u. The process can be carried out diagrammatically, replacing each cross by the diagrams in $-\Sigma(0,p)$, and, as indicated in Figure 7.19, cancellations occur with higher diagrams in $\Sigma(0,p)$. By means of this systematic procedure a diagrammatic expansion can be derived for $(p_0 - p)$ not involving any crosses, and this can then be substituted into the diagrams with crosses in $\Sigma(k, p)$. We will not discuss the development in more detail since our concern will be confined to low-order diagrams.

We can now identify the relations which will be used for evaluating the exponents η and γ. The exponent η was defined in §5.5 (equation (5.163)), and at T_c the Fourier transform of the pair correlation function is of the form $k^{-2+\eta}$. Thus, from (7.274),

$$[\Gamma(k,0)]^{-1} = k^2 - \bar{\Sigma}(k,0) + \bar{\Sigma}(0,0) \sim k^{2-\eta}. \tag{7.275}$$

This relation will be used to determine η.

From (7.273) we deduce that

$$p_0 = p - \bar{\Sigma}(0,p). \tag{7.276}$$

At T_c, $p = 0$; hence we find that

$$p_0 - p_{\sigma c} = p - \bar{\Sigma}(0,p) + \bar{\Sigma}(0,0). \tag{7.277}$$

We know that p_0 is linear in T, and hence that $p_0 - p_{0c}$ is of order τ. Also, near T_c, $p \sim \tau^{\gamma}$, and therefore

$$p - \bar{\Sigma}(0,p) + \bar{\Sigma}(0,0) \sim p^{1/\gamma}. \tag{7.278}$$

This provides a basis for the calculation of γ.

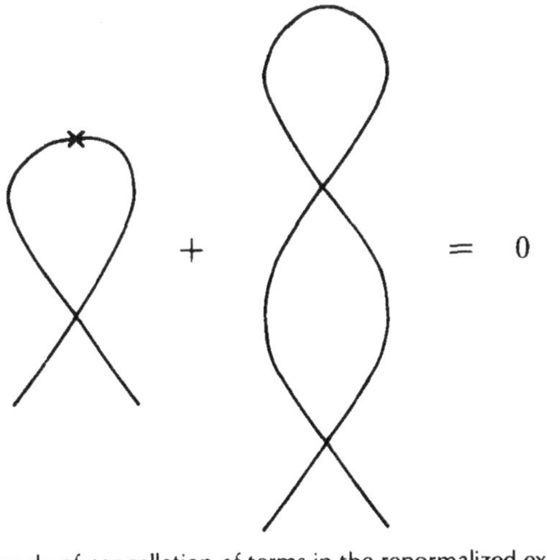

Figure 7.19 An example of cancellation of terms in the renormalized expansion.

7.6.2 The ϵ Expansion

We begin with a few remarks on the nature of the perturbation expansion with which we are concerned, and we start with an analysis of $\Sigma(0,p) - \Sigma(0,0)$ from (7.278). The first term in the expansion corresponds to the first diagram in Figure 7.16, and gives rise to the integral

$$\int_{\mathbf{q}} \frac{\mathrm{d}\mathbf{q}}{q^2 + p} + \int_{\mathbf{q}} \frac{\mathrm{d}\mathbf{q}}{q^2} = p \int_{\mathbf{q}} \frac{\mathrm{d}\mathbf{q}}{q^2(q^2 + p)}. \tag{7.279}$$

Using (7.138) this becomes

$$pC_d \int_0^1 \frac{q^{d-1}\,\mathrm{d}q}{q^2(q^2 + p)}. \tag{7.280}$$

We note that if $d > 4$ this converges at the lower limit for $p = 0$, but diverges if the upper limit is taken to infinity; hence the cut-off is significant. If $d < 4$, however, there is convergence at the upper limit, and the cut-off is of no significance but there is divergence at $p = 0$. In the marginal case $d = 4$ there are logarithmic divergencies at both limits.

As we move to higher-order terms, at each stage a factor u is introduced and a 4-vertex, each vertex being labelled with a separate momentum variable $\mathbf{q}_i$. Two internal lines are introduced to join the new vertices and form a connected diagram, and there are three momentum-conserving conditions. In totality therefore there is one new $\mathbf{q}_i$ variable of integration and two $(\mathbf{q}_i^2 + p)^{-1}$ factors. Typically we are involved (see e.g. equation (7.259)) in a new factor

$$\int_0^1 \frac{q^{d-1}\,\mathrm{d}q}{(q^2 + p)^2} \tag{7.281}$$

and for $d < 4$, the upper limit can be moved to infinity. It is useful in evaluating integrals like (7.281) to make use of the formula

$$\int_0^\infty y^{a-1}(1 + y)^{-a-b}\,\mathrm{d}y = \frac{\Gamma(a)\Gamma(b)}{\Gamma(a + b)} \tag{7.282}$$

derived from the standard formula for the beta function (Sneddon 1961) using the transformation $x = y/(1 + y)$. We then find, for the integral in (7.281), the value

$$p^{-\epsilon}\Gamma(d)\Gamma(2 - d). \tag{7.283}$$

The parameter of the perturbation expansion is therefore $up^{-\epsilon/2}$; if we consider the expansion for $\Sigma(k,0) - \Sigma(0,0)$ the parameter will be $uk^{-\epsilon}$.

Because of the divergence of this parameter when $p \to 0$ $(\epsilon > 0)$ we are dealing with an infrared divergence; a detailed investigation of the divergent terms enables the critical exponents to be expanded as a series in ϵ. However, a background knowledge of the RG as described in the earlier sections of this chapter is essential for a proper understanding of the working of the ϵ expansion.

It should first be pointed out that this expansion does not base itself on the solution for $d = 4$. The RG showed that there are two relevant fixed points: the Gaussian fixed point which dominates when $d > 4$, and the Heisenberg–Ising fixed point when $d < 4$. When $d = 4$ the fixed points coalesce, and this gives rise to a complex situation with logarithmic terms (see e.g. the exact solution for the spherical model, equation (5.153)). But when d differs even very slightly from 4 (e.g. 3.99 in the Wilson–Fisher (1972) terminology) the Gaussian fixed point can be ignored, and attention confined to the Heisenberg–Ising fixed point to which standard scaling behaviour applies. For this situation the RG has established the existence of expansions for critical exponents in powers of ϵ (e.g. (7.204)).

Our aim is to use the perturbation expansions (7.275) and (7.278) to derive these ϵ expansions.

In (7.278) we can take a factor p out of both sides, and the LHS will then yield an expansion of the form

$$1 + up^{-\epsilon/2}\Psi_1(\epsilon) + u^2p^{-\epsilon}\Psi_2(\epsilon) + \ldots + u^s p^{-s\epsilon/2}\Psi_s(\epsilon) + \ldots . \tag{7.284}$$

If we could sum this series to all orders we would obtain a function of (p,u,ϵ) of the form

$$c_1 p^{\gamma^{-1}-1} + c_2 p^{\gamma^{-1}-1+\lambda} \tag{7.285}$$

where the exponents depend on ϵ but are independent of u because of universality; the second term arises from the corrections to scaling discussed in §7.5.7, λ being related to the correction exponent θ. But since the exponents are independent of u, Wilson selected a particular value of u, $u_0(\epsilon)$, for which the irrelevant eigenvalue of the matrix $\mathbf{L}$ of §7.3 makes no contribution, so that the correction term in (7.285) is absent. He showed that the RG allows such a choice to be made (for a detailed exposition see Ma (1976a, (pp. 309–14)). To calculate $u_0(\epsilon)$ the terms in (7.284) are expanded in powers of $\ln p$

$$p^{-s\epsilon/2} = 1 - (s\epsilon/2)\ln p + \frac{1}{2!}(s\epsilon/2)^2(\ln p)^2 \ldots \tag{7.286}$$

and the resulting series is compared with

$$c_1 p^{\gamma^{-1}-1} = c_1\left[1 + (\gamma^{-1} - 1)\ln p + \frac{1}{2!}(\gamma^{-1} - 1)^2(\ln p)^2 + \ldots\right]. \tag{7.287}$$

From this $u_0(\epsilon)$ and $(1/\gamma - 1)$ can be calculated in principle in powers of ϵ. But we will shortly discuss a more direct method for calculating $u_0(\epsilon)$ devised by Wilson.

Let us examine in more detail the evaluation of the integrals arising in expansion (7.278). The integral associated with the term in u and the diagram of Figure 7.15(b) is given by equation (7.280), and can be evaluated exactly. Writing

$$I_1 = \int_0^1 \frac{q^{d-1}\,\mathrm{d}q}{q^2(q^2+p)} \tag{7.288}$$

and transforming to the variable $y = q^2p^{-1}$, we obtain

$$I_1 = \frac{1}{2}p^{-\epsilon/2}\int_0^{p^{-1}} \frac{y^{-\epsilon/2}\,\mathrm{d}y}{1+y} = \frac{1}{2}p^{-\epsilon/2}\left(\int_0^{\infty} \frac{y^{-\epsilon/2}\,\mathrm{d}y}{1+y}t - \int_{p^{-1}}^{\infty} \frac{y^{-\epsilon/2}\,\mathrm{d}y}{1+y}\right). \tag{7.289}$$

The first integral on the RHS is evaluated by using formula (7.282) to give

$$\frac{\Gamma(1-\frac{1}{2}\epsilon)\Gamma(\frac{1}{2}\epsilon)}{\Gamma(1)} = \frac{\pi}{\sin \pi\epsilon/2}. \tag{7.290}$$

For the second integral, since p^{-1} is large, we expand the integral in the form

$$y^{-\epsilon/2-1}(1 - y^{-1} + y^{-2}\ldots) \tag{7.291}$$

and integrate. Collecting the terms together we finally obtain

$$I_1 \simeq \frac{\pi/2}{\sin \pi\epsilon/2}p^{-\epsilon/2} - \frac{1}{\epsilon}. \tag{7.292}$$

If we had taken the upper cut-off of the integral as Λ instead of 1 the second term of (7.292) would be changed to $\Lambda^{-\epsilon}/\epsilon$. Expanding in powers of ϵ we find that

$$I_1 \simeq \ln \Lambda + \frac{\pi^2}{24}\epsilon - \frac{1}{2}\ln p - \frac{\epsilon}{2}\ln \Lambda + \frac{\epsilon}{8}(\ln p)^2 \ldots. \tag{7.293}$$

To complete our calculation we must insert the combinatorial factor for the diagram of Figure 7.15(b) which arises from the permutation of vertices and the different possible spin configurations. Proceeding as in §7.5.3 we find the factor $4(n+2)$. We then obtain for (7.284)

$$1 + 4(n+2)u_0(\epsilon)C_d\left(\ln \Lambda + \frac{\pi^2}{24}\epsilon - \frac{1}{2}\ln p - \frac{\epsilon}{2}\ln \Lambda + \frac{\epsilon}{8}(\ln p)^2\ldots\right) \tag{7.294}$$

and comparison with (7.287) gives

$$\begin{aligned} \gamma^{-1} - 1 &= -2(n+2)C_d\, u_0(\epsilon) \\ \gamma^{-1} &= 1 - 2(n+2)C_d\, u_0(\epsilon). \end{aligned} \tag{7.295}$$

We have anticipated the result that $u_0(\epsilon)$ is of order ϵ which will be derived shortly; to this order the value of γ is independent of the cut-off Λ, a property which is maintained for higher orders in ϵ.

To calculate η we must evaluate k-dependent integrals, the first of which corresponds to the diagram of Figure 7.15(d). This is of the second order in

$u_0(\epsilon)$, and the appropriate integral, I_2, is the renormalized form of (7.265),

$$\int_{\mathbf{q}_1}\int_{\mathbf{q}_2} G(\mathbf{q}_2)G(\mathbf{q}_1+\mathbf{q}_2)[G(\mathbf{k}+\mathbf{q}_1)-G(\mathbf{q}_1)] \tag{7.296}$$

where, since $p = 0$, $G(\mathbf{q}) = 1/\mathbf{q}^2$. We shall evaluate (7.296) in two stages.

We first define the d-dimensional integral

$$J(\mathbf{q}_1) = \int_{\mathbf{q}_2} G(\mathbf{q}_2)G(\mathbf{q}_1+\mathbf{q}_2) = \int_{\mathbf{q}_2} \frac{1}{q_2^2(\mathbf{q}_1+\mathbf{q}_2)^2} \tag{7.297}$$

and evaluate it with the help of the device used in equation (4.54), i.e. we shall replace y^{-1} by

$$\int_0^\infty \exp(-ty)\,\mathrm{d}t.$$

The application of this device to Feynman integrals is described by Brezin, Le Guillou and Zinn-Justin (1976, pp. 139–41) in their review of field theory and critical phenomena. The integral (7.297) is thus transformed into

$$\begin{aligned} J(\mathbf{q}_1) &= \int_0^\infty \mathrm{d}t_1 \int_0^\infty \mathrm{d}t_2 \int \mathrm{d}^d\mathbf{q}_2 \exp(-t_1\mathbf{q}_2^2)\exp[-t_2(\mathbf{q}_1+\mathbf{q}_2)^2] \\ &= \int_0^\infty \mathrm{d}t_1 \int_0^\infty \mathrm{d}t_2 \int \mathrm{d}^d\mathbf{q}_2 \exp\{-[(t_1+t_2)\mathbf{q}_2^2 + 2t_2\mathbf{q}_1\mathbf{q}_{2+t_2}\mathbf{q}_1^2]\} \\ &= \int_0^\infty \mathrm{d}t_1 \int_0^\infty \mathrm{d}t_2 \frac{\pi^{d/2}}{(t_1+t_2)^{d/2}} \exp\left[-\left(\frac{t_1t_2}{t_1+t_2}\right)\mathbf{q}_1^2\right]. \end{aligned} \tag{7.298}$$

The transformation

$$t_1 + t_2 = v, \quad t_1 = wv \tag{7.299}$$

decouples the variables of integration, and we find that

$$\begin{aligned} J(\mathbf{q}_1) &= \pi^{d/2}\int_0^\infty \mathrm{d}v \int_0^1 \mathrm{d}w v^{1-d/2}\exp[w(1-w)vq_1^2] \\ &= \pi^{d/2}q_1^{-\epsilon}\Gamma(\epsilon/2)B(1-\epsilon/2,\,1-\epsilon/2) \\ &= \frac{\pi C_d}{2\sin\pi\epsilon/2}(1-\epsilon/2)B(1-\epsilon/2,\,1-\epsilon/2)q_1^{-\epsilon}. \end{aligned} \tag{7.300}$$

We now note that the integral (7.296) can be written as

$$\int_{\mathbf{q}_1} J(\mathbf{q}_1)\left(\frac{1}{(\mathbf{k}+\mathbf{q}_1)^2} - \frac{1}{q_1^2}\right) \tag{7.301}$$

and we must therefore focus attention on

$$\int_{\mathbf{q}_1}\left(\frac{q_1^{-\epsilon}}{(\mathbf{q}_1+\mathbf{k})^2} - \frac{q_1^{-\epsilon}}{\mathbf{q}_1^2}\right). \tag{7.302}$$

This integral is more difficult to evaluate exactly, and we shall follow the method of Wallace (1973) in identifying the logarithmic terms which are of relevance to the calculation of η. We observe that the RHS of (7.275) can be expanded in the form

$$k^2\left(1 - \eta \ln k + \frac{\eta^2}{2!} (\ln k)^2 \ldots\right) \tag{7.303}$$

so that η is the coefficient of the $k^2 \ln k$ term in our integral.

Introducing d-dimensional spherical polar coordinates as before (equation (4.51)) (7.302) transforms to

$$C_{d-1} \int_0^{\pi} \sin^{d-2} \theta \, d\theta \int_0^{\Lambda} q_1^{3-2\epsilon}\left(\frac{1}{(\mathbf{q}_1 + \mathbf{k})^2} - \frac{1}{\mathbf{q}_1^2}\right) dq_1. \tag{7.304}$$

In the second integral $q_1^{-2\epsilon}$ will be expanded as

$$1 - 2\epsilon \ln q_1 + 2\epsilon^2(\ln q_1)^2 \ldots \tag{7.305}$$

and (7.304) will be written as

$$J_1 + \epsilon J_2 \tag{7.306}$$

where

$$\begin{aligned} J_1 &= C_{d-1} \int_0^{\pi} \sin^{d-2} \theta \, d\theta \int_0^{\Lambda} q_1^3\left(\frac{1}{(\mathbf{q}_1 + \mathbf{k})^2} - \frac{1}{\mathbf{q}_1^2}\right) dq_1 \\ J_2 &= C_{d-1} \int_0^{\pi} \sin^{d-2} \theta \, d\theta \int_0^{\Lambda} q_1^3[2 \ln q_1 - 2\epsilon(\ln q_1)^2 \ldots]\left(\frac{1}{(\mathbf{q}_1 + \mathbf{k})^2}\right) dq_1. \end{aligned} \tag{7.307}$$

We first focus attention on J_1. For the q_1 integral we use a scale transformation

$$\mathbf{q}_1 = k\mathbf{q}_1' \tag{7.308}$$

and obtain

$$k^2 \int_0^{\Lambda/k} q_1'^3\left(\frac{1}{(\mathbf{q}_1' + \mathbf{n})^2} - \frac{1}{q_1'^2}\right) dq_1' \tag{7.309}$$

$\mathbf{n}$ being a unit vector. In the range $0 \leq q_1' \leq 1$ the integral is finite and no logarithmic term arises. In the range $1 \leq q_1' \leq \Lambda/k$, $(\mathbf{q}_1' + \mathbf{k})^{-2}$ is expanded as

$$(1 + 2q_1' \cos \theta + q_1'^2)^{-1} = q_1'^{-2}\left[1 - \frac{2}{q_1'} \cos \theta - \frac{1}{q_1'^2} + \left(\frac{2}{q_1'} \cos \theta + \frac{1}{q_1'}\right)^2 \ldots\right]. \tag{7.310}$$

It is only the q'^{-4} terms coupled with the q'^3 factor in (7.309) which give rise to a $\ln k$ term at the upper limit of integration. We thus obtain a $(4 \cos^2 \theta - 1)$ factor to be combined with the $\sin^{d-2} \theta$ factor in the angular integral in J_1. We then find that the coefficient of $-k^2(\ln k)$ in J_1 is

$$C_{d-1}\left(\frac{4}{d}-1\right)\pi^{1/2}\frac{\Gamma((d-1)/2)}{\Gamma(d/2)}=\left(\frac{4}{d}-1\right)C_d. \qquad (7.311)$$

Combining (7.311) with (7.300) we obtain finally for the coefficient of $-k^2 \ln k$

$$\frac{C_d^2\,\pi\epsilon/2}{\sin \pi\epsilon/2}\,\frac{\Gamma(2-\epsilon/2)\Gamma(1-\epsilon/2)}{(4-\epsilon)\Gamma(2-\epsilon)}. \qquad (7.312)$$

Turning to J_2 in (7.306) and (7.307), the same method of expansion could be used to evaluate logarithmic terms, but this is unnecessary in our approximation because of the ϵ factor multiplying J_2.

The combinatorial weight factor for the diagram is found to be $32(n+2)$. Hence we deduce from (7.303) and (7.312) that

$$\eta \simeq 8(n+2)C_d^2[u_0(\epsilon)]^2. \qquad (7.313)$$

As we have pointed out previously, it would be possible in principle to calculate $u_0(\epsilon)$ by matching higher terms in the expansions (7.287) or (7.303). However, since the exponents are dependent on two parameters only, if a third exponent can be calculated independently, $u_0(\epsilon)$ will automatically be determined. Wilson (1972) showed how to calculate the exponent Δ which is determined by the four-point spin correlation function. We have already noted this connection for the Ising model in §6.2.1, and the argument is readily generalized to the n-vector Heisenberg model, with Hamiltonian

$$\mathscr{H} = -J\sum_{ij}\sum_{\alpha=1}^{n} s_{\mathbf{l}_i}^{\alpha}s_{\mathbf{l}_j}^{\alpha} - mH\sum_i s_{\mathbf{l}_i}^{1}. \qquad (7.314)$$

Taking the fourth derivative w.r.t. H we find that

$$\frac{\partial^4}{\partial H^4}(\ln Z) = \sum \langle s_{\mathbf{l}_1}s_{\mathbf{l}_2}s_{\mathbf{l}_3}s_{\mathbf{l}_4}\rangle_{\text{cumulant}} \qquad (7.315)$$

the sum being taken over all lattice sites $\mathbf{l}_1, \mathbf{l}_2, \mathbf{l}_3, \mathbf{l}_4$.

Here the four-point function is defined in an analogous manner to the pair correlation function (7.260). Taking Fourier transforms as before we are led to

$$\begin{aligned} G_c^{(4)}(\mathbf{k}_1,\mathbf{k}_2,\mathbf{k}_3,\mathbf{k}_4) &= \langle \sigma_{\mathbf{k}_1}\sigma_{\mathbf{k}_2}\sigma_{\mathbf{k}_3}\sigma_{\mathbf{k}_4}\rangle_{\text{cumulant}} \\ &= \frac{\int_{\mathbf{q}} \sigma_{\mathbf{k}_1}\sigma_{\mathbf{k}_2}\sigma_{\mathbf{k}_3}\sigma_{\mathbf{k}_4}\exp[-(\mathscr{H}^{\mathrm{G}}+\mathscr{H}_{\mathrm{I}})]}{\int_{\mathbf{q}}\exp[-(\mathscr{H}^{\mathrm{G}}+\mathscr{H}_{\mathrm{I}})]} \end{aligned} \qquad (7.316)$$

where the translational invariance now leads to the condition

$$\mathbf{k}_1+\mathbf{k}_2+\mathbf{k}_3+\mathbf{k}_4=0. \qquad (7.317)$$

From equation (7.316) $G^{(4)}$ can be developed as an expansion in a series of connected diagrams with four special vertices corresponding to moments $\mathbf{k}_1$, $\mathbf{k}_2$, $\mathbf{k}_3$, $\mathbf{k}_4$ which satisfies condition (7.317). There is no connected diagram in

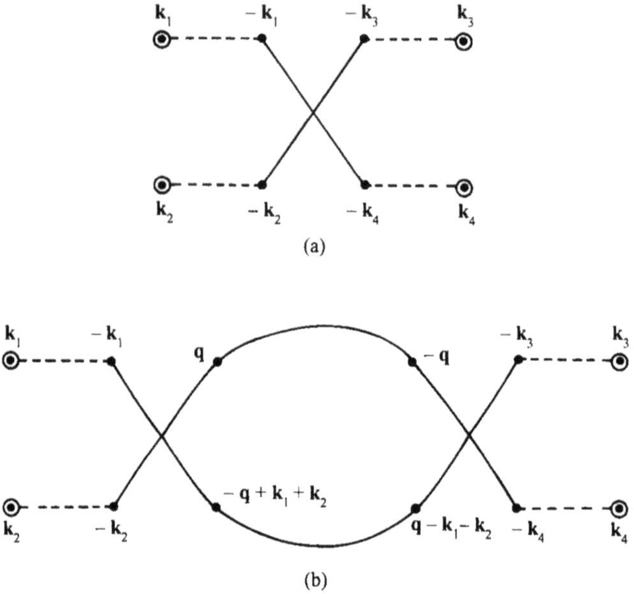

Figure 7.20 Zero-order and first-order diagrams in the expansion of the four-point correlation function.

zero order, and the diagrams corresponding to $u_0(\epsilon)$ and $u_0(\epsilon)^2$ are shown in Figures 7.20(a) and 7.20(b). The rules for associating integrals with diagrams are as before, and therefore these two terms are represented by

$$u_0\, G(\mathbf{k}_1)G(\mathbf{k}_2)G(\mathbf{k}_3)G(\mathbf{k}_4) \tag{7.318}$$

and

$$u_0^2\, G(\mathbf{k}_1)G(\mathbf{k}_2)G(\mathbf{k}_3)G(\mathbf{k}_4) \int_{\mathbf{q}} G(\mathbf{q})G(\mathbf{k}_1 + \mathbf{k}_2 - \mathbf{q}). \tag{7.319}$$

The first diagram has a combinatorial factor 4! associated with it, and the second diagram $4! \times 4(n + 8)$. By the definition of the Fourier transform, the sum over all space lattice points $\mathbf{l}_1, \mathbf{l}_2, \mathbf{l}_3, \mathbf{l}_4$ is obtained from $G^{(4)}$ by allowing all the $\mathbf{k}_i$ to tend to zero. It is convenient to divide out the $G(\mathbf{k}_1)G(\mathbf{k}_2)G(\mathbf{k}_3)G(\mathbf{k}_4)$ factor, and to concentrate on

$$u_{\mathbf{R}}(p) = \operatorname*{Lt}_{\mathbf{k}_i \to 0} \frac{G^{(4)}(\mathbf{k}_1, \mathbf{k}_2, \mathbf{k}_3, \mathbf{k}_4)}{G(\mathbf{k}_1)G(\mathbf{k}_2)G(\mathbf{k}_3)G(\mathbf{k}_4)}. \tag{7.320}$$

This has critical behaviour (see equations (6.17) and (6.18))

$$\tau^{-\gamma - 2\Delta}/\tau^{-4\gamma} = \tau^{3\gamma - 2\Delta} \sim p^{3 - 2\Delta/\gamma} \tag{7.321}$$

and from the exponent relations given in Table 6.1, it is easy to show that the exponent of p in (7.321) is equal to

$$(\epsilon - 2\eta)/(2 - \eta). \tag{7.322}$$

To the first order in ϵ therefore

$$u_R(p) \sim p^{\epsilon/2} \simeq 1 + (\epsilon/2) \ln p. \tag{7.323}$$

The integral in (7.319) reduces to

$$K_1 = \int_q G(\mathbf{q})^2 = C_d \int_0^1 \frac{q^{d-1}\, dq}{(p + q^2)^2}. \tag{7.324}$$

Proceeding as with (7.288), and using (7.282), we derive the value

$$K_1 = \frac{1}{2} C_d p^{-\epsilon/2} \Gamma\left(2 - \frac{1}{2}\epsilon\right)\Gamma\left(\frac{1}{2}\epsilon\right) \sim \frac{C_d}{\epsilon}\left(1 - \frac{\epsilon}{2} \ln p\right). \tag{7.325}$$

Hence, from the diagrammatic expansion,

$$u_R(p) \sim 24u_0[1 + 2(n + 8)u_0 C_d \ln p]. \tag{7.326}$$

Comparing (7.323) and (7.326)

$$u_0(\epsilon)C_d \sim \epsilon/4(n + 8) + O(\epsilon^2). \tag{7.327}$$

Substituting in (7.295) and (7.313) we obtain

$$\gamma \sim 1 + \frac{(n + 2)\epsilon}{2(n + 8)} + O(\epsilon^2) \tag{7.328}$$

and

$$\eta \sim \frac{(n + 2)\epsilon^2}{2(n + 8)^2} + O(\epsilon^3). \tag{7.329}$$

To the first order we have been able to evaluate the integrals with relative ease. For higher-order terms the determination of the combinatorial factors and the values of the integrals present more difficulties. A number of results of this kind derived by Nickel (1974) were issued as a preprint; some of them are quoted in the review article by Wallace (1976), where higher-order terms of the ϵ expansion are derived.

7.6.3 The $1/n$ Expansion

We mentioned in §1.6 that the spherical model can be considered as the limit of the classical Heisenberg model as the spin dimension tends to infinity ($1/n = 0$). This opens up the possibility of an expansion in $1/n$ using the spherical model as the starting base. In practice these expansions are not comparable in significance with the ϵ expansions, but they can be derived with little extra effort, and they illustrate important points of principle. The perturbation parameter u_0 is now taken to be of order $1/n$. Our treatment is again based on the unpublished lecture notes of Wallace (1973).

Let us first consider the evaluation of $u_R(p)$. For large n the first two terms corresponding to the diagrams in Figures 7.20(a) and 7.20(b) are of the same order. But we must now consider the set of 'bubble' diagrams of the type illustrated in Figure 7.21, since each additional bubble contributes a factor of order n (as in (7.320) the external momenta are set to 0). If we sum all these diagrams we obtain

$$u_R(p) = u_0 - 4nK_1u_0^2 + (4nK_1)^2u_0^3 \ldots$$
$$= \frac{u_0}{1 + 4nK_1u_0} \sim \frac{1}{4nK_1} \simeq \frac{\epsilon}{4nC_d} p^{\epsilon/2} \tag{7.330}$$

since from (7.325) K_1 is large compared with 1 for small p. We note that the result we have obtained is independent of u_0 (universality) and is of the correct form given by (7.323).

Turning to the exponent γ, we require, as before, the first diagram of Figure 7.16, which from (7.280) and (7.292), with the configurational factor $4(n + 2)$, gives

$$\frac{2u_0(n + 2)\pi C_d}{\sin \pi\epsilon/2} p^{1-\epsilon/2}. \tag{7.331}$$

One might have expected to take into account the bubble diagrams of the type shown in Figure 7.22, but they cancel with diagrams with crosses as illustrated in Figure 7.19. When we equate (7.331) with $C_1p^{1/\gamma}$, we find that

$$\gamma \simeq \frac{2}{2 - \epsilon}. \tag{7.332}$$

This is the zero-order term. A calculation of the term of order $1/n$ is given in the review article by Ma (1976b).

For the exponent η the first k-dependent integral corresponding to Figure 7.15(d) must be evaluated, but we must again take into account the bubble

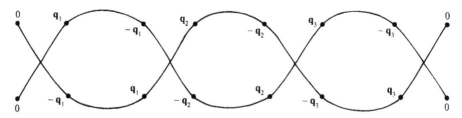

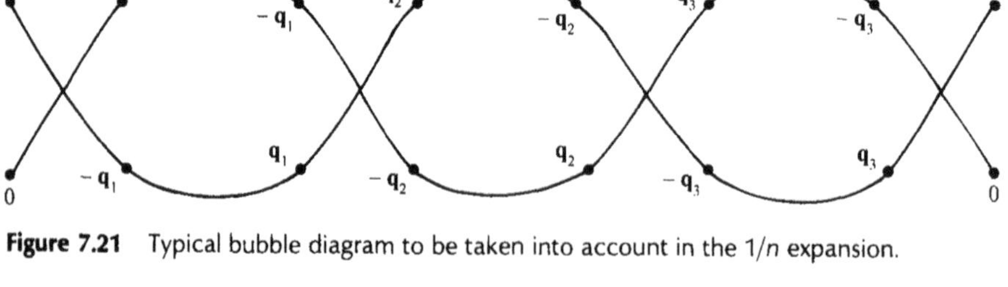

Figure 7.21 Typical bubble diagram to be taken into account in the $1/n$ expansion.

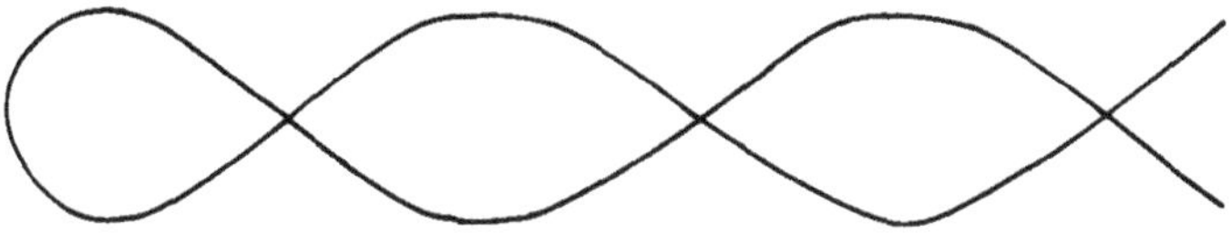

Figure 7.22 Bubble diagram in the $1/n$ expansion which can be ignored because of cancellation.

diagrams shown in Figure 7.23. Each bubble involves an integral $J(\mathbf{q}_1)$ of (7.297) whose value is given by (7.300). The sum of the bubble diagram integrals is

$$32u_0^2(n+2)J(\mathbf{q}_1)\{1 - 4u_0(n+2)J(\mathbf{q}_1) + [4u_0(n+2)J(\mathbf{q}_1)]^2 \ldots\}$$
$$= \frac{32u_0^2(n+2)J(\mathbf{q}_1)}{1 + 4u_0(n+2)J(\mathbf{q}_1)} = 8u_0\left(1 - \frac{1}{1 + 4u_0(n+2)J(\mathbf{q}_1)}\right). \quad (7.333)$$

To obtain the final integral for the diagram, (7.333) must be multiplied by $[(\mathbf{q}_1 + \mathbf{k})^{-2} - \mathbf{q}_1^{-2}]$ and integrated w.r.t. $\mathbf{q}_1$. As before the result must be compared with $k^{2-\eta}$, so that η is the coefficient of $-k^2 \ln k$ in the expansion of the final integral w.r.t. k.

The first term in the RHS of (7.333) gives rise to the integral

$$\int_{\mathbf{q}_1}\left(\frac{1}{(\mathbf{q}_1 + \mathbf{k})^2} - \frac{1}{\mathbf{q}_1^2}\right). \quad (7.334)$$

The transformation by which we evaluated $J(\mathbf{q}_1)$ in (7.298) can be applied to this integral. It is not necessary to evaluate the integral exactly; one must convince oneself that only powers of k, and no $\ln k$ terms, can arise.

In the second term on the RHS of (7.333) the denominator can be ignored in comparison with $nu_0 J(\mathbf{q}_1)$ since we are concerned with small q_1. The key

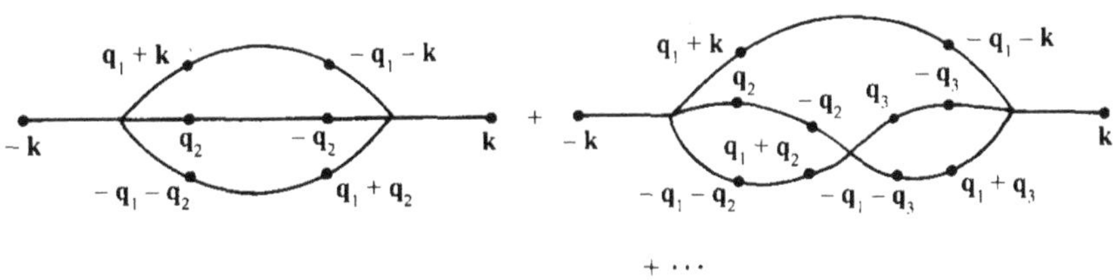

Figure 7.23 Bubble diagrams for the calculation of η in the $1/n$ expansion.

integral is then

$$\int_{\mathbf{q}_q}\left(\frac{q^{\epsilon}}{(\mathbf{q}_1+\mathbf{k})^2}-\frac{q_1^{\epsilon}}{\mathbf{q}_1^2}\right) \tag{7.335}$$

which is converted, by d-dimensional polar coordinates, into

$$C_{d-1}\int_0^{\pi}\sin^{d-2}\theta\,d\theta\int_0^{\Lambda}q_1^3\left(\frac{1}{(\mathbf{q}_1+\mathbf{k})^2}-\frac{1}{\mathbf{q}_1^2}\right)d\mathbf{q}. \tag{7.336}$$

This is precisely the integral J_1 in (7.307) which has already been dealt with. However, since $J(\mathbf{q}_1)$ now appears in the denominator, the combination of beta and gamma functions is different. Dividing (7.311) by the coefficient of $q_1^{-\epsilon}$ in (7.300) as required by (7.333), we find that

$$\eta=\frac{\epsilon^2}{n}\frac{\sin\pi\epsilon/2}{\pi\epsilon/2}\frac{\Gamma(2-\epsilon)}{\Gamma(3-\epsilon/2)\Gamma(1-\epsilon/2)}. \tag{7.337}$$

This term is of order $1/n$, and we note, as before, that u_0 has disappeared from the exponent.

7.7 Field Theory Approach

The ϵ expansion was derived on the assumption that it would apply to small values of ϵ – hence the graphic title of the pioneering paper by Wilson and Fisher (1972) 'Critical exponents in 3.99 dimensions'. Fortunately, and perhaps surprisingly, the first two terms of the expansion with $\epsilon = 1$ gave reasonable agreement with the values of critical exponents derived by three-dimensional series expansions. Despite great technical difficulties, additional terms were derived, and the series seemed to be asymptotic rather than convergent. Padé approximants were used, but gave no clear indication of improvement. The small but persistent differences between the results of the series and ϵ expansions were attributed to the paucity of terms in the latter expansion.

The next major step forward arose from the direct application of field theoretical methods to the problem. The method of renormalization had triumphantly solved the problem of infinities in quantum electrodynamics, and an enormous theoretical effort had gone into refining the approach and applying it to other problems in quantum field theory. An appropriate change of language effected the transformation to statistical mechanics, and made available a large reservoir of techniques and information which could be applied to critical phenomena.

To do full justice to the field theory approach would require a great deal of space, and an expertise in the area which the present author does not possess. Nevertheless, since it played such a vital role in clearing up the discrepancies, it is desirable to indicate the major features of the approach, and to record the

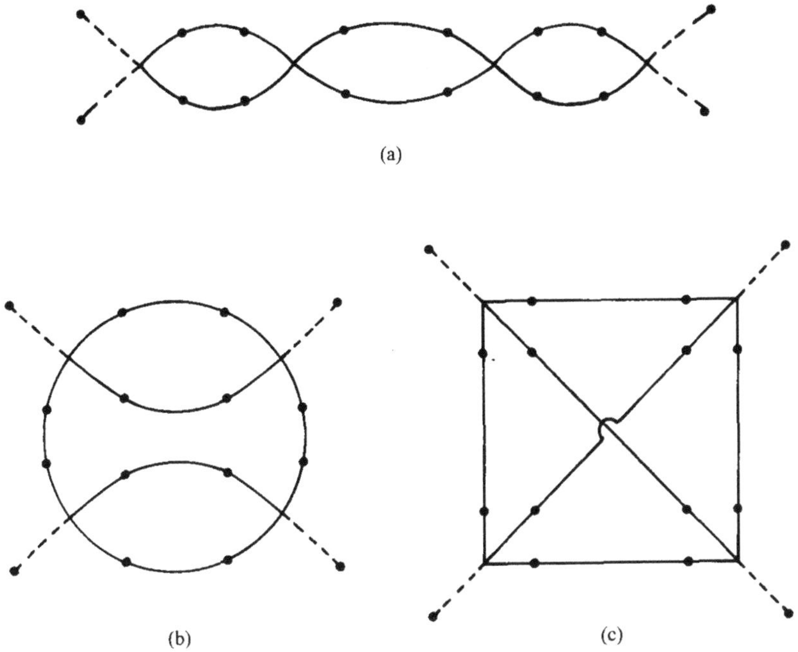

Figure 7.24 Diagrams arising in the fourth-order term of the four-point correlation function (after Larkin and Khmelnitskii 1969).

results derived. Our treatment will be based largely on the review article by Brezin, Le Guillou and Zinn-Justin (1976).

7.7.1 Exact Critical Exponents in Four Dimensions

In §7.2.4 we used the Ginzburg criterion to show that critical exponents attain their classical values when $d = 4$. However, in §7.6.1 we indicated that the coalescence of two fixed points gives rise to logarithmic factors. In 1969 Larkin and Khmelnitskii showed that a field theoretical treatment enables the exponents of these logarithmic factors to be calculated exactly. The above authors were concerned with uniaxial ferroelectrics in three dimensions, but the mathematical problem is identical with the n-vector model in four dimensions.

We have already considered the diagrammatic expansion of the four-point correlation function in the previous section, and the first two terms are given by (7.318) and (7.319) corresponding to the diagrams in Figures 7.20(a) and 7.20(b). When we put $d = 4$ it is easy to see that the integral in (7.324) is $\sim \ln p$.

The diagrams in Figure 7.24 are typical of those that enter at the fourth term. It is not too difficult to show that (a) and (b) are $\sim (\ln p)^3$ whereas (c) is only $\sim(\ln p)$. In fact at any order the maximum power of the logarithm comes from diagrams which can be reduced to the first term by successive contractions of pairs of lines (Figure 7.25). For this group of diagrams Larkin and Khmelnitskii were able to deduce an integral equation which they solved exactly. They adopted a similar procedure for a second independent correlation function involving the energy. The approximation which takes only these diagrams into account and ignores all others is called the *parquet* approximation; the diagrams ignored give rise to a correction $\sim(\ln p)^{-1}$.

From these solutions they deduced the following pattern of critical behavior:

$$\begin{aligned}
&\text{Initial susceptibility} \quad \chi_0 \sim \tau^{-1}(\ln \tau)^{(n+2)/(n+8)} \\
&\text{Spontaneous magnetization} \quad m_0 \sim (\tau')^{1/2}(\ln \tau')^{3/(n+8)} \\
&\text{Specific heat} \quad c_H \sim (\ln \tau)^{(4-n)/(n+8)} \quad (n < 4) \\
&\qquad\qquad\qquad c_H \sim \ln(\ln \tau) \qquad\quad (n = 4).
\end{aligned} \tag{7.338}$$

It is interesting that this was the first occasion on which the numbers $(n + 8)$, $(n + 2)$, so characteristic of Wilson's expansions, made their appearance. The critical equation of state in four dimensions will be considered in the next section.

7.7.2 Expansions in Powers of the Coupling Constant

7.7.2.1 Notation

In field theory the lattice structure is abandoned and the discrete variable **l** is replaced by a continuum variable **x** in d-dimensional space. We shall deal with an Ising spin system ($n = 1$) with a spin function at each point noted by $s(\mathbf{x})$. To conform with standard field theoretical notation we shall in the following sections replace p by m^2 (m denoting mass) and u by $g/4!$. Instead of (7.257)

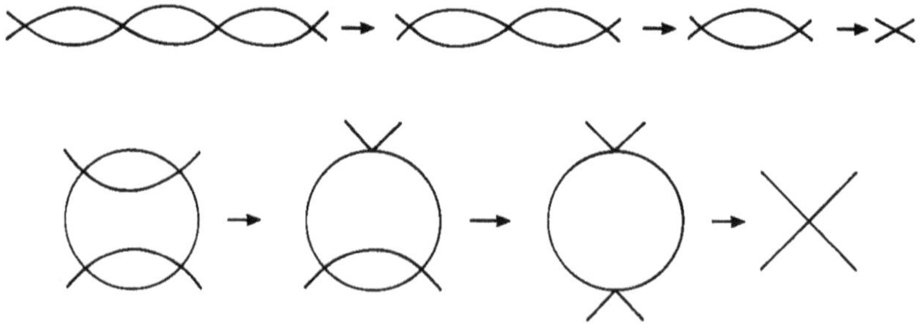

Figure 7.25 Parquet diagrams which can be collapsed by contracting points of lines.

and (7.258) the Hamiltonian in real space is written:

$$\mathscr{H}(\mathbf{x}) = \mathscr{H}^{G} + \mathscr{H}_{I} - H(\mathbf{x})s(\mathbf{x}) = [(\nabla s)^2 + m_0^2 s^2] + \frac{g_0}{4!} s^4 - Hs. \tag{7.339}$$

The summation over lattice points is replaced by $\int \mathscr{H}(\mathbf{x})\, d\mathbf{x}$. The theory can readily be extended to the n-vector model with Hamiltonian

$$\mathscr{H}(\mathbf{x}) = \frac{1}{2} \sum_{\alpha=1}^{n} [\nabla s^\alpha \nabla s^\alpha + m_0^2 (s^\alpha)^2] + \frac{g_0}{4!} \left(\sum_{\alpha=1}^{n} (s^\alpha)^2 \right)^2 . \tag{7.339'}$$

We retain the symbol $\mathbf{q}$ for a momentum which occurs as a variable of integration, and $\mathbf{k}$ for an external momentum.

The N-point cumulant correlation function is defined as in §5.2.2 by

$$G_c^{(N)}(\mathbf{x}_1, \mathbf{x}_2, \ldots, \mathbf{x}_N) = \langle s(\mathbf{x}_1) s(\mathbf{x}_2) \ldots s(\mathbf{x}_N) \rangle_c$$

$$= \frac{\delta}{\delta H_1} \frac{\delta}{\delta H_2} \cdots \frac{\delta}{\delta H_N} [\ln(\text{p.f.})] \tag{7.340}$$

and it is represented diagrammatically as a connected diagram expansion. $G_c^{(2)}$ has previously been denoted by Γ (see e.g. (7.272)). If the independent magnetic field variable $H(\mathbf{x})$ is changed by a Legendre transformation to the magnetization $M(\mathbf{x})$, then only multiply connected graphs enter, as in the theory of condensation of gases (§5.6.1). In momentum space the diagrammatic expansion then contains only graphs which cannot be separated into two parts by cutting a single line. The functions which replace the $G_c^{(N)}$ after this transformation are called vertex functions, and are denoted by $\Gamma^{(N)}(\mathbf{x}_1, \mathbf{x}_2, \ldots, \mathbf{x}_N)$. They are simply related to the $G_c^{(N)}$. In momentum space the first relation is

$$\Gamma^{(2)}(\mathbf{k}, -\mathbf{k}) = [G_c^{(2)}(\mathbf{k}, -\mathbf{k})]^{-1} \tag{7.341}$$

and for a system with magnetic symmetry with which we are concerned

$$\Gamma^{(4)}(\mathbf{k}_1, \mathbf{k}_2, \mathbf{k}_3, \mathbf{k}_4) = \frac{G_c^{(4)}(\mathbf{k}_1, \mathbf{k}_2, \mathbf{k}_3, \mathbf{k}_4)}{G_c^{(2)}(\mathbf{k}_1) G_c^{(2)}(\mathbf{k}_2) G_c^{(2)}(\mathbf{k}_3) G_c^{(2)}(\mathbf{k}_4)}. \tag{7.342}$$

This has been used in our previous work in (7.320).

7.7.2.2 Renormalization

We developed a perturbation expansion in powers of g_0 in §7.6, but since a high-momentum cut-off was then present, the problem of convergence at the upper limit did not arise. If this cut-off is removed integrals corresponding to diagrams of type Figure 7.15(b) will usually diverge if $d \leq 4$, and the expansion will be meaningless. The aim of renormalization theory is to choose new variables which absorb all of the divergencies, so that the expansion in terms of the new *renormalized* variables is finite and meaningful. Formally we arrange to cut off the integrals at some value Λ so that we can undertake our manipulations with convergent integrals. We then allow Λ to tend to infinity and hope that the new expansion coefficients remain finite. A convenient form of this *regularization* is to replace the propagator $G_0(q) = (q^2 + m_0^2)^{-1}$ by $G_0^{(\Lambda)}(q)$ given by

$$G_0^{(\Lambda)}(q) = \int_0^\infty dx[1 - \rho(\Lambda^2 x)]\exp[-x(q^2 + m_0^2)] \tag{7.343}$$

where $\rho(z)$ vanishes for large z, and is sufficiently close to 1 for small z for the relevant integrals to converge. Note that m_0 and q have the dimension of Λ.

The mechanism of renormalization will be illustrated by the specific example of four dimensions. It is not difficult to show by simple power counting that in this case if E is the number of external legs in a diagram, the power of q in the corresponding integral is $3 - E$. Hence there is quadratic divergence for $E = 2$, logarithmic divergence when $E = 4$, and superficial convergence when $E > 4$ (superficial, since subdiagrams may give divergent contributions). These results are illustrated by equations (7.319) and (7.324) corresponding to the diagram of Figure 7.20(b).

We start with mass renormalization which has already been covered in §7.6.1. The original expansion in m_0^2 is given by (7.267), and contains integrals corresponding to diagrams in Figure 7.16 which are quadratically divergent. The renormalized mass m_1 is given by (7.268), which in our new notation we write in the form

$$m_1^2 = \Gamma^{(2)}(0; m_0, g_0). \tag{7.344}$$

The expansion in the renormalized variable is given by (7.274), and it will be seen that diagrams with $k = 0$ like the first two in Figure 7.16 are eliminated. However, for the third diagram, although the quadratic divergence is eliminated a logarithmic divergence remains; this will be dealt with shortly.

The renormalization of g_0 follows the same pattern but instead of (7.344) the four-point vertex function is used,

$$g_1 = \Gamma^{(4)}(0,0,0,0;\ m_0, g_0). \tag{7.345}$$

The diagrams involved in $\Gamma^{(4)}$ are logarithmically divergent, and this divergence is eliminated by the renormalization (7.345).

Returning now to the residual logarithmic divergence, it should be noted that the integral behaves as $k^2 \ln \Lambda$ for large Λ and this suggests the normalization of $\Gamma^{(2)}(k)$ by an appropriate factor. Define

$$\Gamma_R^{(2)}(k) = \Gamma^{(2)}(k) \Bigg/ \left(\frac{\partial}{\partial k^2}\Gamma^{(2)}(k)\right)_{k^2=0} \tag{7.346}$$

so that by definition

$$\frac{\partial}{\partial k^2}\Gamma_R^{(2)}(k) = 1. \tag{7.347}$$

This change can be realized by a renormalization of the spin

$$s(\mathbf{x}) = Z_0^{1/2} s_{\mathrm{R}}(\mathbf{x})$$
$$Z_0^{-1} = \left(\frac{\partial}{\partial k^2} \Gamma^{(2)}(k)\right)_{k^2=0}. \tag{7.348}$$

In this change all the vertex functions are modified

$$\Gamma_{\mathrm{R}}^{(N)}(\mathbf{k}_1, \mathbf{k}_2, \ldots, \mathbf{k}_N) = Z_0^{N/2} \Gamma^{(N)}(\mathbf{k}_1, \mathbf{k}_2, \ldots, \mathbf{k}_N). \tag{7.349}$$

Finally (7.344) and (7.345) are replaced by

$$m^2 = \Gamma_{\mathrm{R}}^{(2)}(0)$$
$$g = \Gamma_{\mathrm{R}}^{(4)}(0,0,0,0). \tag{7.350}$$

We now proceed in the same manner as in §7.6.1, but the details are much more involved. Instead of (7.339) we write

$$\mathscr{H}^{\mathrm{G}} = \frac{1}{2}[(\nabla s_{\mathrm{R}})^2 + m^2 s_{\mathrm{R}}^2]$$
$$\mathscr{H}_1 = \frac{g}{4!} s_{\mathrm{R}}^4 + \frac{1}{2}(Z_0 m_0^2 - m^2) s_{\mathrm{R}}^2 + \frac{1}{2}(Z_0 - 1)(\nabla s_{\mathrm{R}})^2 + \frac{1}{4!} g(Z_{(4)} - 1) s_{\mathrm{R}}^4 \tag{7.351}$$

where the last three terms in the interaction are power series in g,

$$Z_0 - 1 = \sum_2^\infty A_n g^n$$
$$Z_0 m_0^2 - m^2 = \Lambda^2 \sum_1^\infty B_n g^n \tag{7.352}$$
$$g(Z_{(4)} - 1) \equiv g_0 Z_0^2 - g = \sum_2^\infty C_n g^n.$$

The A_n, B_n, C_n are dimensionless functions of m/Λ, and are determined order by order in powers of g by the three renormalization conditions (7.350) and (7.347). Equations (7.352) enable us to transform from g_0 to g and from m_0 to m.

The central result of renormalization theory is that the $\Gamma_{\mathrm{R}}^{(N)}(\mathbf{k}_1, \mathbf{k}_2, \ldots, \mathbf{k}_N; m,g,\Lambda)$ calculated in this manner have expansions with finite coefficients when Λ goes to infinity. Moreover, this limit is independent of the particular cut-off function $\rho(z)$ in (7.343).

7.7.2.3 *The Callan–Symanzik equation*

In 1970 Callan and Symanzik independently put forward a simple idea in field theory which proved very fruitful in its application to critical phenomena. Since $\Gamma_{\mathrm{R}}(\mathbf{k}_i; m,g,\Lambda)$ is independent of Λ when $\Lambda \to \infty$ its derivative must tend to zero in this limit. Hence we can write

$$\Lambda \frac{\mathrm{d}}{\mathrm{d}\Lambda} [\Gamma_{\mathrm{R}}^{(N)}(\mathbf{k}_i;m,g,\Lambda)]_{g,m} \simeq 0. \tag{7.353}$$

Let us now use (7.352) to relate g to the original parameter g_0 whilst retaining the renormalized mass m as a variable. We then have from (7.349)

$$\Lambda \frac{\mathrm{d}}{\mathrm{d}\Lambda} \left\{ \left[Z_0\left(g_0, \frac{m}{\Lambda}\right) \right]^{N/2} \Gamma^{(N)}(\mathbf{k}_i;\, m,g_0,\Lambda) \right\}_{g,m} = 0. \tag{7.354}$$

This readily yields

$$\left[\Lambda \frac{\partial}{\partial \Lambda} + W\left(g_0, \frac{m}{\Lambda}\right) \frac{\partial}{\partial g_0} - \frac{N}{2}\, \eta\left(g_0, \frac{m}{\Lambda}\right) \right] \Gamma^{(N)}(\mathbf{k}_i;\, g_0,m,\Lambda) = 0. \tag{7.355}$$

Here

$$W\left(g_0, \frac{m}{\Lambda}\right) = \Lambda \left(\frac{\partial g_0}{\partial \Lambda} \right)_{g,m} \tag{7.356}$$

and

$$\eta\left(g_0, \frac{m}{\Lambda}\right) = -\Lambda \frac{\mathrm{d}}{\mathrm{d}\Lambda} \left[\ln Z_0\left(g_0, \frac{m}{\Lambda}\right) \right]_{g,m} \tag{7.357}$$

both of which can be derived from (7.352).

7.7.2.4 *Application to critical phenomena*

So far the discussion has related to the elimination of divergent integrals at the upper limit of high momenta with the help of a large cut-off Λ. The adaptation of this theory to critical phenomena in d dimensions ($d \leq 4$), where attention is focused on low momenta and infrared divergence, is due largely to Brezin, Le Guillou and Zinn-Justin (1973). We have already introduced the Landau–Ginzburg–Wilson Hamiltonian in §7.2.3; let us write it in the form

$$\beta \mathscr{H} = \int \mathscr{H}(\mathbf{y})\, \mathrm{d}\mathbf{y} \tag{7.358}$$

with

$$\mathscr{H}(\mathbf{y}) = \frac{1}{2} [c_0(\nabla s)^2 + a_0 s^2] + \frac{b_0}{4!} s^4 \tag{7.359}$$

where $\mathbf{y}$ is a d-dimensional vector and the parameters a_0, b_0, c_0 are smooth non-singular functions of temperature near T_c. This Hamiltonian was used in Wilson's development described previously and has a momentum cut-off of order $1/a$, where a is the lattice spacing.

Critical phenomena are related to distances large compared with the lattice spacing, and it is convenient to introduce a change of scale

$$\begin{aligned} \mathbf{y} &= \Lambda \mathbf{x} \\ s(\mathbf{y}) &= \zeta s(\mathbf{x}) \end{aligned} \tag{7.360}$$

leading to a transformed Hamiltonian

$$\mathscr{H}(\mathbf{x}) = \frac{1}{2}\,[(\nabla s)^2 + m_0^2\, s^2] + \frac{u_0}{4!}\, s^4. \tag{7.361}$$

Here $\zeta = c_0^{1/2}\Lambda^{1-d/2}$ is chosen to make the coefficient of $(\nabla s)^2$ equal to unity, and we then find that

$$\begin{aligned} m_0^2 &= \frac{a_0}{c_0}\,\Lambda^2 \\ u_0 &= \frac{b_0}{c_0^2}\,\Lambda^{4-d}. \end{aligned} \tag{7.362}$$

As before the generalization to an n-vector Hamiltonian is straightforward:

$$\mathscr{H}(\mathbf{x}) = \frac{1}{2}\sum_{\alpha=1}^{n}\,[\nabla s^\alpha \nabla s^\alpha + m_0^2 (s^\alpha)^2] + \frac{u_0}{4!}\left(\sum_{\alpha=1}^{n}(s^\alpha)^2\right)^2. \tag{7.363}$$

Λ is assumed to be large compared with the lattice spacing but small compared with the correlation length which becomes infinite at T_c. We have therefore transformed to a model with a large cut-off Λ to which the previously described renormalization theory can be applied. m has the dimension of Λ, an inverse length, and will later be identified with κ (of Chapter 5), the inverse of the correlation length. The critical temperature is characterized by $m = 0$, and we are interested in exploring the behaviour of the model for small m.

An important difference from the previous model is the coupling constant u_0 in (7.362), which being of order Λ^ϵ is large when $\epsilon > 0$. It is convenient therefore to define a dimensionless coupling constant

$$g_0 = u_0\,\Lambda^{-\epsilon} \tag{7.364}$$

and the second renormalization condition of (7.350) becomes

$$\Gamma_{\mathbf{R}}^{(4)}(0,0,0,0;\, m,g) = m^\epsilon g. \tag{7.365}$$

We first examine the behaviour at $T = T_c$ ($m = 0$). The Callan–Symanzik equation (7.355) then assumes the simplified form

$$\left(\Lambda\,\frac{\partial}{\partial n} + W(g_0)\,\frac{\partial}{\partial g_0} - \frac{N}{2}\,\eta(g_0)\right)\Gamma^{(N)}(\mathbf{k}_i\,; g_0\,, n) = 0. \tag{7.366}$$

This equation can readily be integrated. The term in $\eta(g_0)$ can be removed to give

$$\Gamma^{(N)}(g_0,\Lambda) = \Phi(g_0,\Lambda)\exp\left(\frac{N}{2}\int^{g_0}\frac{\eta(g')}{W(g')}\right)\mathrm{d}g' \tag{7.367}$$

where $\Phi(g_0,\Lambda)$ satisfies

$$\left(\Lambda\frac{\partial}{\partial\Lambda} + W\frac{\partial}{\partial g_0}\right)\Phi(g_0,\Lambda) = 0. \tag{7.368}$$

This linear partial differential equation can be integrated to show that Φ is an arbitrary function of a single variable

$$\Phi = \Phi\left(\ln\Lambda - \int^{g_0}\frac{\mathrm{d}g'}{W(g')}\right). \tag{7.369}$$

A more transparent way of using this result is to show that $\Gamma^{(N)}(g_0,\lambda\Lambda)$ corresponding to a change of cut-off, i.e. a change of lattice spacing, can be simply related to $\Gamma^{(N)}(g_0(\lambda),\Lambda)$ with an appropriate choice of $g_0(\lambda)$. We would be applying an RG transformation of the type introduced by Wilson, and would expect it to lead to a fixed point.

In forming the ratio of $\Gamma^{(N)}(g_0,\lambda\Lambda)$ to $\Gamma^{(N)}(g_0(\lambda),\Lambda)$ using (7.367) and (7.369) we see that there are two factors. The factor involving $\eta(g)$ is equal to

$$\exp\left(\frac{N}{2}\int_{g_0(\lambda)}^{g_0}\frac{\eta(g')}{W(g')}\right)\mathrm{d}g'. \tag{7.370}$$

If we define $g_0(\lambda)$ by the equation

$$\lambda\frac{\mathrm{d}g_0}{\mathrm{d}\lambda} = -W[g_0(\lambda)] \quad (g_0(1) = g_0) \tag{7.371}$$

and use an appropriate change of variable, the integral in (7.370) becomes

$$\int_1^{\lambda}\frac{\mathrm{d}\sigma}{\sigma}\eta[g_0(\sigma)]. \tag{7.372}$$

But since we now have, from (7.371),

$$\ln\lambda = -\int_{g_0}^{g_0(\lambda)}\frac{\mathrm{d}g'}{W(g')} \tag{7.373}$$

the other factor is unity. Hence we can write

$$\Gamma^{(N)}(g_0,\lambda\Lambda) = \Gamma^{(N)}(g_0(\lambda),\Lambda)\exp\left(\frac{N}{2}\int_1^{\lambda}\frac{\mathrm{d}\sigma}{\sigma}\eta[g_0(\sigma)]\right). \tag{7.374}$$

The RG transformation generated by $\Lambda\rightarrow\lambda\Lambda$ is controlled by equation (7.371), and its evolution depends on the form of $W(g_0)$. From (7.352) we see that $W(0) = 0$, and the initial behaviour determined by the early Feynman integrals shows that

$$W(g_0,\epsilon) = -\epsilon g_0 + \frac{3g_0^2}{16\pi^2} + O(g_0^3,\epsilon g_0^2). \tag{7.375}$$

Assume that the behaviour of $W(g_0)$ is of the form depicted in Figure 7.26 (this can be checked from higher-order integrals), and that there is a value g_0^* satisfying $W(g_0^*) = 0$ with

$$W'(g_0^*) = \omega > 0. \tag{7.376}$$

The fixed point $g_0 = 0$ is unstable and, given a small initial value of g_0, equation (7.371) tells us that $g_0(\lambda)$ will move away until it reaches the stable fixed point g_0^*. Linearizing about g_0^* we find from (7.376) that the asymptotic behaviour of $g_0(\lambda)$ is given by

$$g_0(\lambda) - g_0^* \sim \lambda^{-\omega}. \tag{7.377}$$

Moreover, if $\Gamma^{(N)}(\mathbf{k}_i;g_0^*)$ and $\eta(g_0^*)$ are finite we have from (7.374)

$$\Gamma^{(N)}(\mathbf{k}_i;g_0,\lambda\Lambda) \sim \Gamma^{(N)}(\mathbf{k}_i;g_0^*,\Lambda)\lambda^{N\eta/2} \tag{7.378}$$

as $\lambda \to \infty$, where

$$\eta = \eta(g_0^*). \tag{7.379}$$

Let us now see what change is induced in the N-point correlation function $\langle s(\mathbf{y}_1)s(\mathbf{y}_2)\ldots s(\mathbf{y}_N)\rangle$ by the transformation $\mathbf{y} = \lambda\mathbf{x}$. Using an adapted version of (7.360) we see that it is multiplied by $\zeta^N(\lambda\Lambda)^d$. Hence on taking Fourier

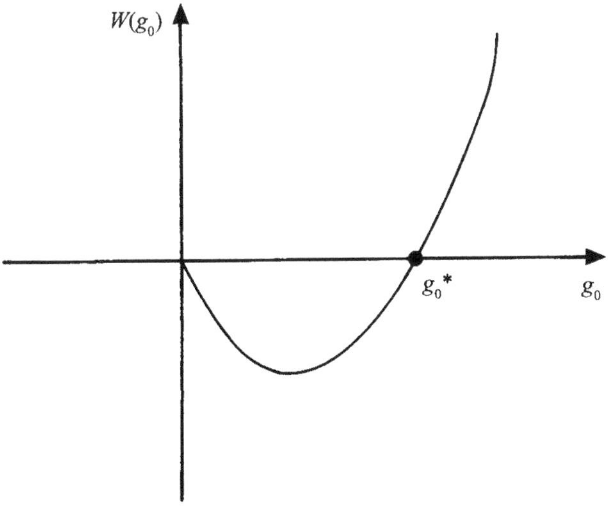

Figure 7.26 Shape of $W(g_0)$ below four dimensions.

transforms we find that

$$\Gamma^{(N)}(\mathbf{k}_i/\lambda;\, g_0,\Lambda) = \lambda^{-d+1/2N(d-2)}\Gamma^{(N)}(\mathbf{k}_i;\, g_0,\lambda\Lambda). \tag{7.380}$$

Thus from (7.378) we derive the asymptotic relation for low momenta

$$\Gamma^{(N)}(\mathbf{k}_i/\lambda;\, g_0,\Lambda) \sim \lambda^{1/2N(d-2+\eta)-d}\Gamma^{(N)}(\mathbf{k}_i;\, g_0^*,\Lambda). \tag{7.381}$$

Two important conclusions can be drawn from (7.381): the power-law behaviour of the correlation functions for low momenta with exponent $-\frac{1}{2}N(d-2+\eta)+d$; and the universality of the correlation functions in this region, i.e. the behaviour does not depend on the initial value of g_0. Also when $N=2$ the behaviour of the pair correlation function is given by

$$G^{(2)}(k) = [\Gamma^{(2)}(k)]^{-1} \sim 1/k^{2-\eta} \tag{7.382}$$

and this identifies $\eta(g_0^*)$ with the critical exponent η.

Turning now to the identification of m, we use equation (5.176), which is rewritten in terms of our present notation as

$$\Gamma^{(2)}(\mathbf{k}) = \chi_0(w)^{-1} + L_2(w)k^2a^2 + L_4(w)k^4a^4 + \dots . \tag{7.383}$$

The renormalized $\Gamma_R^{(2)}(\mathbf{k})$ is defined by (7.346), from which, using (7.350), we find that

$$m^2 = \frac{\chi_0(w)^{-1}}{L_2(w)} \sim \frac{\tau^\gamma}{\tau^{\gamma-2\nu}} = \tau^{2\nu} \tag{7.384}$$

using (7.178).

To proceed to critical behaviour for T near to T_c the Callan–Symanzik equation must be applied to the renormalized $\Gamma_R^{(N)}(\mathbf{k},m,g)$. After some manipulation (Brezin, Le Guillou and Zinn-Justin 1976) the following equation is derived for the asymptotic behaviour (large k/m) of $\Gamma_R^{(N)}$.

$$\left(m\frac{\partial}{\partial m} + W(g)\frac{\partial}{\partial g} - \frac{N}{2}\eta(g)\right)\Gamma_R^{(N)} = 0. \tag{7.385}$$

Using the same procedure as before a fixed point g^* is identified satisfying $W(g^*)=0$, and critical behaviour is determined by $\Gamma_R^{(N)}(\mathbf{k}_i;\, m,g^*)$. This is now temperature dependent and hence describes the critical *region*; a second critical exponent $\nu(g^*)$ can be evaluated as a power series in g^*, and from η and ν all other critical exponents can be derived in accordance with the relations given in Table 6.1. A relation analogous to (7.377) is deduced,

$$g(\lambda) - g^* \sim \lambda^\omega \tag{7.386}$$

and ω is identified with the correction term to scaling.

With this basis all of the scaling and thermodynamic properties discussed in Chapter 6 can be obtained. The derivation is not easy, and familiarity with different aspects of field theory is almost an essential requisite. But the tools available are very powerful.

In order to obtain numerical results, double power series in (g,ϵ) are developed for $W(g)$, and from the equation $W(g) = 0$ a power series in ϵ is derived. From this critical exponents can be calculated as ϵ expansions, and they agree with (7.14) and (7.15) obtained by Wilson's direct method described in §7.6.2. But the field theoretical approach can draw on all the experience gained in renormalization calculations and expansions can be developed in a more systematic and satisfactory manner.

The power of the approach is well illustrated by the calculation of the equation of state in four dimensions where the exponents of the logarithmic terms can be obtained exactly. In this case the fixed point is at $g = 0$, and corrections to scaling become the focus of attention. For details of the derivation we refer to Brezin, Le Guillou and Zinn-Justin (1976), but we quote the interesting final result

$$H(m,\tau) \propto m^3\left(\frac{\tau}{m^2}\,|\ln m|^{[(n+2)/(n+8)]} + \frac{1}{(n+8)}|\ln m|^{-1}(1 + \text{constant}\,|\ln m|^{-1}\right). \tag{7.387}$$

From this the critical behaviour of magnetization, susceptibility, etc., given previously in (7.338), can readily be deduced.

We have already noted at the end of §7.1 small but significant differences between numerical values of critical exponents derived from ϵ expansions and those derived from series expansions in Chapter 5. These were disturbing, but it was generally assumed that the ϵ expansion series had not been developed to sufficient length to provide an accuracy comparable with that of series expansions. Series expansions workers maintained a hope that the exponents would be rational fractions in three dimensions as in two dimensions. The ϵ expansions did not contradict such a possibility.

7.7.2.5 *Expansions in three dimensions*

The crucial step which enabled the field theoretical approach to undertake accurate calculations of critical exponents in three dimensions arose from a suggestion of Parisi (1973). It had been assumed (see Brezin, Le Guillou and Zinn-Justin 1976) that the Callan–Symanzik equation could be used only when both g and ϵ were varied, otherwise the right-hand-side of equations like (7.353) and (7.385) would not be zero. Parisi proposed that the procedure should be followed with a fixed value of d equal to 3, and his proposal was implemented by two separate groups, Baker *et al.* (1976, 1978) and Le Guillou and Zinn-Justin (1977, 1980).

The detailed calculations follow the same pattern as before, using expansions in powers of the renormalized coupling constant g, and the Callan–Symanzik equation (7.385). Expansions for $W(g)$ and $\eta(g)$ are derived from renormalization theory in three dimensions, and the fixed point g^* is calculated numerically from the equation $W(g^*) = 0$. The value of critical exponent η is given by $\eta(g^*)$, and the exponent ω which represents the correction to scaling is given by $W'(g^*)$. Further development of the theory introduces a function $\gamma^{-1}(g)$ which can likewise be expanded as series in powers of g, and whose value at g^* gives the critical exponent γ^{-1}.

The essential preliminary task is the evaluation of the appropriate Feynman integrals in three dimensions to as high an order as possible. Both groups made use of the calculations of Nickel (unpublished) who succeeded in extending the power series for $W(g)$ as far as g^7. The coefficients are functions of n. Typical series for the Ising case ($n = 1$) are

$$\begin{aligned} W(g) &= -g + g^2 - 0.422\,496\,6g^2 + 0.351\,069\,6g^4 \\ &\quad - 0.376\,526\,8g^5 + 0.495\,547\,5g^6 - 0.749\,689g^7 \\ \eta(g) &= 0.010\,973\,94g^2 + 0.000\,914\,22g^3 + 0.017\,962g^4 \\ &\quad - 0.000\,653\,7g^5 + 0.013\,878g^6 \\ \gamma^{-1}(g) &= 1 - \frac{g}{6} + \frac{g^2}{27} - 0.023\,069\,6g^3 + 0.019\,886\,8g^4 \\ &\quad - 0.022\,459\,5g^5 + 0.030\,367\,9g^6 . \end{aligned} \tag{7.388}$$

The purpose of renormalization is to eliminate coefficients which become infinite as the cut-off $\Lambda \to \infty$. But when this has been successfully achieved, it is well known that the resulting series in powers of g are asymptotic and do not have a non-zero radius of convergence. In this respect the present expansions differ from those discussed in §5.2 which are standard power series expansions. Problems arose then when the region of convergence of the series did not extend to the singularity associated with critical behaviour. A powerful tool discussed in §5.2.3, the Padé approximant, served as a mechanism for analytical continuation of the function defined by the power series.

In the case of any asymptotic series it is useful first to apply a Borel transformation (Whittaker and Watson 1927). For a series

$$A(g) = \sum_r a_r g^r \tag{7.389}$$

the Borel transform $B(g)$ is defined by

$$B(g) = \sum_r \frac{a_r g^r}{r!} . \tag{7.390}$$

If $B(g)$ is convergent, $A(g)$ can be defined as an analytic function for all values of g,

$$A(g) = \int_0^\infty \exp(-t) B(tg)\, dt . \tag{7.391}$$

Table 7.3

	$n = 1$	$n = 2$	$n = 3$
g^*	1.416 ± 0.005	1.406 ± 0.004	1.391 ± 0.004
ω	0.79 ± 0.03	0.78 ± 0.025	0.78 ± 0.02
γ	1.241 ± 0.0020	1.316 ± 0.0025	1.386 ± 0.0040
ν	0.630 ± 0.0015	0.669 ± 0.0020	0.705 ± 0.0030
η	0.031 ± 0.004	0.033 ± 0.004	0.033 ± 0.004
β	0.325 ± 0.0015	0.3455 ± 0.0020	0.3645 ± 0.0025
α	0.110 ± 0.0045	-0.007 ± 0.006	-0.115 ± 0.009
$\theta = \omega\nu$	0.498 ± 0.020	0.522 ± 0.018	0.550 ± 0.016

For example, if $a_r = (-1)^r r!$, $B(g) = (1 + g)^{-1}$.

In standard mathematical analysis a_r would have to be known for all r in order to define $B(g)$. In our particular case only a finite number of terms are available, but the Padé approximant can be used to fit these terms by an analytic function defined over the whole plane, and hence to provide an approximation to $B(g)$. $A(g)$ can then be calculated by (7.391). (This process has been termed the Padé Borel approximation (Baker 1975).)

An important additional piece of information became available which enabled the accuracy of the approximation to $B(g)$ to be improved considerably. Lipatov (1977) showed how to calculate the asymptotic behaviour of a_r in series in the coupling parameter g of the type we are considering. In our particular case Brezin, Le Guillou and Zinn-Justin (1977) found the form

$$cr!(-a)^r r^b \tag{7.392}$$

where the constants a, b, c could be determined. This gives the location, nature and residue of the singularity of $B(g)$ closest to the origin. For details of the subsequent calculation see Le Guillou and Zinn-Justin (1980).

Although there were initial disagreements between the calculations of the two groups, these were ironed out to reach a reasonable consensus. Table 7.3 lists the critical values of exponents as given in the latest paper (Le Guillou and Zinn-Justin 1980). We shall see later that error bars are optimistic.

7.8 Removal of the Discrepancy

Let us recall the values of the critical exponents derived for the spin $\frac{1}{2}$ Ising model, and summarized in §6.2.3. These had been obtained by careful analysis of long series expansions for a variety of lattices from the loose-packed diamond to the close-packed FCC. The results could be approximated by rational fractions, and the hope was maintained that these might prove to be exact, by analogy with two dimensions. The thermodynamic exponents were given by $\gamma = 5/4 = 1.250$, $\beta = 5/16 = 0.3125$, $\alpha = 1/8 = 0.125$, and these differ from the values given in column 1 of Table 7.1 by amounts well above the error bounds quoted in both analyses.

The correlation exponents given in Chapter 5, $\nu = 9/14 = 0.643$ (equation (5.179)) and $\eta = 1/18 = 0.056$ (equation (5.188)) presented a more serious challenge, since in addition to their deviation from the values given in Table 7.2, they did not satisfy the strong scaling (or hyperscaling) relation

$$d\nu = \gamma + 2\beta. \tag{7.393}$$

This relation is an essential constituent of the RG treatment.

A major fault in the analysis of the series expansion results had been the assumption that the singularities corresponding to the critical point are of Darboux form, and that the corrections to scaling are therefore analytic. We have seen in §7.5.7 that the RG predicts a confluent singularity having a leading correction term to scaling, and its exponent is found to be about 0.5. It was therefore necessary to reanalyze the series data allowing for the possibility of such a confluent exponent.

The first attempt at such a reanalysis (Saul, Wortis and Jasnow 1975) dealt with the high-temperature susceptibility of the Ising model of spin s. Clear evidence was found of the presence of a confluent singularity term with exponent θ, comparable with the RG estimate. However, the data seemed to indicate that the amplitude of this term vanished when $s = \frac{1}{2}$, and the authors confirmed the value of 1.25 for exponent γ. Further work by Camp and van Dyke (1975), and subsequent analysis of series for the second moment of the correlation function (Camp *et al.* 1976) supported this result. The latter authors concluded that 'the validity of hyperscaling remains problematic'.

A remarkable exploitation by Sati McKenzie (1975) of a method pioneered by Rapaport (1973) (using the embedding approach of §5.2.1) enabled the high-temperature susceptibility series for the FCC lattice to be extended to 15 terms. McKenzie (1979) noted that her results could be well fitted to $\gamma = 1.241$ if a non-zero confluent singularity term were included with $\theta = 0.496$. But Gaunt and Sykes (1979) using the same data and additional data for the SC and diamond lattices had found support for $\gamma = 1.250$.

The clarification of this confused situation and the removal of the discrepancy are due largely to Nickel (1981, 1982) who developed a new approach to the analysis of series expansions for the Ising model of spin s (Nickel and Sharpe 1979). This model was defined in (4.73) and we shall write the p.f. for the model in the form

$$Z = \exp\left(K \sum_{\langle ij \rangle} s_i s_j + h \sum_i s_i\right) \qquad (K = \beta J, \quad h = \beta \mu_0 H) \tag{7.394}$$

where s_i $(= s_{zi}/s)$ can take the values $-1, -1 + 1/s, \ldots, 1 - 1/s, 1$.

Nickel and Sharpe looked for a parameter which would be analogous to the renormalized coupling parameter g in the field theoretical model. To do this they used the sums of the cumulant averages

$$\sum \langle s_1 s_2 \ldots s_{2N} \rangle_c = \sum G_c^{(2N)} = \left(\frac{\partial^{2N}}{\partial h^{2N}} (\ln Z) \right)_{h=0} \tag{7.395}$$

which diverge at T_c as $\tau^{-\gamma - 2N\Delta}$ (equation (6.25)).

They focused attention on the dimensionless rato

$$R_N = \frac{\sum \langle s_1 \ldots s_{2N} \rangle_c}{(\sum \langle s_1 s_2 \rangle_c)^N} = \frac{\sum G_c^{(2N)}}{(\sum G_c^{(2)})^N} \tag{7.396}$$

and assumed that in the critical region the lattice spacing, a, would have no significance and the only length scale of relevance would be the coherence length ξ (§6.3). This must now be given a precise definition in terms of the second moment of the pair correlation μ_2 (equation (5.173)),

$$\left(\frac{\xi}{a} \right)^2 = \frac{(2d)^{-1} \mu_2}{\chi_0} \tag{7.397}$$

since it needs to be expanded as a high-temperature series.

Nickel and Sharpe argued that since each cumulant average is extensive, the dependence on volume is $R_{N+1} \propto V^{-N}$, and since the only length scale near T_c is ξ, we must have

$$R_{N+1} \sim \frac{k_N \xi^{Nd}}{V^N} \tag{7.398}$$

where k_N is a numerical constant. Hence $R_2 V/\xi^d$ is dimensionless, and this was used for the renormalized coupling constant

$$G = -A_0 V \frac{\xi^{-d} (\sum G_c^{(4)})}{\chi_0^2} \tag{7.399}$$

where A_0 is a convenient numerical constant. Referring back to (7.365) we note the definition of the renormalized coupling constant g for the field theoretical model,

$$g = m^{-\epsilon} \Gamma_R^{(4)}(0,0,0,0) = m^{d-4} Z_0^2 \Gamma^{(4)}(0,0,0,0) \sim \xi^{4-d} \frac{\Gamma^{(4)}(0,0,0,0)}{[L_2(w)]^2} . \tag{7.400}$$

Using (7.342) and (7.384) it is easy to see that (7.399) and (7.400) are equivalent.

Moreover, from (7.399) as $\tau \to 0$, G behaves as $\tau^{\gamma + d\nu - 2\Delta}$. It can be shown rigorously (Schroder 1976) that

$$\gamma + d\nu - 2\Delta \geq 0; \tag{7.401}$$

the values given at the beginning of this section imply that for the Ising model this is an inequality ($\gamma + d\nu - 2\Delta \simeq 0.04$) and hence that $G \to 0$ as $T \to 0$. But hyperscaling implies an equality, and that $G \to G^*$ (non-zero) as $\tau \to 0$.

Nickel and Sharpe inverted (7.399) to obtain ξ^2 as a function of G, but since ξ^2 is of order K for small K it was necessary to introduce a new variable

$$y = G^{-2/d} \tag{7.402}$$

and to derive a series in the form

$$(\xi/a)^2 = \sum_{n=1} a_n y^n. \tag{7.403}$$

They made use of high-temperature series expansions for the susceptibility and its second derivative in zero field for the SC, BCC and FCC lattices (Essam and Hunter 1968, Moore, Jasnow and Wortis 1969) to deal with the $\sum G_c^{(4)}$ and χ_0^2 factors and achieve the inversion. If hyperscaling is satisfied the coherence length should diverge as $y \to y^*$, a finite value. In addition, if the divergence is of power-law form $\xi \sim (y^* - y)^{-1/\omega}$, the inverse logarithmic derivative

$$Y(y) \equiv \left(\frac{\mathrm{d}}{\mathrm{d}y} \ln(\xi/a)^2\right) = \frac{1}{2}\,\omega(y^* - y) \tag{7.404}$$

should have a zero at $y = y^*$.

In fact they did find such a zero close to the value obtained for g in the field theory model, and they showed that the behaviour of $Y(y)$ parallels the behaviour of $W(g)$ in (7.388), with the exponent ω having the same significance as a correction to scaling in both. This result provided clear and unequivocal evidence of hyperscaling for the spin $\frac{1}{2}$ Ising model. It was now necessary to examine how the standard ratio and Padé approximant methods of estimating critical exponents are modified in the presence of a confluent singularity, and to explain how anomalous results might be obtained.

As a preliminary, a *tour de force* calculation by Nickel using the Wortis linked cluster method of §5.2.2 added six new terms to the high-temperature expansions χ_0 and ξ for the BCC lattice, and brought the total number of terms available to 21. A simplifying feature of the BCC lattice was exploited; its generating function for random walks (free graphs) is the product of three one-dimensional generating functions (see e.g. Domb 1960).

Consider first the ratio method for a series $\sum a_n w^n$ (equation (5.95)). Denoting a_n/a_{n-1} by ρ_n, Domb and Sykes (1961) noted that convergence could be accelerated by plotting $\sigma_n = n\rho_n - (n-1)\sigma_{n-1}$ since this eliminated the $1/n$ term. If the function we are dealing with is of the form

$$f(w) = A\left(1 - \frac{w}{w_c}\right)^{-\gamma} \times \left[1 + A_1\left(1 - \frac{w}{w_c}\right) + A_2\left(1 - \frac{w}{w_c}\right)^2 + \ldots + A_\theta\left(1 - \frac{w}{w_c}\right)^\theta + \ldots\right] \tag{7.405}$$

near the singularity w_c, then

$$\sigma_n = n\rho_n - (n-1)\rho_{n-1}$$
$$= \frac{1}{w_c}\left(1 + (\gamma - 1)A_1 n^{-2} + \ldots + \theta^2 A_\theta \frac{\Gamma(\gamma)}{\Gamma(\gamma - \theta)} n^{-(\theta+1)} + \ldots\right). \tag{7.406}$$

Estimates of γ can be obtained from the normalized slope

$$\tau_n = \frac{\sigma_n^{-1}(\rho_n - \rho_{n-1})}{n^{-1} - (n-1)^{-1}}$$
$$\simeq (\gamma - 1)(1 - 2A_1 n^{-1} + \ldots) - \theta(1+\theta)A_\theta \frac{\Gamma(\theta)}{\Gamma(\gamma - \theta)} n^{-\theta} \tag{7.407}$$

or the extrapolated slope

$$n\tau_n - (n-1)\tau_{n-1} \simeq (\gamma - 1)$$
$$\times \left(1 - [2A_1(\gamma - 3) - 6A_2(\gamma - 2) + 3A_1^2(\gamma - 1)]n^{-2} + \ldots \right.$$
$$\left. - \theta(1 - \theta^2)A_\theta \frac{\Gamma(\gamma)}{\Gamma(\gamma - \theta)} n^{-\theta} + \ldots\right) \tag{7.408}$$

in which the $O(n^{-1})$ correction terms have again been eliminated.

Whilst (7.407) and (7.408) can give a reasonable first approximation γ_a to γ, the slowly varying function $n^{-\theta}$, with θ expected to be $\sim\frac{1}{2}$, makes accurate assessment difficult. Nickel took two additional steps to improve the accuracy of the estimate. First he derived series for the function f^{1/γ_a} for which the analytic correction terms in A_1, A_2 ... are nearly eliminated. Second, to reduce 'noise' in the early terms of the series he applied an Euler transformation $y = w/(1 + \alpha w/w_c)$ with an appropriate choice of α; this effectively averages over several consecutive terms (for details see Nickel (1982)). Applying the method to 15 terms of the FCC lattice expansion and 21 terms of the BCC lattice expansions, Nickel concluded that reasonable fits would be obtained with $\gamma = 1.237 \pm 0.005$ and $2\nu = 1.259 \pm 0.005$. The value of θ was more uncertain: $\theta \sim 0.5$ was reasonable but $\theta \sim 0.6$ could also be entertained.

Turning now to the Padé approximant, Nickel (1981) noted that with the functional form (7.405), the logarithmic derivative will be of the form

$$\frac{d}{dw}(\ln f) \simeq \frac{\gamma}{w_c - w} - \frac{\theta A_\theta}{w_c}\left(1 - \frac{w}{w_c}\right)^{\theta - 1} + \ldots. \tag{7.409}$$

If $A_\theta \neq 0$ the Padé approximant to (7.409) will attempt to represent the branch cut $(1 - w/w_c)^{\theta-1}$ by a sequence of poles. A substantial number of terms will be needed to establish these poles, and for shorter series the residue from the pole near w_c will contain a contribution from the cut. He found that for the FCC lattice, approximants based on the first 12 terms seemed to be well converged at $\gamma \simeq 1.248$. But approximants based on 13, 14 and 15 terms were qualitatively different in that the single real-axis pole had now split into two poles suggestive of the branch cut given by (7.409). For the BCC lattice the same pattern of behaviour was observed with the characteristic break occurring at 16 or 17 terms.

Nickel concluded that below some critical number of series terms D log Padé approximants do not have the resolution necessary to distinguish the presence of confluent correction terms, and will yield artificially stable but wrong estimates. Using a more sophisticated analysis he was led to the estimates $\gamma = 1.239 \pm 0.002$, $\theta = 0.53 \pm 0.15$ for the Ising model. These are consistent with the above estimates derived by the ratio method, and with the estimates given in the previous section for the field theoretical model.

In a more recent paper published in a volume commemorating Michael Fisher's 60th birthday, Nickel (1991) reviewed subsequent developments in relation to this topic. An important contribution by Fisher himself was the application of partial differential approximants to the problem. Partial differential approximants were introduced (Fisher 1977, Fisher and Kerr 1977, Fisher and Chen 1982) as a mechanism for exploring the behaviour of double power series, identifying their singular points, and determining singular behaviour in the vicinity of these singularities. Using models dependent on a parameter which can interpolate between the pure Ising model and the Gaussian model it was possible to find special parameter values where corrections to scaling vanish. Typical results (Chen, Fisher and Nickel 1982, Fisher and Chen 1985) are $\gamma = 1.2385 \pm 0.0015$, 1.2395 ± 0.0004, $\nu = 0.632 \pm 0.001$, $\eta = 0.039 \pm 0.004$, $\theta = 0.54 \pm 0.05$.

The direct ϵ expansion was extended as far as ϵ^5 in an impressive calculation by Gorishny, Larin and Tkachov (1984), and Nickel felt that it should be treated as an independent approach to be analyzed on its own merits. In regard to the relation between values obtained from high-temperature series and from expansions in the renormalized coupling parameter g, Nickel felt that the situation was reversed relative to what it was in 1980, in that the latter are now more suspect; a possible reason for this is the presence of confluent singularities associated with the correction to scaling terms. But one could be happy that all three approaches are consistent with the values $\gamma = 1.238$, $\nu = 0.630$, $\eta = 0.0355$.

A question which naturally arises now is the origin of the good agreement between experiment and the initial exponent values reported in §5.7. In fact, if we look back at these experimental results, we find that the error bars are often sufficient to include the more accurate values derived above. Also, it is difficult to ensure in experimental measurements that one is sufficiently close to T_c to measure the true exponent value rather than an effective value. On the whole the agreement between experiment and the latest theoretical work can be regarded as satisfactory.

Notation for Chapter 7

a lattice spacing, b reduction factor in correlation length during the RG transformation, $N \to N' = N/b^d$.
$\mathbf{l}$, $\mathbf{m}$ lattice points, $\sigma_{\mathbf{l}}$ continuous variable lattice spin at $\mathbf{l}$, $\hat{\sigma}_{\mathbf{q}}$ its Fourier transform. .
Z_{I} p.f. for Ising model, $\mathbf{R}(\mathscr{H})$ RG transformation.
c, u, c_0, u_0 constants in weighting factor $W(\sigma)$ (7.6), (7.12) and (7.33).

Block spin transformation (§7.2.2)

$\mathbf{r} \to \mathbf{r}' = \mathbf{r}/b$, $N \to N' = N/b^d$, $\sigma_\alpha \to \sigma_{\alpha'} = \sigma_\alpha/c$.
Free energy $f(h',\tau') = b^d f(h,\tau)$.
Pair correlation $\Gamma(\mathbf{r}',h',\tau') = c^{-2}\Gamma(\mathbf{r},h,\tau)$.
p, v, f coefficients in expansion of local free energy (7.31).
Landau–Ginzburg–Wilson Hamiltonian (7.37) (using (7.32)).

Real-space renormalization

$\mathbf{K} = (K^1,K^2,K^3\ldots)$ coupling constants, $\mathbf{K}^*$ fixed point, $\mathbf{K}_0$ starting point, $\mathbf{K}_n$ nth iteration.
$Z(\mathbf{K})$ p.f., $\psi(\mathbf{K})$ free energy, $\xi(\mathbf{K})$ correlation length.
Near $\mathbf{K}^*$, $\mathbf{K}_n = \mathbf{K}^* + \mathbf{k}_n$, $\mathbf{L}$ matrix, $\mathbf{k}_{n+1} = \mathbf{L}\mathbf{k}_n$.
Λ_i eigenvalues, $\mathbf{l}_i$ eigenvectors of $\mathbf{L}$.
$u_l^{(0)}$ scaling fields (coefficients in expansion (7.67)); superscript 0 is dropped.
y_T, y_H, y_g scaling exponents (7.73).

Momentum-space renormalization

$s_{\mathbf{l}}$ Ising spin, $\sigma_{\mathbf{q}}$ its Fourier transform.
$s_{\mathbf{l}}^\alpha$ component of classical vector spin, $\sigma_{\mathbf{q}}^\alpha$ its Fourier transform.
v_0 volume of unit cell.
Constant $\beta = 1/kT$ is absorbed into Hamiltonian $\mathscr{H}(\exp(-\mathscr{H}))$.
sc, K constant in Gaussian Hamiltonian (7.135).
$\tilde{p} = c - 2dK$, $p = \tilde{p}/K$, $C_d = 2\pi^{d/2}/\Gamma(d/2)$.
A term $-u\,\Sigma_{\mathbf{l}} s_{\mathbf{l}}^4$ is added for the s^4 model.
RG transformation $\mathbf{q}' = b\mathbf{q}$, $\sigma_{\mathbf{q}} = \zeta\sigma_{\mathbf{q}'}$.
u_2', u_4' coefficients in expansion of $\mathscr{H}'$ (7.152).
For properties of Gaussian integrals see (7.158), (7.159) and (7.160).
$A(1/b, p)$ integral defined in (7.168), $C(1/b, p)$ in (7.174).
For n-vector model Hamiltonian $s_{\mathbf{l}}^2$ is replaced by

$$\sum_{\alpha=1}^{n} s_{\mathbf{l}}^\alpha s_{\mathbf{l}}^\alpha \quad \text{and} \quad s_{\mathbf{l}}^4 \text{ by } \left(\sum_{\alpha=1}^{n} s_{\mathbf{l}}^\alpha s_{\mathbf{l}}^\alpha\right)^2.$$

u, v are parameters in Hamiltonian giving rise to fixed points of different symmetry (7.205).
A_2, A_σ coefficients in Hamiltonian for long-range interactions (7.220).
Interaction energy for dipolar forces given in (7.231).
g in (7.234) is a standard atomic constant; g, h in (7.240) are parameters in the Hamiltonian for dipolar interactions.
m, **M** are matrices defined in (7.242a), (7.242b).

Perturbation expansions (§7.6)

q is a momentum variable of integration, **k** is a fixed momentum, $\Gamma(\mathbf{k})$ pair correlation (7.262).
$G_0(k) = (p_0 + k^2)^{-1}$ is the bare propagator.
$G(k) = p + k^2$ is the renormalized propagator.
$\Sigma(\mathbf{k},p_0)$ sum of diagrams in Figure 7.16.
$\bar{\Sigma}(\mathbf{k},p)$ sum of modified diagrams in Figure 7.18.
Λ upper cut-off of (divergent) integrals.
$G_c^{(4)}(\mathbf{k}_1,\mathbf{k}_2,\mathbf{k}_3,\mathbf{k}_4)$ four-point correlation function (cumulant).
$u_R(p)$ renormalized coupling constant defined in (7.320).
Integrals $J(\mathbf{q}_1)$ defined in (7.297), J_1 and J_2 in (7.307), K_1 in (7.324).

Field theory approach (§7.7.2)

m_0 bare mass, g_0 bare coupling constant (7.339).
m_1 initial renormalized mass (7.344), g_1 initial renormalized coupling constant (7.345).
m final renormalized mass, g final renormalized coupling constant (7.350).
$G_c^{(N)}(\mathbf{x}_1, \mathbf{x}_2, \ldots, \mathbf{x}_N)$ N-point correlation function (cumulant) (7.340).
$\Gamma^{(N)}(\mathbf{x}_1, \mathbf{x}_2, \ldots, \mathbf{x}_N)$ N-point vertex function (e.g. (7.342)).
$W(g_0, m/\Lambda)$ defined in (7.356), $\eta(g_0, m/\Lambda)$ in (7.357).
g_0^* fixed point, $W(g_0^*) = 0$, $W'(g_0^*) = \omega$.
Borel transformation defined in (7.389) and (7.390).

References

Aharony, A. (1973) *Phys. Rev. B* **8**, 3349.
Aharony, A. (1976) DG **6**, Ch. 6.
Aharony, A. and Fisher, M. E. (1973) *Phys. Rev. B* **8**, 3323.
Baker, G. A. (1975) *Essentials of Padé Approximants*, New York and London: Academic Press.
Baker, G. A., Nickel, B. G., Green, M. S. and Meiron, D. I. (1976) *Phys. Rev. Lett.* **36**, 1351.
Baker, G. A., Nickel, B. G. and Meiron, D. I. (1978) *Phys. Rev. B* **17**, 1365.

BREZIN, E., LE GUILLOU, J. C. and ZINN-JUSTIN, J. (1973) *Phys. Rev. D* **8**, 434, 2418.

BREZIN, E., LE GUILLOU, J. C. and ZINN-JUSTIN, J. (1976) DG **6**, Ch. 3.

BREZIN, E., LE GUILLOU, J. C. and ZINN-JUSTIN, J. (1977) *Phys. Rev. D* **15**, 1544, 1558.

CALLAN, C. G. (1970) *Phys. Rev. D* **2**, 1541.

CAMP, W. J. and VAN DYKE, J. P. (1975) *Phys. Rev. B* **11**, 2579.

CAMP, W. J., SAUL, D. M., VAN DYKE, J. P. and WORTIS, M. (1976) *Phys. Rev. B* **14**, 3990.

CHEN, J. H., FISHER, M. E. and NICKEL, B. G. (1982) *Phys. Rev. Lett.* **48**, 630.

DOMB, C. (1960) *Adv. Phys.* **9**, 344 (Appendix 2).

DOMB, C. (1974) DG **3**, Ch. 1.

DOMB, C. and SYKES, M. F. (1961) *J. Math. Phys.* **2**, 61.

DYSON, F. J. (1949) *Phys. Rev.* **75**, 1736.

ESSAM, J. W. and HUNTER, D. L. (1968) *J. Phys. C: Solid State Phys.* **1**, 392.

EWALD, P. P. (1921) *Ann. Phys., Leipzig* **64**, 253.

FEYNMAN, R. P. and HIBBS, A. R. (1965) *Quantum Mechanics and Path Integrals*, New York: McGraw-Hill.

FISHER, M. E. (1974) *Rev. Mod. Phys.* **46**, 597.

FISHER, M. E. (1977) *Physica B–C* **86–88**, 590.

FISHER, M. E. (1983) *Critical Phenomena*, Lecture Notes in Physics No. 186, ed. F. J. W. HAHNS, Berlin: Springer, p. 117.

FISHER, M. E. and CHEN, J. H. (1982) Cargèse Lectures *Phase Transitions*, ed. M. LEVY, J. C. LE GUILLOU and J. ZINN-JUSTIN, New York: Plenum Press.

FISHER, M. E. and CHEN, J. H. (1985) *J. Phys.* **46**, 1645.

FISHER, M. E. and KERR, R. M. (1977) *Phys. Rev. Lett.* **39**, 667.

FISHER, M. E., MA, S. K. and NICKEL, B. G. (1972) *Phys. Rev. Lett.* **29**, 917.

GAUNT, D. S. and SYKES, M. F. (1979) *J. Phys. A: Math. Gen.* **12**, L25.

GINZBURG, V. L. (1960) *Sov. Phys. Solid State* **2**, 1824 (translated from (1960) *Fiz. Tverd. Tela* **2**, 2031).

GINZBURG, V. L. (1978) *Key Problems of Physics and Astrophysics*, Moscow: Mir.

GINZBURG, V. L. and LANDAU, L. D. (1950) *Sov. Phys. JETP* **20**, 1064.

GORISHNY, S. G., LARIN, S. A. and TKACHOV, F. V. (1984) *Phys. Lett. A* **101**, 120.

GRIFFITHS, R. B. (1970) *Phys. Rev. Lett.* **24**, 1479.

HILEY, B. J. and JOYCE, G. S. (1965) *Proc. Phys. Soc.* **85**, 493.

LARKIN, A. I. and KHMELNTTSKII, D. E. (1969) *Sov. Phys. JETP* **29**, 1123 (translated from (1969) *Zh. Eksp. Teor. Fiz.* **56**, 2087).

LE GUILLOU, J. C. and ZINN-JUSTIN, J. (1977) *Phys. Rev. Lett.* **39**, 95.

LE GUILLOU, J. C. and ZINN-JUSTIN, J. (1980) *Phys. Rev. B* **21**, 3976.

LEVY, P. M. and STINCHCOMBE, R. B. (1968) *J. Phys. C: Solid State Phys.* **1**, 1584.

LIPATOV, L. N. (1977) *Sov. Phys. JETP* **45**, 216 (translated from (1977) *Zh. Eksp. Teor. Fiz.* **72**, 411).

LUTTINGER, J. M. and TISZA, L. (1946) *Phys. Rev.* **70**, 954

MA, S. K. (1976a) *Modern Theory of Critical Phenomena*, Reading, PA: Benjamin.

MA, S. K. (1976b) DG **6**, Ch. 4.

MCKENZIE, S. (1975) *J. Phys. A: Math. Gen.* **8**, L102.

MCKENZIE, S. (1979) *J. Phys. A: Math. Gen.* **12**, L185.

MOORE, M. A., JASNOW, D. and WORTIS, M. (1969) *Phys. Rev. Lett.* **22**, 940.
NICKEL, B. G. (1974) Unpublished preprint on Feynman integrals.
NICKEL, B. G. (1981) *Physica* **106A**, 48 (*Stat. Phys.* **14**).
NICKEL, B. G. (1982) *Proc. Cargèse Summer Institute on Phase Transitions*, ed. J. C. LE GUILLOU and J. ZINN-JUSTIN, New York: Plenum Press, p. 291.
NICKEL, B. G. (1991) *Physica A* **177**, 189 (Volume in Commemoration of Michael Fisher's 60th Birthday).
NICKEL, B. G. and SHARPE, B. (1979) *J. Phys. A: Math. Gen.* **12**, 1819.
NIEMEIJER, Th. and VAN LEEUWEN, J. M. J. (1973) *Phys. Rev. Lett.* **31**, 1411.
NIEMEIJER, Th. and VAN LEEUWEN, J. M. J. (1974) *Physica* **71**, 17.
NIEMEIJER, Th. and VAN LEEUWEN, J. M. J. (1976) DG **6**, Ch. 7.
NIJBOER, B. R. A. and DE WETTE, F. W. (1958) *Physica* **24**, 422.
ONSAGER, L. (1939) *J. Phys. Chem.* **43**, 189.
PARISI, G. (1973) Lecture to Cargèse Summer School, unpublished.
RAPAPORT, D. C. (1973) *Phys. Lett.* **44A**, 327.
SAUER, J. A. (1940) *Phys. Rev.* **57**, 142.
SAUL, D. M., WORTIS, M. and JASNOW, D. (1975) *Phys. Rev. B* **11**, 2541.
SCHRODER, R. (1976) *Phys. Rev. B* **14**, 172.
SNEDDON, I. N. (1961) *Special Functions of Mathematical Physics and Chemistry*, Edinburgh: Oliver & Boyd.
STANLEY, H. E. (1968) *Phys. Rev.* **176**, 718
SYMANZIK, N. (1970) *Commun. Math. Phys.* **18**, 227.
THOMPSON, C. J. (1978) *Contemp. Phys.* **19**, 203.
THOULESS, D. J. (1969) *Phys. Rev.* **181**, 954.
VAKS, V. G., LARKIN, A. I. and PIKIN, S. A. (1966) *Sov. Phys. JETP* **24**, 240 (translated from (1966) *Zh. Eksp. Teor. Fiz.* **51**, 361).
WALLACE, D. (1973) Unpublished lecture notes given at Southampton University.
WALLACE, D. (1976) DG **6**, Ch. 5.
WEGNER, F. J. (1972) *Phys. Rev. B* **5**, 4529; **6**, 1891.
WHITTAKER, E. T. and WATSON, G. N. (1927) *Modern Analysis*, 4th edition, Cambridge, pp. 140, 154.
WILSON, K. G. (1971) *Phys. Rev. B* **4**, 3171, 3184.
WILSON, K. G. (1972) *Phys. Rev. Lett.* **28**, 548.
WILSON, K. G. (1974) *Physica* **73**, 119 (van der Waals Centennial Conference, 1973).
WILSON, K. G. (1975a) *Adv. Math.* **16**, 176.
WILSON, K. G. (1975b) *Rev. Mod. Phys.* **47**, 773.
WILSON, K. G. (1976) DG **6**, Ch. 1.
WILSON, K. G. (1979) *Sci. Am.* **241**, 158.
WILSON, K. G. (1983) *Rev. Mod. Phys.* **55**, 583.
WILSON, K. G. and FISHER, M. E. (1972) *Phys. Rev. Lett.* **28**, 240.
WILSON, K. G. and KOGUT, J. (1974) *Phys. Rep.* **12C**, 77, 95–6.
ZIMAN, J. M. (1964) *Principles of the Theory of Solids*, Cambridge.

Appendix: Related Topics

Whilst this monograph is aimed at the non-specialist, I have tried to provide key references which enable a student or research worker to find more specialized information if he needs it. In this spirit, I wish to mention briefly in this appendix a number of topics which have not been discussed in the main body of the text, but which bear a close physical or mathematical relationship to our main theme.

A tool which has developed in scope and power during the past two decades is that of Monte Carlo simulation. Model systems containing a relatively small number of interacting spins (perhaps 1000) are activated kinetically and the equilibrium, or non-equilibrium, property of interest is estimated for the system. By varying the size of the system and extrapolating it is possible to estimate bulk properties. The Monte Carlo approach received a great impetus with the development of the finite size scaling techniques described in §6.6.3. The potentialities of Monte Carlo methods grew as computers developed higher speed and increased storage capacity.

In relation to the basic problems with which we have been concerned in this monograph, evaluating critical exponents precisely, establishing scaling and universality, and investigating the validity of hyperscaling for simple models, Monte Carlo methods could not initially compare in accuracy with series and RG methods. But there are more sophisticated models like spin glasses (see e.g. Binder and Young 1986), for which the latter methods provided little information, and in which Monte Carlo simulations played a central role. Also the combination of Monte Carlo methods with RG transformations substantially increased the accuracy obtainable, which now approaches that of series expansions (see e.g. Pawley *et al.* 1984).

A general introduction to the use of Monte Carlo methods for spin systems is given in a review article by Binder (1976); more recent developments are described in Binder and Heerman (1988) and Binder (1992). For a comparison between series and Monte Carlo methods see Adler (1995, 1996).

Exact solutions have played a role of special importance in critical phenomena. We have seen in Chapter 5 how Onsager's solution revealed the defects of the classical theories, and served as a test for new techniques of calculation. For about 20 years after Onsager's work there were no new solutions for realistic interactions which differed in a substantial way from that of Onsager, but interesting alternative methods were used for deriving Onsager's results (see e.g. McCoy and Wu 1973).

In 1967 Lieb produced a variety of new solutions for two-dimensional ferroelectric models (see Lieb and Wu 1972) which followed a completely different pattern from the Ising model.

The models originated in attempts to explain ferroelectric and antiferroelectric phase transitions in hydrogen-bonded crystals like potassium dihydrogen phosphate, KH_2PO_4. The equilibrium positions of the hydrogen atoms are near one or other end of the bonds of the tetrahedral lattice connecting the phosphate groups, and a 'state' of the crystal is characterized by a specification of the hydrogen positions. If a hydrogen atom near a vertex is denoted by an arrow pointing along the bond to the vertex, and a hydrogen atom distant from the vertex by an arrow pointing away from the vertex, there will be 16 possible vertex configurations. Onsager (1939) suggested that the ferroelectric transition in KH_2PO_4 is connected with an ordering of these vertex configurations, and an appropriate model assigns a suitable energy to each vertex configuration. For a particular vertex configuration, specific sets of the other configurations will be compatible with it as neighbours along the bonds of the lattice, and a p.f. must be constructed which takes all possible configurations into account and gives them their correct energy weighting.

It can be shown that the above general 16-vertex problem is equivalent to an Ising model with two, three and four spin interactions in an external magnetic field. Lieb's exact solutions correspond to particular cases in which 10 of the 16 vertex configurations are disallowed (six vertex models).

Lieb was followed by Baxter (1971, 1972), whose solution of a more complex ferroelectric model (the eight-vertex model) gave rise to exponents which vary continuously with the strength of the interaction. This was a major challenge to the concept of universality, since the model is equivalent to an Ising model with a two and four spin interaction. The issue was resolved by Kadanoff and Wegner (1971), who demonstrated that a special symmetry of the model gives rise to the continuously varying exponents; this specific feature would not be expected to occur in normal physical systems.

Baxter subsequently derived a number of other new solutions (see Baxter 1982), and as a by-product extended significantly exact series expansions for two-dimensional models which he was unable to solve exactly (Baxter and Enting 1979).

Closely related to the properties of the Ising model are the properties of self-avoiding walks (SAWs) on lattices. An SAW on a lattice is a random walk subject to the condition that no lattice site may be visited more than once in the walk. The term SAW was first used by Hammersley and Morton (1954), although it is not clear who first introduced the model (Domb 1990). An SAW is a realistic model of a polymer chain molecule which takes into account the *excluded volume effect*, i.e. the finite size of the chain which prevents any region of space from being occupied more than once. The standard random walk, which was used as a first model, does not have this property. It does not take much effort to convince oneself that the standard mathematical techniques, which are so powerful in elaborating the properties of random walks, are inapplicable to SAWs (see e.g. Domb 1969).

Properties of interest of SAWs are the number of walks c_N of length N, the number of self-avoiding polygons u_N (i.e. walks which are self-avoiding but close on the last step), the mean square end-to-end length $\langle R_N^2 \rangle$, and the probability distribution of end-to-end length, $f(\mathbf{u})\, d\mathbf{u}$ ($\mathbf{u} = R/\langle R_N^2 \rangle^{1/2}$).

Quite early on, an analogy was noted between the configurational problems posed by the SAW model and those arising in the development of high-temperature series expansions for the Ising model (Fisher and Sykes 1959, Domb and Sykes 1961a). We have seen that the critical behaviour of thermodynamic quantities for this model are determined by the asymptotic behaviour as a function of N of the coefficients of power series discussed in §5.2 and §5.5. When the embedding approach is used the dominant contribution to the magnetic susceptibility series is c_N, and to the series for the p.f. u_N; $\langle R_N^2 \rangle$ contributes to the pair correlation function series, and $f(\mathbf{u})$ parallels the scaling functions discussed in §6.2 (Domb 1971).

The precise connection between the SAW model and magnetic models was revealed by de Gennes (1972) who pointed out that the SAW configurations correspond to a ferromagnetic model of spin dimension n when n is allowed to take the value zero. The RG calculation of critical exponents described previously can be used immediately for the calculation of SAW exponents.

A theoretical model which burst into prominence after being ignored for nearly 20 years is the Potts model (Potts 1952; the origin of the model is discussed in Domb (1974)). It represents a generalization of the Ising model, having Q orientations but only two different energies of interaction. Although the critical point can be located exactly, no exact solution analogous to Onsager has been put forward to date. However, a good deal of exact information is available on critical behaviour and critical exponents (see e.g. Nienhaus 1987). A number of physical systems have been identified for which the Potts model serves as a reasonable representation, and the difference in symmetry between the n-vector magnetic model and the Q-component Potts model gives rise to substantial differences in critical behaviour. An excellent review of the Potts model starting from first principles has been given by Wu (1982).

In §4.8 we introduced the model of solid solution formed from A and B atoms on a lattice, where the interactions favour separation into A and B phases at low temperatures. We showed that the mathematical problem in forming a p.f. for this solid solution is identical with the Ising model of ferromagnetism. Let us now consider a purely random mixture of non-interacting A and B atoms on the lattice, and starting with pure A phase let us gradually introduce B atoms at random in steadily increasing concentration p. At first the B atoms will form isolated cluster islands (atoms which are nearest neighbour are regarded as belonging to the same cluster), but at a sufficiently high concentration there will be clusters of B atoms spanning the whole crystal. It can be shown that the transition from isolated islands to a spanning cluster takes place sharply at a critical concentration p_c. If the A atoms are non-magnetic and the B atoms magnetic, the formation of a spanning cluster is associated with the onset of ferromagnetism (Elliott *et al.* 1960).

The above is termed a *site percolation process.* The term *percolation process* was first introduced by Broadbent and Hammersley (1957) to describe the flow of a fluid in a random *medium* in contrast to a diffusion process in which randomness is associated with the particles of the *fluid.* Consider a network, which for convenience we take to be a crystal lattice, with the bonds of the lattice serving as connecting channels; each channel has a finite probability $(1 - p)$ of being blocked. Fluid is introduced at a particular point, and one wishes to calculate the probabilities of other points in the network being wet or dry. In particular one is interested in the probability that the fluid will spread to infinity, and Broadbent and Hammersley were able to establish the existence of a critical probability p_c above which a fluid starting at one point of the lattice would not spread to infinity.

Percolation processes have a variety of applications in different fields, the most immediate application in physics being to semiconductors. The bonds of an electric network have a finite probability p of being conducting, and $(1 - p)$ of being non-conducting. What is the probability of finding a conducting path between two different points of the network?

Percolation processes involving randomness in the bonds of a lattice are described as *bond percolation processes.* The relationship between bond and site percolation processes was clarified by Fisher (1961) who showed that every bond process is equivalent to a site process on a different lattice which he called the *covering lattice.*

During the 1960s the world of physics showed little interest in percolation phenomena, and only a handful of workers contributed to the literature. There were clear analogies with the magnetic critical phenomena described in Chapter 5. For example, the behaviour of $P(p)$, the number of sites belonging to the infinite cluster, closely parallels the behaviour of the spontaneous magnetization (e.g. Sykes, Glen and Gaunt 1974); and the probability that two sites belong to a single cluster, the *pair connectedness*, parallels the pair correlation function (see e.g. Essam 1972). Sykes and Essam (1964) conjectured the exact values of p_c for a number of two-dimensional lattices, recalling the Kramers and Wannier conjecture (1941) for the Ising model. Mean field solutions had their analogue in closed-form solutions for tree and cactus structures (Fisher and Essam 1961). Series expansions could be developed in powers of p (Domb and Sykes 1961b) and the techniques used for the exploration of critical behaviour in magnets could be applied to them.

The origin of this analogy was revealed when Kasteleyn and Fortuin (1969) showed that the bond percolation model is identical with the Q-component Potts model with $Q = 1$. Hence the general theoretical description provided by the RG applies to percolation but the symmetry is different, and the appropriate universality class is not in the n-vector framework. Toulouse (1974) showed that for percolation a mean field solution is attained only when the space dimension d becomes 6; hence an ϵ expansion is not a very useful tool for calculating critical behaviour in three dimensions.

As an interesting aside Temperley and Lieb (1971) connected the percolation model (as well as a number of other models in lattice statistics) to a mathematical problem formulated by Whitney in 1932. They pointed out that the determination by Sykes and Essam of the exact value of p_c for two-dimensional lattices was a rediscovery of a result obtained by Whitney.

Useful general references on percolation theory are Stauffer (1985) and Stauffer and Aharony (1991).

In connection with site percolation we referred to a model of a mixture of non-magnetic A atoms with magnetic B atoms, but we were then concerned only with the configurations at zero temperature. The problem becomes much more involved when we introduce the temperature as a new variable. If we start with a crystal of magnetic B atoms with a Curie temperature $T_c(1)$, and introduce A atoms of concentration p, the Curie temperature $T_c(1 - p)$ will decrease as p increases until it falls to zero at the critical percolation concentration p_c. The detailed form of the $T_c(1 - p)$ curve, as well as other thermodynamic properties of this *dilute magnetic* system, have attracted the attention of many research workers; for a comprehensive review see Stinchcombe (1983).

Experimental work on the λ-point of liquid helium was discussed in §1.5. Essential to the interaction which produces the transition are the Bose–Einstein statistics which govern the behaviour of ^{4}He atoms; no such transition is observed in liquid ^{3}He whose atoms obey Fermi–Dirac statistics. Let us now consider liquid ^{4}He – with an admixture of ^{3}He atoms. At low concentrations the effect of the ^{3}He atoms is to weaken the λ-point transition, but not to change its character. But when a critical concentration is reached a first-order transition with phase separation takes over.

An analogous magnetic system is an antiferromagnet in the presence of a disordering magnetic field (§5.3.4). Near the Néel temperature the disordering field does not change the nature of the critical point, but at sufficiently low temperatures and high fields the transition from the antiferromagnetic to the paramagnetic phase is first order.

Such transition points had been envisaged by Landau in 1937 who called them 'critical points of a continuous phase transition', and even earlier by

Kohnstamm in 1926 who called them 'critical points of the second order'. Griffiths (1970) provided new insight into the nature of such transition points by introducing, as an additional variable, the field conjugate to the order parameter, and considering the phase transition in three-dimensional space. It then becomes clear that these transition points represent the intersection of three second-order transition lines – hence the proposed term 'tricritical point'.

An account from first principles of tricritical points, together with a survey of the extensive literature, is given in Lawrie and Sarbach (1984).

For an antiferromagnetic make-up of spins with a Heisenberg n-vector interaction Fisher and Nelson (1974) showed that a *bicritical point* can result. This is a point at which a first-order transition which separates two distinct ordered phases splits into two distinct critical lines which separate these phases from a common disordered phase. For certain ranges of phenomenological constants a new phase can enter to replace the first-order transition line, and a *tetracritical point* can result at which four critical lines meet. For a review of such multicritical phenomena see Fisher (1975).

Finally, in relation to finite size and surface effects discussed in §6.6, a great deal of progress has taken place in the past two decades, and reviews by Barber (1983) and Binder (1983) are strongly recommended.

References

ADLER, J. (1995) *Recent Developments in Computer Simulation Studies in Condensed Matter Physics* **8**, ed. D. LANDAU, Berlin: Springer.

ADLER, J. (1996) *Annual Reviews of Computational Physics* **4**, ed. D. STAUFFER, Singapore: World Scientific.

BARBER, M. N. (1983) DL **8**, Ch. 2.

BAXTER, R. J. (1971) *Phys. Rev. Lett.* **26**, 832.

BAXTER, R. J. (1972) *Ann. Phys., NY* **70**, 193.

BAXTER, R. J. (1982) *Exactly Solved Problems in Statistical Mechanics*, New York and London: Academic Press.

BAXTER, R. J. and ENTING, I. G. (1979) *J. Stat. Phys.* **21**, 103.

BINDER, K. (1976) DG **5b**, Ch. 1.

BINDER, K. (1983) DL **8**, Ch. 1.

BINDER, K. (ed.) (1992) *Monte Carlo Methods in Condensed Matter Physics*, Berlin: Springer.

BINDER, K. and HEERMAN, D. W. (1988) *Monte Carlo Methods in Statistical Physics – An Introduction*, Berlin: Springer.

BINDER, K. and YOUNG, A.P. (1986) *Rev. Mod. Phys.* **58**, 801.

BROADBENT, S. R. and HAMMERSLEY, J. M. (1957) *Proc. Cambridge Philos. Soc.* **53**, 629.

DE GENNES, P. G. (1972) *Phys. Lett.* **38A**, 339.

DOMB, C. (1969) *Adv. Chem. Phys.* **15**, 229.

DOMB, C. (1971) *Critical Phenomena in Alloys, Magnets and Superconductors*, ed. R. E. MILLS, E. ASCHER and R. I. JAFFEE, New York: McGraw-Hill, p. 89.

DOMB, C. (1974) *J. Phys. A: Math. Gen.* **7**, 1335.
DOMB, C. (1990) *Disorder in Physical Systems*, ed. G. R. GRIMMETT and D. J. A. WELSH, Oxford, p. 33.
DOMB, C. and SYKES, M. F. (1961a) *J. Math. Phys.* **2**, 63.
DOMB, C. and SYKES, M. F. (1961b) *Phys. Rev.* **122**, 77.
ELLIOTT, R. J., HEAP, B. R., MORGAN, D. J. and RUSHBROOKE, G. S. (1960) *Phys. Rev. Lett.* **5**, 366.
ESSAM, J. W. (1972) DG **2**, Ch. 6.
EWALD, P. P. (1921) *Ann. Phys., Leipzig* **64**, 253.
FISHER, M. E. (1961) *J. Math. Phys.* **2**, 620.
FISHER, M. E. (1975) *Proc. Conf. No. 24 on Magnetism and Magnetic Materials*, New York: AIP, p. 273.
FISHER, M. E. and ESSAM, J. W. (1961) *J. Math. Phys.* **2**, 609.
FISHER, M. E. and NELSON, D. R. (1974) *Phys. Rev. Lett.* **32**, 1350.
FISHER, M. E. and SYKES, M. F. (1959) *Phys. Rev.* **114**, 45.
GRIFFITHS, R. B. (1970) *Phys. Rev. Lett.* **24**, 715.
HAMMERSLEY, J. M. and MORTON, K. W. (1954) *J. R. Stat. Soc.* **B16**, 23.
KADANOFF, L. P. and WEGNER, E. J. (1971) *Phys. Rev. B* **4**, 3989.
KASTELEYN, P. W. and FORTUIN, C. (1969) *J. Phys. Soc. Jpn. (Suppl)* **26**, 21.
KOHNSTAMM, Ph. (1926) *Handbuch der Physik*, ed. H. GEIGER and K. SCHEEL, Vol. 10, Berlin: Springer.
KRAMERS, H. A. and WANNIER, G. H. (1941) *Phys. Rev.* **60**, 252, 263.
LANDAU, L. D. (1937) *Phys. Z. Sow.* **11**, 26, 545.
LAWRIE, I. D. and SARBACH, S. (1984) DL **9**, Ch. 1.
LIEB, E. and WU, F. Y. (1972) DG **1**, Ch. 8.
MCCOY, B. and WU, T. I. (1973) *The Two Dimensional Ising Model*, Cambridge, MA: Harvard.
NIENHAUS, B. (1987) DL **11**, pp. 36–42.
ONSAGER, L. (1939) Discussion of Conference on Dielectrics, New York Academy of Sciences.
PAWLEY, G., SWENDSON, R. H., WALLACE, D. J. and WILSON, K. G. (1984) *Phys. Rev. B* **29**, 4030.
POTTS, R. B. (1952) *Proc. Cambridge Philos. Soc.* **48**, 106.
STAUFFER, D. (1985) *Introduction to Percolation Theory*, London: Taylor & Francis (2nd edition with A. AHARONY, 1991).
STINCHOMBE, R. B. (1983) DL **7**, Ch. 3.
SYKES, M. F. and ESSAM, J. W. (1964) *J. Math. Phys.* **5**, 1117.
SYKES, M. F., GLEN, M. and GAUNT, D. S. (1974) *J. Phys. A: Math. Gen.* **7**, L105.
TEMPERLEY, H. N. V. and LIEB, E. H. (1971) *Proc. R. Soc. A* **322**, 251.
TOULOUSE, G. (1974) *Nuovo Cimento B* **23**, 234.
WHITNEY, H. (1932) *Ann. Math.* **33**, 688.
WU, F. Y. (1982) *Rev. Mod. Phys.* **54**, 235.

Author Index

Subject Index

Milton Keynes UK
Ingram Content Group UK Ltd.
UKHW020319111024
449327UK00040B/1400

9 780367 401382